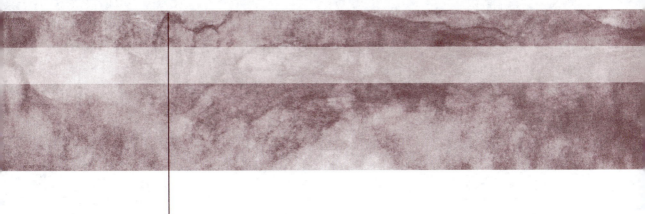

Mathematics of Interest Rates, Insurance, Social Security, and Pensions

Mathematics of Interest Rates, Insurance, Social Security, and Pensions

ROBERT MUKSIAN

Professor of Mathematics
Bryant College Smithfiled, RI

Pearson Education, Inc.
Upper Saddle River, New Jersey 07458

Library of Congress Cataloging-in-Publication Data

Muksian, Robert
　　Mathematics of interest rates, insurance, social security, and pensions/Robert
Muksian.—1st ed.
　　　p.cm.
　　Includes bibliographical references and index.
　　ISBN: 0-13-009425-0
　　1. Interest rates—Mathematical models.　2. Insurance—Mathematical models.
3. Pensions—Mathematical models.　4. Mathematics.　I. Title.

HG1621 .M857　2003
332.8′2′0151—dc21　　　　　　　　　　　　　　　　2002034632

Acquisition Editor: *George Lobell*
Editor-in-Chief: *Sally Yagan*
Vice President/Director of Production and Manufacturing: *David W. Riccardi*
Executive Managing Editor: *Kathleen Schiaparelli*
Senior Managing Editor: *Linda Mihatov Behrens*
Production Editor: *Bob Walters*
Manufacturing Buyer: *Michael Bell*
Manufacturing Manager: *Trudy Pisciotti*
Marketing Assistant: *Rachel Beckman*
Assistant Managing Editor, Math Media Production: *John Matthews*
Editorial Assistant/Supplements Editor: *Jennifer Brady*
Art Director: *Wanda España*
Interior Design: *Heather Scott*
Cover Design: *Jonathan Boylan*
Creative Director: *Carole Anson*
Art Editor: *Thomas Benfatti*
Director of Creative Services: *Paul Belfanti*
Cover Photo: Exterior Entrance: Merryl Lynch Executive Group Somerset, New Jersey. View at Dusk.
　Courtesy of Architect: Spector Group. Photo by George Erml.
Art Studio: *Laserwords*

ⓒ 2003 Pearson Education, Inc.
Pearson Education, Inc.
Upper Saddle River, New Jersey 07458

Printed in the United States of America

10　9　8　7　6　5　4　3　2　1

ISBN: 0-13-009425-0

Pearson Education LTD., *London*
Pearson Education Australia PTY, Limited, *Sydney*
Pearson Education Singapore, Pte. Ltd
Pearson Education North Asia Ltd, *Hong Kong*
Pearson Education Canada, Ltd., *Toronto*
Pearson Educación de Mexico, S.A. de C.V.
Pearson Education–Japan, *Tokyo*
Pearson Education Malaysia, Pte. Ltd

To my grandson, Mark Steven Pitts.
He showed an affinity to mathematics at a very early age.

Contents

Preface

This book is intended for individuals whose career paths may include the need for mathematics of finance, insurance, breakeven analyses, and retirement planning via Social Security and private pensions. It is intended to be a helpful adjunct to business concentrations such as:

- Accounting, since many accounting firms have included retirement planning on a consulting basis.
- Finance and Financial services, in that the lines of separation between banks, insurance companies, and brokerage houses are no longer firmly delineated.
- Management, to be able to "converse" in the mathematics of all segments of a business.
- Marketing services, in order to be "literate" in the vocabulary of finance, insurance, and pensions for professional and personal use.
- Computer Information Systems, in order to know specific mathematical concepts so as to be able to develop appropriate algorithms for solution.
- Economics, in order to extend the use of financial mathematics to insurance and pensions.

The book can also be used for those students whose concentrations are in the Liberal Arts. It will help to make them "literate" in the vocabulary of finance, insurance, and pensions and to be able to utilize the appropriate mathematics for professional and personal use.

The prerequisite preparation that is necessary is at least a solid foundation in high school algebra. Knowledge of the contents of this book can be useful for personal financial planning as well as for the business uses cited above.

Chapter 1 discusses concepts that involve Simple Interest and Simple Discount. Chapter 2 introduces Practical Applications of Simple Interest, including the determination of short-term rates of return on investments. Chapter 3 is a presentation of Compound Interest and includes the determination of time-weighted, compounded rates-of-return on long-term investments. Chapter 4 discusses the concepts of Simple Annuities, where the frequency of payments and the frequency of compounding are the same. In Chapter 5, the concepts of annuities are extended to include deferred annuities, complex annuities (where the annual frequency of payments and the annual frequency of compounding are different), annuities in perpetuity, and annuities where the periodic payments vary geometrically. Additional practical applications of annuities are included in Chapter 6 (Bonds). In order to assist in the optimization of investment port-folios, elements of linear programming are introduced in Chapter 7. After a brief introduction to optimization concepts with two variables, the use of the **Solver** tool in Microsoft Excel is shown for optimization with more than two variables. Advanced topics of Capital Rationing and Working Capital Man-agement are included to show how the **Solver** can be used in advanced topics of finance. Breakeven Models for cost-revenue, supply-demand, and financial considerations such as the effect of commissions on purchases and early retire-ment decisions are discussed in Chapter 8. Chapter 9 introduces concepts of life insurance through life annuities, net annual premiums, and terminal reserves. Elements of the mathematics of Social Security are presented in Chapter 10, and Chapter 11 introduces you to the elements of the mathematics associated with private pensions.

This is a better book because of many useful comments and suggestions received from colleagues and correspondents Kevin Charlwood, Washburn University; Jeffrey Forrest, Slippery Rock University; Joseph Greene, Augusta State University; Kerri McMillan, Clemson University; Yujin Shen, The Richard Stockton College of New Jersey; Thomas Springer, Florida Atlantic University, and Jianzhong Su, University of Texas at Arlington.

I wish to give special thanks to Dr. Hsi Li, Professor of Finance, Dr. Alan Olinksy, Professor of Mathematics, and Dr. Phyllis Schumacher, Professor of Mathematics, all of Bryant College, Smithfield, R. I., for their kind suggestions relative to the material. I also wish to give special thanks to Mr. William Pitts, Vice-President, Strategic Planning, Textron Financial Corporation, Providence, R.I., and to Mr. Paul Merlino, Esq. of Lamoriello and Company, Warwick, R.I for comments relative to private pensions. I would especially thank my daugh-ter, Dr. Robin Muksian Schutt, Associate Professor of English, New England Institute of Technology, Warwick R.I., for commenting on the readability of the manuscript.

Robert Muksian
rmuksian@bryant.edu
Bryant College, Smithfield, RI

Mathematics
of Interest Rates,
Insurance, Social Security,
and Pensions

CHAPTER 1

Simple Interest and Discount

The payment of interest is the cost for borrowing money. If an individual borrows an amount of money from a lender (e.g., bank, finance company, credit union, individual), the amount of the repayment is greater than the amount borrowed. The difference between the two amounts is the cost of borrowing—called interest. If an individual deposits money in a savings institution (i.e., bank, credit union, savings and loan association), that institution has, in effect, borrowed the money from the depositor and pays the depositor interest. This interest is the cost to the savings institution for having use of the depositor's money. If an individual invests money in corporate bonds, the corporation, in effect, has borrowed the money from the investor and pays interest for the use of that money. Since this interest decreases profit, it is a cost to the corporation for having use of the money.

Simple interest is the interest earned only on the borrowed amount. If a corporate bond pays a fixed interest rate annually, that rate and face value (the amount "borrowed") determine the amount of interest of the bond. The interest rate is a percent of the principal amount. The interest charged for automobile financing, for example, is based on the amount of the automobile cost that must be borrowed. Interest charges in business transactions, on the other hand, are based on the amount owed after partial payments are made. The concepts of simple interest are discussed in this chapter and Chapter 2. **Compound interest** is interest that is earned on interest. Simple interest concepts are the foundation for compound interest, which will be discussed in Chapters 3, 4, and 5.

1.1 SIMPLE INTEREST

If money is placed in a savings account to earn interest or if interest must be paid on money that is borrowed, the amount of interest depends upon three factors:

1. The principal,
2. The interest rate, and
3. The length of time.

Note that the interest rate and time must be in compatible units. If interest rates are specified as monthly, time must be the number of months. If interest rates are specified as annual, time must be the number of years. The usual specifications of interest rates are annual.

All stated interest rates in this book are annual rates unless indicated to be different.

Then, letting:

P be the principal, the amount earning interest,

r be the annual interest rate in %,

t be the number of years, and

I be the interest,

simple interest is defined by

$$I = Prt. \tag{1.1-1}$$

EXAMPLE 1.1.1   What is the simple interest on $1,000 at 6% for a) 2 years, and b) 6 months?

Solution   The principal is 1,000 and 6% converts to 0.06.

a) For 2 years, $t = 2$ and from Equation (1.1-1)

$$I = Prt$$

$$= (1,000)(0.06)(2) = \$120.$$

b) For 6 months, $t = \frac{1}{2}$ of 1 year (0.5 years). From Equation (1.1-1)

$$I = Prt$$

$$= (1,000)(0.06)(0.5) = \$30. \quad \blacksquare$$

Interest is earned or paid for the exact amount of time that the principal is outstanding. Frequently, the exact time involves a number of days only, or a number of days in conjunction with a number of months or years. Since it is customary to express interest rates on an annual basis, the number of days must be converted to a fraction of a year. Two methods for this conversion are used, and the respective interests are named ordinary interest and exact interest.

ORDINARY INTEREST

A lending institution may use ordinary interest when interest is paid to the institution. The method assumes a 360-day year. Letting:

N be the exact number of days, and

t_o be ordinary time (hence "ordinary" interest),

$$t_o = \frac{N}{360}. \tag{1.1-2}$$

EXACT INTEREST

A savings institution may use exact interest when the institution pays the interest. The method uses a 365-day year. Letting:

N be the exact number of days, and

t_e be exact time (hence "exact" interest)

$$t_e = \frac{N}{365}.$$ (1.1-3)

EXAMPLE 1.1.2 For a principal of $1,800, a 7% interest rate, and a time of 92 days, find a) the ordinary interest and b) the exact interest.

Solution

a) For ordinary interest, Equation (1.1-1) becomes

$$I = Prt_o,$$

and from Equation (1.1-2),

$$t_o = \frac{92}{360}.$$

Then,

$$I = (1,800)(0.07)\left(\frac{92}{360}\right) = 32.20,$$

and the ordinary interest is $32.20.

b) For exact interest, Equation (1.1-1) becomes

$$I = Prt_e,$$

and from Equation (1.1-3),

$$t_e = \frac{92}{365}.$$

Then,

$$I = (1,800)(0.07)\left(\frac{92}{365}\right) = 31.76,$$

and the exact interest is $31.76. ∎

Given a starting date for earning or paying interest, on what date will a specific number of days occur? Another way of asking the question is how many days are there between two calendar dates? The most convenient solution is to count the number of days in conjunction with a calendar, but in so doing, *the first date is excluded and the last date is included*. Thus, if time begins on March 14, 92 days later occurs on a date as follows:

Month	Days in the Month	Cumulative Days
Starting on March, 14	17	
April	30	47
May	31	78
Ending on June, 14	14	92

Then, June 14 is 92 days following March 14. If available, a table of Julian Dates will facilitate the determination of dates. Such a table is supplied as Table A-1 in the Appendix. The months are listed across the top, and the day of the month is listed down the left. Thus, for March 14, find March at the top and 14 at the left. The intersection of these two factors indicates March 14 to be the 73rd day of the year. Then, 92 days later would be the 165th day (73 + 92) of the year which is June (at the top) 14 (at the left). For the number of days between March 14 and June 14, find June 14 to be the 165th day and March 14 to be the 73rd day. Then, the number of days between the two dates is 92 (165 − 73). For leap years, 1 would be added to each date after February 28.

EXAMPLE 1.1.3   A $1,500 loan is initiated on April 19 and is to be repaid on July 7. What will be the ordinary interest if the interest rate is 10%?

Solution   Since this is ordinary interest, 360 is used for the number of days in a year. From Equations (1.1-1) and (1.1-2),

$$I = Prt_o,$$

and

$$t_o = \frac{N}{360}.$$

Using a table of Julian Dates, Table A-1, July 7 is the 188th day and April 19 is the 109th day. Thus, $N = 188 - 109 = 79$ days, and

$$t_o = \frac{79}{360}.$$

Then,

$$I = (1,500)(0.10)\left(\frac{79}{360}\right) = 32.92,$$

and the ordinary interest is $32.92. ■

EXAMPLE 1.1.4   How many days will there be between August 11 and May 8 of the following year? (Use 365-day years.)

Solution   Since two calendar years are involved, the number of days in each must be determined. August 11 is the 223rd day and the number of days in the first year is 365 − 223 or 142. May 8 is the 128th day of the year and is the number of days in the following year. Then, the number of days between August 11 and May 8 is 142 + 128, or 270 days. To simplify the determination of the number of days between calendar dates in two consecutive years, let D_1 be the number of the date in the first year, and D_2 be the number of the date in the following year. The number of days, N, between the two dates can be determined by

$$N = 365 - (D_1 - D_2). \tag{1.1-4}$$

For the dates above, $N = 365 - (223 - 128) = 365 - 95 = 270$ days. ■

The definition of simple interest, Equation (1.1-1) involves four parameters—I, P, r, t. Given any three of the parameters, the 4th may be determined algebraically. From Equation (1.1-1), $I = Prt$. Given r, t, and I, $P = \frac{I}{rt}$; given P, t, and I, $r = \frac{I}{Pt}$; and given P, r, and I, $t = \frac{I}{Pr}$. These modified forms of Equation (1.1-1) should not be memorized as such. The algebraic manipulations are quite simple and are shown in Example 1.1.5.

EXAMPLE 1.1.5   a) What principal will earn $50 interest in five years at a simple interest rate of 6%? b) What simple interest rate is necessary for $1,000 to earn $90 interest in 18 months? c) How long will it take to earn $110 interest with a principal of $1,000 at a simple interest rate of 5%?

Solution

a) The given information is: $I = 50$, $r = 0.06$, and $t = 5$.
From Equation (1.1-1), $I = Prt$ and substituting the given values gives
$$50 = P(0.06)(5), \text{ or}$$
$$50 = P(0.3).$$

Dividing both sides by 0.3 results in $P = \$166.67$.

b) The given information is: $P = 1,000$, $I = 90$, and $t = 18$ months. From the definition of simple interest, $I = Prt$, and substituting the given values, gives
$$90 = (1,000)(r)\left(\frac{18}{12}\right), \text{ or}$$
$$90 = 1,500r.$$

Dividing both sides by 1,500 gives $r = 0.06$, or 6% per year since the time was indicated as $\frac{18}{12}$ years.

c) The given information is: $I = 110$, $P = 1,000$, and $r = 0.05$.
From the definition of simple interest, $I = Prt$ and substituting the given values yields
$$110 = (1,000)(0.05)t, \text{ or}$$
$$110 = 50t.$$

Dividing both sides by 50 gives $t = 2.2$ years since the interest rate is assumed to be annual. Now, 0.2 years $= (0.2)(360)$ days if it is ordinary time and $(0.2)(365)$ days if it is exact time, or 72 days and 73 days respectively. Thus, $t_o = 2$ years and 72 days or $t_e = 2$ years and 73 days. ∎

If time is indicated as a number of months, then the number of years is obtained by dividing the number of months by 12. That is, 3 months is $\frac{3}{12}$ years and reduces to $\frac{1}{4}$ of a year. A longer period such as 15 months is $\frac{15}{12}$ years, which is $1\frac{1}{4}$ years or one year and three months.

PROBLEM SET 1.1

1. Determine the simple interest on $1,000 at 5% for the following:
a) 4 months, **b)** 7 months, **c)** 9 months, **d)** 15 months, **e)** 27 months, **f)** 3 years, **g)** 5 years, and **h)** 10 years.

2. Determine the simple interest on $2,000 at 8% for 142 days using **a)** ordinary time, and **b)** exact time.

3. If a loan is initiated on June 2, of a 365-day year, determine the due date if the loan is for **a)** 90 days, **b)** 180 days, and **c)** 270 days.

4. Determine the number of days between

a) January 4 and September 8 of a 366-day year.

b) April 29, yyyy and January 12 of the following year. Assume yyyy is a leap year.

c) October 16, yyyy and February 26 of the following year. Assume both years are 365-day years.

d) November 12, yyyy and March 3 of the following year. Assume the "following year" is a leap year.

5. Determine the ordinary interests on a $5,000 loan at 9% for

a) 90 days, **b)** 180 days, and **c)** 270 days.

6. Determine the exact interests on $2,500 at 5% for the number of days determined for each part in Problem 4.

Complete the following for Problems 7–11.

	I	P	r	t
7.	$1,000	-----	7.25%	3 years
8.	$100	$2,000	-----	6 months
9.	$75	$1,000	-----	1 year
10.	$250	$5,000	8.0%	-----
11.	$100	$1,500	10.0%	-----

12. If the simple interest on $5,000 is $1,105 for an 8.5% interest rate, determine the **a)** ordinary and **b)** exact times in years and days.

1.2 FUTURE VALUE AND PRESENT VALUE

The "future amount" and the "present value" of money refer to "growth" of money because money has a time-value. The following are the respective definitions.

1. The **future value** is the amount to which a present value will grow, given the interest rate and time.

2. The **present value** is the amount necessary in order that a specified future value is realized, given the interest rate and time. Time is measured from the original maturity date backwards to the date on which the present value is being determined.

FUTURE VALUE

Given the amount of interest, the future value (e.g., the maturity value of a loan) is the sum of the principal and the interest. Letting:

S be the future amount, and
I be the interest,

the future amount is defined by

$$S = P + I. \tag{1.2-1}$$

In the preceding section, simple interest was defined by $I = Prt$. Substituting Equation (1.1-1) into Equation (1.2-1) gives

$$S = P + Prt.$$

and factoring P gives

$$S = P(1 + rt). \tag{1.2-2}$$

As indicated by Equations (1.2-1) and (1.2-2), interest is added at the end of an interest earning time period. This concept is shown in Figure 1-1.

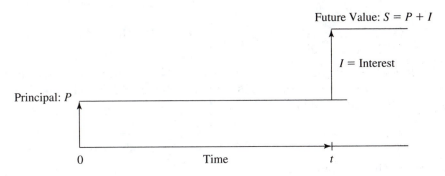

FIGURE 1-1 Future Amount at Simple Interest

Equation (1.2-2) allows the future amount to be determined without first computing the interest. If knowing the amount of the interest is necessary, Equation (1.2-1) may be rearranged to give

$$I = S - P. \tag{1.2-3}$$

The use of Equations (1.1-1) and (1.2-1), in combination, in order to determine a future value is a discretionary matter. However, the form of Equation (1.2-2) is necessary in later chapters; therefore, familiarity with that form for a future amount is recommended.

EXAMPLE 1.2.1   Find the future amount and interest on $1,000 at 6% for three years.

Solution   The future amount will be determined by using both approaches. The given information is $P = 1,000$, $r = 0.06$, and $t = 3$.
From Equation (1.1-1)

$$I = Prt$$
$$= (1,000)(0.06)(3) = 180.$$

Thus, the interest is $180. From Equation (1.2-1),

$$S = P + I$$
$$= 1,000 + 180 = 1,180,$$

and the future amount is $1,180.
From Equation (1.2-2),

$$S = P(1 + rt)$$
$$= (1,000)[1 + (0.06)(3)]$$
$$= (1,000)(1 + 0.18)$$
$$= (1,000)(1.18) = 1,180.$$

The interest is determined by Equation (1.2-3) as
$$I = S - P$$
$$= 1,180 - 1,000 = 180,$$

as above. ∎

EXAMPLE 1.2.2   At an interest rate of 7.25%, how many years are necessary for $2,000 to double?

Solution   Since the $2,000 is to "grow," it is the principal, and the doubled amount is the future amount. Thus, $P = 2,000$, $S = 2(2,000) = 4,000$, and $r = 0.0725$. From the Equation (1.2-3),

$$I = S - P$$
$$= 4,000 - 2,000 = 2,000.$$

Then, from Equation (1.1-1),

$$I = Prt,$$
$$2,000 = (2,000)(0.0725)t, \text{ or}$$
$$2,000 = 145t.$$

Dividing both sides by 145 gives 13.793 years. ∎

In this problem the amount of the interest was not requested directly. Therefore, Equation (1.2-2) could have been used as follows:

$$S = P(1 + rt), \text{ or}$$
$$4,000 = 2,000(1 + 0.0725t).$$

Dividing both sides by 2,000 gives

$$1 + 0.0725t = 2,$$

and subtracting 1 from both sides gives

$$0.0725t = 1.$$

Now, dividing both sides by 0.0725 gives 13.793 years as above. This latter approach becomes a necessity in later chapters; therefore, its use is recommended.

EXAMPLE 1.2.3 An automobile dealer advertises automobile financing with a 5% simple interest rate. What will be the total of payments for financing $10,000 for four years?

Solution This is an income producing activity for the agency in that it is earning interest on $10,000 of principal. Thus, for $r = 5\% = 0.05$ and $t = 4$ years, using Equation (1.2-2),

$$S = P(1 + rt)$$

$$= (10,000)[1 + (0.05)(4)]$$

$$= (10,000)(1.2) = 12,000.$$

Therefore, the total of payments is $12,000. The interest is determined from Equation (1.2-3) as

$$I = S - P.$$

$$= 12,000 - 10,000 = \$2,000. \quad \blacksquare$$

Note: The true interest rate for this type of financing is almost double the stated rate. This will be discussed in detail in Chapter 4.

PRESENT VALUE

In the determination of a future amount, the principal that is saved (or invested) is known. However, when the *future amount* is known, the *present value* problem is to determine the principal that will grow to that specified future amount. The word "present" should not be interpreted to mean the actual present date ("now"). It could be a date earlier than "now," it could be today's date "now," or it could be a future date, which occurs earlier than the date when the future amount is specified. A "time chart" may be helpful for visualizing the time frame in a present value problem. Such a chart is shown in Figure 1-2.

In order to solve for the present value of a specified future amount Equation (1.2-2) may be solved for P. That is, P is the present value of S, given r and t. Equation (1.2-2) is $S = P(1 + rt)$, and dividing both sides by $(1 + rt)$ gives the present value of S as

$$P = \frac{S}{1 + rt}. \tag{1.2-4}$$

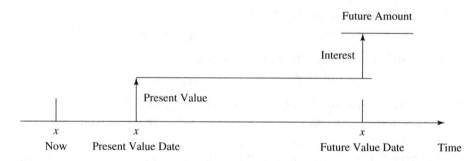

FIGURE 1-2 A Time Chart for the Present Value of Money

EXAMPLE 1.2.4   What is the present value of $1,180 three years from now at a 6% simple interest rate?

Solution   We are seeking the present value of $1,180, given $S = 1,180$, $r = 0.06$, and $t = 3$. From Equation (1.2-4),

$$P = \frac{1,180}{1 + (0.06)(3)} = 1,000.$$

Thus, the present value of $1,180 in three years at 6% simple interest is $1,000, which is the principal in Example 1.2.1. ∎

EXAMPLE 1.2.5   The total amount of a loan to which interest had been added is $15,000. If the interest rate was 5% and the term of the loan was four years, what was the amount of the loan?

Solution   Since $15,000 is to be paid at the end of four years, it is the future amount. Therefore $S = 15,000$. The amount of the loan is the present value of 15,000 and from Equation (1.2-4), the present value is given by

$$P = \frac{15,000}{1 + (0.05)(4)} = 12,500.$$

Therefore, the amount of the loan was $12,500. ∎

EXAMPLE 1.2.6   It is now April 8, yyyy. On January 8, yyyy an investment was made that will mature to $10,000 on October 5, of the same year. How much was invested if the interest rate was 9% simple interest?

Solution   Since the investment was made on January 8 and the future value date is October 5, the table of Julian Dates (Table A-1) shows the time, $t = 270$ days, $S = 10,000$, and $r = 0.09$. Then from Equation (1.2-4)

$$P = \frac{10,000}{1 + (0.09)\left(\dfrac{270}{365}\right)} = 9,375.80,$$

and the present value of $10,000 in 270 days at 9% is $9,375.80. ∎

The key to the solution of present value problems is that the amount specified on the future value date is the future amount. If the future value date and the amount for that date are known, the problem is a present value problem.

PROMISSORY NOTES

One type of financial transaction that involves simple interest is the promissory note wherein one party borrows money and agrees to pay another party a specified sum in return. There are two types of **promissory notes**: interest-bearing and non-interest-bearing notes. **Interest-bearing notes** indicate the amount borrowed, the date of the loan, the interest rate to be charged, and the due date. The maturity value is considered to be the future value. **Non-interest-bearing notes** indicate the date of the debt, the maturity value, and the due date. Neither the principal nor the interest rate appears. If either type of note is liquidated before the maturity date, the amount of the payment is called the **present value**; the interest rate charged for early discharge is often referred to as the **discount rate**, and the date of discharge is called the **discharge date**. For both types of notes, the discharge amount is the present value of the note on the discharge date.

A. NON-INTEREST-BEARING NOTES

Only the initiation date, the maturity date, and maturity value would be indicated on a non-interest-bearing note. If the note is paid prior to the maturity date, the value of the note is the present value. The difference between the maturity value and the present value on the discharge date is called **discount**. A time chart for non-interest-bearing notes is shown in Figure 1-3.

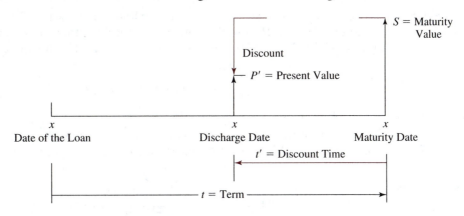

FIGURE 1-3 A Time Chart for Non-Interest-Bearing Notes

B. INTEREST-BEARING NOTES

The time chart for interest bearing notes is the same as Figure 1-3 except that the principal, P, the interest rate, r, and the maturity value, S, are indicated on the note.

EXAMPLE 1.2.7   A color television set costs $575. After agreeing to a down payment of $50, you sign a promissory note to pay $525 cash in 90 days. At the end of 7 days you decide to discharge the note. The dealer informs you that money is worth 10% at this time. How much is necessary to discharge the note?

Solution   This is a non-interest-bearing note since neither the principal nor the interest rate is indicated. Then, $S = 525$ and $r' = 0.10$. Since the note is due in 90 days and was discharged after seven days, $t' = 83$ as may be seen in the time chart below.

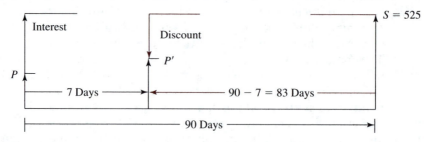

From Equation (1.2-4),

$$P' = \frac{525}{1 + (0.10)\left(\dfrac{83}{365}\right)} = 513.33.$$

Thus, the dealer accepts \$513.33 in full settlement of the note.   ■

EXAMPLE 1.2.8   You sign a promissory note for \$1,000 at 12% interest due in one year. What amount is necessary to discharge the note after nine months if money is worth 15% on the discharge date?

Solution   Since the interest rate and amount of the loan are shown on the note, this is an interest-bearing note. Before the note can be discharged, its maturity value must be known. Then, $P = 1,000$, $r = 0.12$, and $t = 1$ year. From Equation (1.2-2),

$$S = P(1 + rt)$$

$$= 1,000[1 + (0.12)(1)]$$

$$= 1,120.$$

Since the term of the note is one year, and it was discharged after nine months, $r' = 0.15$ and $t' = 12 - 9 = 3$ months as may be seen from the time chart below.

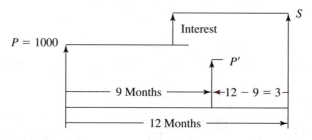

Then, from Equation (1.2-4),

$$P' = \frac{1,120}{1 + (0.15)\left(\dfrac{3}{12}\right)} = 1,079.52.$$

A payment of \$1,079.52 will discharge the note three months before it is due.   ■

PROBLEM SET 1.2

In Problem Set 1.1, the problems were to determine the interest amounts for various principals and interest rates. Using Equation (1.2-1), the future amounts are readily known by adding those interests to the respective principals. In order to "memorize" by repeated use, solve the following problems by using Equations (1.2-2) and (1.2-3).

1. Determine the simple interest on $1,000 at 5% for the following:

  a) 4 months, **b)** 7 months, **c)** 9 months, **d)** 15 months, **e)** 27 months, **f)** 3 years, **g)** 5 years, **h)** 10 years.

2. Determine the simple interest on $2,000 at 8% for 142 days using **a)** ordinary time and **b)** exact time.

3. Determine ordinary interests on a $5,000 loan at 9% for **a)** 90 days, **b)** 180 days, and **c)** 270 days.

4. Determine the exact interests on $2,500 at 5% for the number of days in Problem 3.

5. The interest for financing an automobile is called **add-on** interest because the simple interest for the amount borrowed is added on to the principal. What will be the total amount due for financing an $8,000 automobile for which a $2,500 trade-in allowance was made on the present automobile for a 10% simple interest rate and a term of four years?

6. A bank pays 5.4% simple interest. In how many years will $3,000 become $4,500?

7. For Problem 6, in how many years will the $3,000 triple?

Complete the following table.

	S	r	t	P
8.	$5,000	7%	5 years	
9.	$20,000	5%	20 years	
10.	$6,000	7%	4 years	

11. The total amount paid for financing the purchase of furniture was $4,321.50. If a 12% simple interest rate was charged and the term was two years, what was the purchase price of the furniture?

12. A loan for $5,000 at 12% simple interest is payable in two years. After 18 months, the borrower repaid the loan. What amount was paid to liquidate the note?

13. A non-interest-bearing note for $2,000 is paid 60 days before the due date. If 18% ordinary interest is charged for early payment, what is the amount to be paid?

14. A non-interest-bearing note for $5,000 was dated June 4, yyyy with a due date of June 4 of the following year. It was paid on March 4 of the following year. Using 15% ordinary interest, what was the payment?

15. A loan for $4,000 at 16% simple interest for three years was repaid after one year. At the time of repayment interest rates had increased to 18%. What was the amount of the payment?

16. In Problem 15, what was the actual simple interest rate on the $4,000 for the three years?

1.3 BANK DISCOUNT

When banks compute interest on loans, that interest is usually based on the **maturity value** of the loan and is deducted from the loan, a process known as **bank discount**, or simply **discount**. In this manner interest is paid in advance. The borrower receives the **proceeds** of the loan and repays the maturity value.

MATURITY VALUE

Consider a bank loan from the simple interest concept. Suppose the interest on a $1,000 loan was $100. The borrower would repay $1,100. But, using bank discount, the proceeds of the loan would be $900, the amount the borrower actually receives, and the amount repaid would be $1,000. However, if the borrower needed $1,000, the bank-discount method would be $100 less than needed, and the loan would not be helpful. An amount that is greater than $1,000 would need to be borrowed in order to receive proceeds of $1,000.

Let us calculate the simple interest rate for both approaches using a one-year time reference. From the definition of simple interest $I = Prt$, $t = 1$, and $I = Pr$. Solving for r,

$$r = \frac{I}{P}.$$

For simple interest on a principal of $1,000 and an interest of $100,

$$r = \frac{100}{1,000} = 0.10 \text{ or } 10\%.$$

For bank discount on a loan of $1,000 and an interest of $100, the proceeds would be $900. From a simple interest perspective, a $900 principal required an interest payment of $100 and

$$r = \frac{100}{900} = 0.1111 \text{ or } 11.11\%.$$

It is clear that the bank discount yields a higher simple interest rate than the bank discount rate. Therefore, in order for the borrower to receive proceeds of $1,000 and the bank to earn an 11.11% interest rate, a discount rate would have to be established that would be equivalent to the 11.11%. Letting:

S be the maturity value,

P be the proceeds,

d be the bank discount rate,

D be the discount (interest), and

t be time in years,

bank discount is defined by

$$D = Sdt, \qquad (1.3\text{-}1)$$

and

$$P = S - D. \qquad (1.3\text{-}2)$$

Equation (1.3-1) may be substituted into Equation (1.3-2) to yield

$$P = S - Sdt,$$

and factoring S gives proceeds as

$$P = S(1 - dt). \qquad (1.3\text{-}3)$$

Then, for proceeds of $1,000, an interest rate of 11.11% gives an interest of $111.11 in one year. Substituting into Equation (1.3-3) gives

$$1,000 = 111.11(1 - d).$$

Dividing both sides by 111.11 gives

$$0.9000 = 1 - d,$$

or

$$d = 1 - 0.9000 = 0.10.$$

For this situation a discount rate of 10% would be equivalent to a simple interest rate of 11.11%.

EXAMPLE 1.3.1   What are the proceeds on a loan if $5,200 is to be repaid in five months and the simple discount rate is 8%?

Solution   It is given that $S = 5,200$, $t = \frac{5}{12}$, and $d = 0.08$. From Equation (1.3-3)

$$P = S(1 - dt)$$

$$= 5,200 \left[1 - (0.08) \left(\frac{5}{12} \right) \right]$$

$$= 5,200(1 - 0.03333) = 5,026.68,$$

and the proceeds of the loan are $5,026.68. ■

EXAMPLE 1.3.2   The government borrows money periodically by selling Treasury bills for short periods of time. For these circumstances, the proceeds to the government are the purchase prices of the bills. At a discount rate of 7.25%, what will be the purchase price of a $10,000 Treasury bill for 90 days using ordinary interest?

Solution   It is given that $S = 10,000$, $d = 0.0725$, and $t = 90$. From Equation (1.3-3),

$$P = S(1 - dt)$$

$$= 10,000\left[1 - (0.0725)\left(\frac{90}{360}\right)\right]$$

$$= 10,000[1 - 0.018125] = 9,818.75.$$

Therefore, the purchaser would pay \$9,818.75 for the bill and receive \$10,000 in 90 days. ■

SHORT-TERM LOANS

There may be many reasons for the government, a manufacturer, or a merchant to borrow money for short periods of time. The cash position of a business may be such that there are insufficient funds in the checking account to meet the payroll. The business will then borrow money for a short term in the expectation that cash will flow into the business before the next payroll date, thereby enabling the business to repay the loan and have cash available for the next payroll.

Another reason for incurring short-term loans is to take advantage of cash discounts that are offered to merchants by manufacturers in order to induce the merchants to pay merchandise invoices in a timely manner. If the invoice is paid by a certain number of days after the invoice date, the cash discount amount may be deducted from the merchandise amount on the invoice. Borrowing to take advantage of a cash discount is done on the last date where the cash discount is applicable—usually 10 days. If the net is due in 30 days and there is an expectation that cash will flow into the business during the 20-day period from the date of the loan, only 20 days interest will be paid on the loan. If r is the cash discount rate on an invoice, d is a bank discount rate on a loan, and n is the number of days of the loan, it is desirable to arrange a short-term loan for taking advantage of a cash discount, if the following condition is met:

$$d \le \frac{r}{n}(360) \qquad\qquad\qquad (1.3\text{-}4)$$

for ordinary time, or

$$d \le \frac{r}{n}(365) \qquad\qquad\qquad (1.3\text{-}5)$$

for exact time.

For example, a 20-day loan to take advantage of a 2% cash discount could "tolerate" a bank discount rate as high as 36% using a 360-day year or as high as 36.5% using a 365-day year. Rates as high as these may be considered **usurious** (imposing an interest rate that exceeds a legal maximum), but the point should be well taken that borrowing at extremely high interest rates can still enable a savings of cash discount. If a business does have the cash to pay the net amount of an invoice and does not take advantage of any cash discount on the last possible day or must use that cash for other purposes, such as payroll, that business would have, in effect, "borrowed" the cash discount from the wholesaler or retailer at the high rates indicated above. However, "most of the time" a business will not have sufficient cash available to pay net amounts on invoices. In a high inflationary economy, bank discount rates would be in the 15% to 21% range, and a savings would ensue by borrowing the cash discount. The following examples will illustrate this concept.

In order to encourage timely payment of bills, manufacturers may include incentive terms on invoices such that the receiver of the goods may take a cash discount if the receiver makes the payment by a certain date. One representation of the terms could be $r\%/n$ days, net/30 days. This means that if the merchandise total is paid within n days of the date of the invoice, the merchandise cost may be reduced by $r\%$ and the balance may be remitted as payment in full. Other forms of terms could be $r\%/n$ days R.O.G., where R.O.G. means "receipt of goods," and $r\%/n$ days E.O.M., where E.O.M means "end of month."

EXAMPLE 1.3.3   The merchandise cost on an invoice is $40,000. The cash discount terms of the invoice are 2%/10 days, net/30 days. In order to take advantage of the cash discount, the business borrows sufficient money on the 10th day and pays the invoice expecting that sufficient funds will be available to repay the loan on the 30th day. If the bank discount rate is 10%, what will be the actual cash discount?

Solution   If the total merchandise cost is thought of as the maturity value of a debt at time 1, the discounted cost as the present value of the debt at time 0, and cash discount rate as r (replacing d), then Equation (1.3-3), gives the net cost as
$$P = (1 - 0.02)(40,000)$$
$$= (0.98)(40,000) = 39,200.$$

Then, a payment of $39,200 by the 10th day will satisfy the invoice.

From Equation (1.3-1), replacing d with r, the cash discount, D, is
$$D = rS$$
$$= (0.02)(40,000) = 800.$$

In order for the merchant to take advantage of the $800 cash discount, the amount that must be borrowed is such that the proceeds equal $39,200. From Equation (1.3-3),
$$P = S(1 - dt).$$

For a discount rate of 10% and a term of 20 days,
$$39,200 = S\left[1 - (0.10)\left(\frac{20}{360}\right)\right]$$
$$= S(1 - 0.0056) = (0.9944)S.$$

Dividing both sides by 0.9944 gives 39,420.76. At the specified discount rate, a loan of $39,420.76 will yield proceeds of $39,200, which may be used to pay the cash discounted amount of the invoice, thereby saving $800 from the invoice amount. Twenty days later, the $39,420.76 is repaid to the bank. Equation (1.3-2)

gives the bank discount as
$$D = S - P$$
$$= 39{,}420.76 - 39{,}200 = 220.76.$$

The actual cash discount that the business realizes is the difference between the cash discount of the invoice and the bank discount. Thus,
$$\text{Actual } D = 800 - 220.76 = \$579.24. \quad \blacksquare$$

EXAMPLE 1.3.4 For proceeds of \$39,200 and a cash discount of \$800, what will be the actual savings if the bank discount rate is 24% for 20 days?

Solution Utilizing Equation (1.3-3)
$$39{,}200 = S\left[1 - (0.24)\left(\frac{20}{360}\right)\right]$$
$$= S(1 - 0.0133333) = (0.986667)S.$$

Dividing both sides by 0.986667 gives \$39,728.72. The interest cost will be \$528.72 (or $39{,}728.72 - 39{,}200$), and the actual savings will be \$271.28. $\blacksquare$

REDISCOUNTING NOTES

If a bank extends itself too greatly by granting a large number of large loans, it may have a cash shortfall for meeting its own obligations. If this occurs, the bank may sell some of the notes to another bank. The second bank will discount the already discounted note. Thus, the original note is **rediscounted**. The first bank does not incur an absolute loss of money in the rediscounting process. It simply earns less interest.

EXAMPLE 1.3.5 In Example 1.3.3, it was shown that at a discount rate of 10% for 20 days, a loan of \$39,420.76 was necessary to yield proceeds of \$39,200. Suppose the bank, which made the loan, sells the note to a second bank 10 days after the loan was made. If the second bank charges the first bank a 9% discount rate, what are proceeds for the first bank, and what are the respective discounts?

Solution A time diagram of the transactions is shown below. The subscripts 1 and 2 refer to the first and second banks, respectively. In Example 1.3.3, it was shown that the discount for the first bank was \$220.76. Thus $D_1 = 220.76$. When, the first bank sells the note, it receives proceeds from the second bank based on the maturity value of the loan. Then, from Equation (1.3-3),
$$P_1 = S(1 - dt)$$
$$= (39{,}420.76)\left[1 - (0.09)\left(\frac{10}{360}\right)\right]$$
$$= (39{,}420.76)(1 - 0.0025) = 39{,}322.21.$$

The first bank receives \$39,322.21 from the second bank, and its net discount is
$$I_1 = P_2 - P_1$$
$$= 39{,}322.21 - 39{,}200 = 122.21.$$

Thus, the first bank earns $122.21 interest instead of $220.76 on the original loan. Since the second bank now owns the note, it will receive the maturity value from the business, which borrowed the money originally. Its discount is the difference between the maturity value and the proceeds it gave to the first bank. Thus,

$$D_2 = S - P_1$$

$$= 39{,}420.76 - 39{,}322.21 = 98.55,$$

and the second bank earns $98.55 interest on the transaction. Notice that the sum of the two discounts $(122.21 + 98.55)$ is equal to the original discount of $220.76. ∎

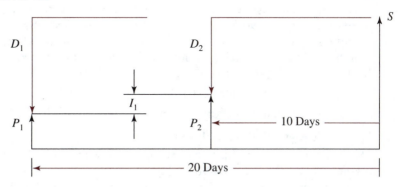

EQUIVALENT DISCOUNT-SIMPLE INTEREST RATES

Given a desired simple interest rate, the necessary discount rate is determined by

$$d = \frac{r}{1 + rt} \times 100 \text{ for } \%, \tag{1.3-6}$$

and given a discount rate, the equivalent simple interest rate is determined by

$$r = \frac{d}{1 - dt} \times 100 \text{ for } \%. \tag{1.3-7}$$

Equations (1.3-6) and (1.3-7) are derived from the following identity. The derivation is left as exercises for the student

$$(1 + rt)(1 - dt) = 1. \tag{1.3-8}$$

PROBLEM SET 1.3

Determine the proceeds and discount for problems 1–8.

	Maturity Value	Discount Rate	Time
1.	$10,000	9%	2 years
2.	$5,000	10%	1 year
3.	$2,500	12%	9 months
4.	$1,500	11%	6 months

	Maturity Value	Discount Rate	Time
5.	$10,000	10%	90 days
6.	$5,000	13%	60 days
7.	$2,500	12%	30 days
8.	$10,000	12%	20 days

9. If the U.S. Government pays 7.5% ordinary discount on 180-day Treasury bills, what is the cost to buy a $10,000 Treasury bill?

10. You purchase furniture from "Honest John." The total cost of the furniture is $2,632.85. Honest John asks for a $300 down payment and agrees that if the balance is paid in 90 days, there will be no interest charge. You sign such a promissory note. Thirty days later, you receive a letter from a bank informing you that it has purchased the note from Honest John. If the bank charged Honest John a 12% ordinary discount rate,

    a) How much did Honest John receive?

    b) How much must you pay the bank?

    c) How much interest does the bank earn?

11. A trade-discounted invoice was in the amount of $20,000. The terms on the invoice were 3%/10 days, net/30 days.

    a) Determine the least amount to be borrowed in order to save the cash discount if the discount rate is 21%.

    b) What will be the actual cash discount that is realized?

12. Merchandise cost on an invoice is $50,400. The cash discount terms of the invoice are 3%/10 days, net/30 days. How much must be borrowed in order to save the cash discount if a discount rate of 18% is charged?

Problems 13–20 refer to Problems 1–8 respectively. Find the proceeds after *rediscounting* and the actual discount realized by the first discount process after the rediscount.

	Rediscount Rate	Time Before Maturity
13.	9%	1 year
14.	8%	9 months
15.	10%	6 months
16.	9%	2 months
17.	8%	30 days
18.	10%	30 days
19.	10%	20 days
20.	12%	10 days

21. The State Trust Company granted a loan of $40,000, due in 90 days at a discount rate of 10%. After 30 days had elapsed, it sold the note to the National Bank, which charges other banks a 6% discount rate.

a) What were the proceeds to the State Trust Company?

b) What was the net discount that State Trust earned on the original loan?

c) What was the discount earned by the National Bank?

In Problems 22–26 determine the equivalent discount rates for the following simple interest rates and one-year terms.

22. 7% **23.** 8% **24.** 9% **25.** 12% **26.** 18%

In Problems 27–31 determine the equivalent simple interest rates for the following discount rates and one-year terms.

27. 9% **28.** 10% **29.** 12% **30.** 15% **31.** 18%

The U.S. Treasury issues Treasury bills for terms of 91 days, 180 days, and 365 days. The discount rate, which is used, is based on a 360-day year. For each of the discount rates in Problems 32–36, determine the cost per $1,000 (which will be the proceeds to the government) and the equivalent ordinary simple interest rate for 180-day Treasury bills.

32. 6% **33.** 7% **34.** 8% **35.** 9% **36.** 10%

37. Derive Equations (1.3-5) and (1.3-6) from Equation (1.3-7).

CHAPTER 2

Practical Applications Using Simple Interest

2.1 EQUATIONS OF VALUE

EQUATED AMOUNT

Situations occur where businesses borrow money for relatively short periods of time with subsequent borrowings occurring prior to the repayment of an earlier debt. Situations also occur when a debt is about to mature, and the business will not have the funds to repay the debt. In order to maintain credit worthiness with the lending institution, the business may seek to replace some or all of the outstanding debts.

Consider the situation shown in Figure 2-1.

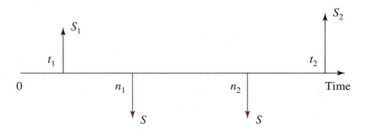

FIGURE 2-1 Consolidation with Known Payment Dates

As shown in the diagram, two debts, S_1 and S_2, are due at the respective times, t_1 and t_2. These are to be replaced by two new, equal debts, S, at times n_1 and n_2. If the interest rate is $r\%$ simple and we choose time 0 as a comparison date, then the present values of the new debts must be equal to the present values of the old debts. Mathematically,

$$\frac{S}{1+rn_1} + \frac{S}{1+rn_2} = \frac{S_1}{1+rt_1} + \frac{S_2}{1+rt_2},$$

and the only unknown is S. In general, for m old debts and k new debts, if time 0, "now," is chosen as the comparison date,

$$\sum_{j=1}^{k} \frac{S}{1 + rn_j} = \sum_{i=1}^{m} \frac{S_i}{1 + rt_i}. \qquad (2.1\text{-}1)$$

If we chose time t_2 as the comparison date, the future values of the new debts must equal the future values of the old debts. Mathematically,

$$S(1 + r[t_2 - n_1]) + S(1 + r[t_2 - n_2]) = S_1(1 + r[t_2 - t_1]) + S_2$$

and, again, the only unknown is S. In general, if the date of the last debt is used as the comparison date,

$$\sum_{j=1}^{k} S(1 + r[t_k - n_j]) = \sum_{i=1}^{m} S_i(1 + r[t_k - t_i]). \qquad (2.1\text{-}2)$$

Equations (2.1-1) and (2.1-2) represent the extreme points in time when a comparison date may be established. It is also possible to select a comparison date between time 0 and the date of the last debt. For debts prior to the comparison date, the future value in the forward direction to the comparison date would be used, and for debt dates subsequent to the comparison date, the present values to the comparison date would be used. In each of the above, it was assumed that one or more equal new debts would replace the old debts. The new debts need not be equal, but the relationship, such as a constant times S, between the first replacement debt and subsequent replacement debts must be established.

EXAMPLE 2.1.1   A loan of $1,000 is due in one month, and a loan of $5,000 is due in ten months. These loans are to be replaced by two loans of S each payable in three months and six months respectively. If the interest rate is 12% simple interest, find S for

(a) the comparison date at time 0, and

(b) the comparison date ten months from time 0.

Solution   A time diagram for this situation is shown below.

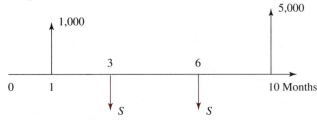

a) For time 0 as the comparison date, the present value of the new loans is set equal to the present value of the old loans. Then, Equation (2.1-1) gives

$$\frac{S}{1+\dfrac{12}{100}\left(\dfrac{3}{12}\right)}+\frac{S}{1+\dfrac{12}{100}\left(\dfrac{6}{12}\right)}=\frac{1{,}000}{1+\dfrac{12}{100}\left(\dfrac{1}{12}\right)}+\frac{5{,}000}{1+\dfrac{12}{100}\left(\dfrac{10}{12}\right)}.$$

Solving for S gives $S = 2{,}891.73$. Then \$2,891.73 payable in three months and six months is equivalent to the original loans at a simple interest rate of 12%.

b) For the comparison date in 10 months, Equation (2.1-2) gives

$$S\left(1+\frac{12}{100}\left[\frac{10-3}{12}\right]\right)+S\left(1+\frac{12}{100}\left[\frac{10-6}{12}\right]\right)$$
$$=1{,}000\left(1+\frac{12}{100}\left[\frac{10-1}{12}\right]\right)+5{,}000.$$

Solving for S gives \$2,886.26. ■

Note that the new loan amounts using the future as the comparison date are less than the new loan amounts using the present as the comparison date. Therefore, the lender would be wise to choose "now" as the comparison date. As a practical matter, "now" is the transaction date, and the lender is actually lending the borrower an amount of money sufficient to liquidate the old loans from its records "now." The amount of "new" money lent is the present value, on the transaction date, of the new loans. Then, for part (a), two interest-bearing notes could be created. The first would indicate a loan of \$2,807.50 plus 12% simple interest with a maturity value of \$2,891.73 due in three months. The second would indicate a loan of \$2,728.05 plus 12% simple interest with a maturity value of \$2,891.73 due in six months.

EXAMPLE 2.1.2 Assume today is April 15. A loan of \$5,000 was due on March 15, and a loan of \$10,000 is due on December 15. The borrower wishes to replace the past-due and future loans by a single loan payable on September 15. If the lender charges 12% simple interest and uses a 365-day year, what is the maturity value of the new loan?

Solution The time diagram for this situation is shown below. Since the loan structure is to be revised "today" on April 15, the \$5,000 loan has incurred additional interest from March 15 to April and its future value on April 15 will be needed.

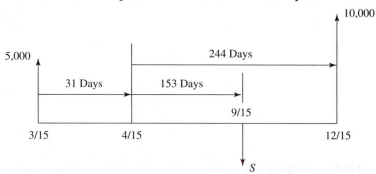

The numbers of days between the dates were determined using Table A-1. Using April 15 as the comparison, the $5,000 loan must be brought forward to April 15, and the present value of the $10,000 loan that is due on April 15 must be determined. Then, using a combination of Equations (2.1-1) and (2.1-2), we obtain

$$\frac{S}{1 + \frac{12}{100}\left[\frac{153}{365}\right]} = 5{,}000\left(1 + \frac{12}{100}\left[\frac{31}{365}\right]\right) + \frac{10{,}000}{1 + \frac{12}{100}\left[\frac{244}{365}\right]}.$$

Solving for S gives $S = \$15{,}028.06$. Substituting 15,028.06 into the left side of the equation above gives the present value of S on April 15 as 14,308.34. The borrower could sign a promissory note for $14,308.34 plus 12% simple interest with a maturity value of $15,208.06 due on September 15, or the borrower could sign a noninterest-bearing note for the maturity value due on September 15. Effectively, the lender charged 12% simple interest on the past-due amount of $5,000 and lent the borrower $5,050.96 to liquidate that loan, and the lender also lent the borrower $9,257.38 to liquidate the $10,000 loan. ■

The practical situation is the use of the date of the new consolidation transaction as the comparison date. Then, the present value of all new debts would be equal to the present value of all old debts. A more general statement would be that **the "current" value of all old debts is equal to the "current" value of all new debts on the transaction date, "now."** For any old debts that are past due, the current value would be the future amount from the due date to the comparison date.

Using the concept of bank discount, recall from Equation (1.3-3) that the present value (i.e., proceeds) is given by

$$P = S(1 - dt).$$

With this concept and using the transaction date "now" as the comparison date, the equation of value would be

$$\sum_{j=1}^{k} S(1 - dn_j) = \sum_{i=1}^{m} S_i(1 - dt_i). \tag{2.1-3}$$

where, k equals the number of new debts, n equals the time to the jth new debt, m equals the number of old debts, and t equals the time to the ith old debt, the same as in Equations (2.1-1) and (2.1-2).

EXAMPLE 2.1.3 Assume today is April 15. A loan of $5,000 was due on March 15, and a loan of $10,000 is due on December 15. The borrower wishes to replace the past-due and future loans with a single loan payable on September 15. If the lender charges a 12% simple discount rate and uses a 360-day year, what is the maturity value of the new loan on September 15?

Solution The time diagram for this situation is shown below.

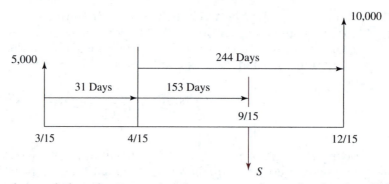

The numbers of days between the dates were determined using Table A-1. Since the $5,000 loan is past due by 31 days, its "maturity" value must be determined as of 4/15. Modification of the appropriate term of Equation (2.1-3) to determine S as

$$S = \frac{P}{1 - dt},$$

we get the following equation of value,

$$S\left(1 - \frac{12}{100}\left[\frac{153}{360}\right]\right) = \frac{5,000}{1 - \frac{12}{100}\left[\frac{31}{360}\right]} + 10,000\left(1 - \frac{12}{100}\left[\frac{244}{360}\right]\right).$$

This reduces to

$$0.949S = 5,052.20 + 9,186.67 = 14,238.87,$$

and solving for S gives

$$S = \$15,004.08. \quad \blacksquare$$

EXACT TIME

A business or an individual may desire to determine a date when a specified amount, such as the sum of all existing debts, could be used to liquidate all of the debts. Consider the time diagram in Figure 2-2.

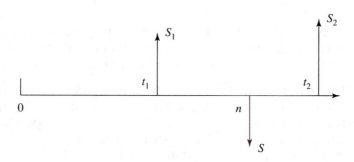

FIGURE 2-2 Consolidation with Unknown Payment Date

We are seeking the value of n for which

$$\frac{S}{1+rn} = \sum_{i=1}^{m} \frac{S_i}{1+rt_i}, \tag{2.1-4}$$

or the value of n for which

$$S(1-dn) = \sum_{i=1}^{m} S_i(1-dt_i). \tag{2.1-5}$$

EXAMPLE 2.1.4 A loan of \$1,000 is due in 1 month, and a loan of \$5,000 is due in 10 months. When will a single payment of \$6,000 liquidate the debts? The interest rate is 12% simple interest.

Solution Using Equation (2.1-4),

$$\frac{6,000}{1+\frac{12}{100}[n]} = \frac{1,000}{1+\frac{12}{100}\left[\frac{1}{12}\right]} + \frac{5,000}{1+\frac{12}{100}\left[\frac{10}{12}\right]},$$

from which it is determined that $n = 0.6992$ years, or 8.4 months. ∎

EQUATED TIME

Equation (2.1-4) gives the exact time when a specified amount would liquidate a specified number of debts. However, if an approximation of the date when a single payment will liquidate a series of debts is sufficient, the *method of equated time* may be used. This approach is the weighted-average time each debt is outstanding. Thus, from the time diagram above,

$$Sn = \sum_{i=1}^{m} S_i t_i. \tag{2.1-6}$$

The value of n that is determined by Equation (2.1-6) will always be greater than the exact value of n. Using Equation (2.1-6) on Example 2.1.4 gives

$$6,000n = 1,000(1) + 5,000(10) = 51,000$$

from which n is determined to be 8.5 months.

One of the reasons that equations of value are used extensively in business is because cash flows may not always be timely for the liquidation of debts. They afford a way to restructure existing debt obligations in order to maintain credit worthiness.

PROBLEM SET 2.1

In the following problems, if time is given in months, use the number of months divided by 12 for the fraction of a year. If dates are given in days, use the number of days divided by 365 for the fraction of the year when the interest rate is given as a simple interest rate, and the number of days divided by 360 when the interest rate is given as a simple discount rate. Use Table A-1 in order to determine the number of the day of the year, and assume that the year "yy" is a leap year, and the year "yz" is the following year.

1. A loan of $2,000 is due in five months, and a loan of $6,000 is due in ten months. The simple interest rate is 10%. What single new loan transacted today, principal and maturity value, will replace these two loans by a single loan due in seven months?

2. A loan of $1,000 is due in one month, and a loan of $5,000 is due in nine months. The simple interest rate is 12%. What single new loan transacted today, principal and maturity value, will replace these loans in five months?

3. A loan of $2,000 is due in five months, and a loan of $6,000 is due in 10 months. The simple discount rate is 10%. What single new loan transacted today, principal and maturity value, will replace these two loans by a single loan due in seven months?

4. A loan of $1,000 is due in one month, and a loan of $5,000 is due in nine months. The simple discount rate is 12%. What single new loan transacted today, principal and maturity value, will replace these loans in five months?

5. A loan of $2,000 is due in two months, and a loan of $8,000 is due in 12 months. These loans are to be repaid by two equal payments due in five months and nine months respectively. If the simple interest rate is 15%, what will be the size of each payment?

6. A loan of $13,000 is due in two months, and a loan of $20,000 is due in 12 months. These loans are to be repaid by two equal payments due in five months and nine months respectively. If the simple interest rate is 12%, what will be the size of each payment?

7. A loan of $13,000 is due in two months, and a loan of $20,000 is due in 12 months. These loans are to be repaid by two equal payments due in five months and nine months respectively. If the simple discount rate is 15%, what will be the size of each payment?

8. A loan of $12,000 is due in two months, and a loan of $18,000 is due in 12 months. These loans are to be repaid by two equal payments due in five months and nine months respectively. If the simple discount rate is 12%, what will be the size of each payment?

9. A loan of $2,000 is due on 8/12/yy, and a loan of $6,000 on 2/18/yz. If the simple interest rate is 10%, what single new loan, principal and maturity value, transacted on 3/15/yy and due on 12/10/yy will replace these two loans?

10. A loan of $1,000 is due on 4/30/yy, and a loan of $5,000 is due on 12/26/yy. The simple interest rate is 12%. What single new loan, principal and maturity value, transacted on 3/31/yy and due on 9/27/yy will replace these two loans?

11. A loan of $2,000 is due on 8/12/yy, and a loan of $8,000 is due on 2/28/yz. The simple discount rate is 10%. What single new loan, principal and maturity value, transacted on 3/15/yy and due on 12/10/yy will replace these two loans?

12. A loan of $4,000 is due on 4/30/yy, and a loan of $9,000 is due on 12/26/yy. The simple discount rate is 12%. What single new loan, principal and maturity value, transacted on 3/31/yy and due on 9/27/yy will replace these two loans?

13. A loan of $12,000 is due on 5/31/yy, and a loan of $18,000 is due on 4/1/yz. These loans are to be repaid by two equal payments due on 9/28/yy and 12/27/yy respectively. If the simple interest rate is 15%, what will be the size of each payment? Comparison date: 4/1/yy.

14. A loan of $12,000 is due on 6/13/yy, and a loan of $18,000 is due on 4/14/yz. These loans are to be repaid by two equal payments due on 10/11/yy and 1/9/yz respectively. If the simple interest rate is 12%, what will be the size of each payment? Comparison date: 4/14/yy.

15. A loan of $12,000 is due on 5/31/yy, and a loan of $20,000 is due on 3/27/yz. These loans are to be repaid by two equal payments due on 8/29/yy and 12/27/yy respectively. If the simple discount rate is 15%, what will be the size of each payment? Comparison date: 4/1/yy.

16. A loan of $12,000 is due on 6/13/yy, and a loan of $20,000 is due on 4/9/yz. These loans are to be repaid by two equal payments due on 9/11/yy and 1/9/yz respectively. If the simple discount rate is 12%, what will be the size of each payment? Comparison date: 4/14/yy.

17. A loan of $2,000 is due in five months, and a loan of $6,000 is due in ten months. If the simple interest rate is 10%, when will a single payment of $8,000 replace these two loans?

18. A loan of $1,000 is due in one month, and a loan of $5,000 is due in nine months. If the simple interest rate is 12%, when will a single payment of $6,000 replace these loans?

19. A loan of $2,000 is due in five months, and a loan of $6,000 is due in 10 months. If the simple interest rate is 10%, when will a single payment of $8,000 replace these two loans using the method of equated time?

20. A loan of $1,000 is due in one month, and a loan of $5,000 is due in nine months. If the simple interest rate is 12%, when will a single payment of $6,000 replace these loans using the method of equated time?

21. A loan of $12,000 is due on 3/15/yy, and a loan of $20,000 is due on 1/9/yz. If the simple discount rate is 15%, on what date will a single payment of $32,000 replace these loans using

   a) exact time and

   b) the method of equated time?

   Comparison date: 1/10/yy.

22. A loan of $12,000 is due on 3/15/yy, and a loan of $20,000 is due on 1/9/yz. If the simple discount rate is 12%, on what date will a single payment of $32,000 replace these loans using

a) exact time and

b) the method of equated time?

Comparison date: 1/10/yy.

23. A loan of $5,000 is due on 4/30/yy, a loan of $10,000 is due on 5/30/yy, and a loan of $25,000 is due on 10/27/yy. The simple discount rate is 15%. The borrower proposes to repay these loans with two payments. The first payment is due on 5/30/yy. The second payment, which is to be twice the first payment, is due on 8/28/yy. What is the size of each payment if the transaction date is 3/1/yy?

24. A loan of $5,000 is due in two months, a loan of $10,000 is due in three months, and a loan of $20,000 is due in nine months. The simple interest rate is 15%. The borrower proposes to repay these loans with two payments. The first payment is due in three months. The second payment, which is to be twice the first payment, is due in six months. What is the size of each payment?

2.2 ADD-ON INTEREST AND APPROXIMATE TRUE INTEREST RATE

ADD-ON INTEREST

There are certain methods of financing purchases where the future amount of an initial balance is divided by the number of payments in order to determine the equal periodic payment. This type of financing is called **add-on interest** because the cost for financing a purchase is added to the initial balance to obtain the future amount. Lending institutions and merchants who, themselves, finance customers' purchases use add-on interest for such items as furniture and automobiles. Because the balance due on a loan is "due" only until a payment is received, and because the interest charge is determined as if the entire loan was payable at one time in the future, the quoted interest rate in the past was not the true interest rate. The Truth-in-Lending Act of 1973 specified that all loan transactions required a disclosure statement to be given to the borrower showing the true annual interest rate, the annual percentage rate—APR, within $\frac{1}{4}$% of the exact value.

Using the concepts of simple interest, the interest is

$$I = B_o rt, \tag{2.2-1}$$

where B_o is the initial balance (amount to be financed). The future amount is

$$S = B_o + I. \tag{2.2-2}$$

Usually, add-on interest-type financing requires monthly payments. Therefore, the number of payments (N) is

$$N = 12t, \tag{2.2-3}$$

and the monthly payment (P) is

$$P = \frac{S}{N}. \tag{2.2-4}$$

The total of "true" interest, handling, postage, and all other fees is often called the **finance charge**.

EXAMPLE 2.2.1   A young, married couple wished to purchase furniture that cost $3,000. John, of "Honest John's Furniture," told them that because they were just getting started, he would accept $600 as a down payment and finance the remainder at "only 5% for three years."

(a) What is the interest on the financing?

(b) What is the monthly payment?

(c) What is the total cost of the furniture?

Solution   The amount to be financed is the cost of the furniture less the down payment. Then

$$B_o = 3,000 - 600 = 2,400.$$

a) From Equation (2.2-1),

$$I = B_o rt$$

$$= (2,400)(0.05)(3) = 360.$$

b) From Equation (2.2-2),

$$S = B_o + I$$

$$= 2,400 + 360 = 2,760,$$

or $2,760. For 3-year financing, the number of monthly payments is, from Equation (2.2-3)

$$N = 12t$$

$$= (12)(3) = 36,$$

and from Equation (2.2-4),

$$P = \frac{2,760}{36} = 76.67.$$

Thus, the monthly payment is $76.67.

c) The total cost of the furniture is the sum of the down payment and S. Then,

$$\text{Total Cost} = 600 + 2,760 = \$3,360,$$

or, equivalently, it is the sum of the original cost plus the interest. That is,

$$\text{Total Cost} = 3,000 + 360 = \$3,360. \quad \blacksquare$$

Later in this section it will be shown that because the balance of the loan decreases after each payment is made, the true annual interest rate that is being charged is approximately 9.7%—*not the 5% that was quoted.*

EXAMPLE 2.2.2   Suppose the young couple of Example 2.2.1 were told that they could finance the furniture for four years at a "service charge" of $4 per week. What will be the finance charge rate they are paying?

Solution   At $4 per week, the yearly finance charge is $208 and for four years equals $832. Then, from Equation (2.2-1)

$$I = B_o rt,$$

or

$$832 = (2,400)(4)r = 9,600r.$$

Dividing both sides by 9,600 gives 0.0867 or 8.67%. The simple interest rate that is being applied to the entire $2,400 is 8.67%; however, since the balance due would be decreasing with each payment, the true annual finance charge rate would be approximately 17% (as will be shown later in this section). ■

THE RULE OF 78

When add-on interest is used in financing, the monthly payment is part interest and part reduction of the balance due. However, the interest portion is not spread uniformly over the term of the loan. That is, if there are to be 36 payments, the interest portion of each payment is not 1-36th of the total interest. The method that is often used for apportioning the interest is called the **Rule of 78**.

There are twelve months in a year, and the sum of digits that identify the months is 78 ($= 1 + 2 + 3 + \cdots + 11 + 12$). For a loan of one year, there would be twelve payments and the interest portion of the first payment would be 12-78^{ths} of the total interest. The second payment would include 11-78^{ths} of the total interest and so forth until the twelfth payment would include 1-78th of the interest. Similarly, to finance a loan for two years, the first payment would be 24-300^{ths} of the total interest because the sum of the digits from 1 through 24 (2 years $= 24$ months) is 300. The sum-of-the-years' digits, S_d, may be determined by

$$S_d = \frac{N(N+1)}{2},$$

where, $N =$ the total number of payments. Thus, for 1 year

$$S_d = \frac{(12)(13)}{2} = 78,$$

and for 2 years

$$S_d = \frac{(24)(25)}{2} = 300.$$

Suppose that after n payments the borrower desires to liquidate the remaining balance. How much of the total interest will the borrower have paid? In order to determine the single payment, it will be easier to analyze the problem by determining how much does NOT have to be paid. Figure 2-3 shows the situation, where B_n is the balance after n payments according to the Rule of 78.

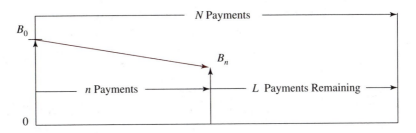

FIGURE 2-3 Rule of 78 Repayments

It may be seen from Figure 2-3 that the number of remaining payments after making n payments is

$$L = N - n \tag{2.2-5}$$

and the interest for the sum-of-the-years' digits that does NOT have to be paid is

$$S_L = \frac{L(L+1)}{2}. \tag{2.2-6}$$

Defining a **rebate factor**, R_F as the fraction of the interest that does NOT have to be paid,

$$R_F = \frac{L(L+1)}{N(N+1)}, \tag{2.2-7}$$

the liquidation payment, P_f, is the total of the remaining payments (PL) less the interest rebate $(R_F I)$,

$$P_f = PL - R_F I. \tag{2.2-8}$$

Any final liquidation of a debt will include the original debt plus the interest to the date of liquidation. Therefore, a **rebate** is interest that does not have to be paid.

EXAMPLE 2.2.3   Suppose the interest on a note, using add-on interest, is $200 for 2-year financing. After 10 payments have been made, the debt is liquidated in a single payment. What is the rebate on the interest?

Solution   For two years, $N = 24$, and after 10 payments, $L = 14$. Then, from Equation (2.2-7),

$$R_F = \frac{L(L+1)}{N(N+1)}$$

$$= \frac{(14)(15)}{(24)(25)} = 0.35,$$

or 35%. The interest rebate $= (0.35)(200) = \$70$. Notice that when the interest rebate is 35% of the interest at time n, 65% of the interest has already been paid in less than half the term (N). For terms of 2, 3, and 4 years, we expect 50% of the interest will have been paid by approximately 30% of the term, 75% of the interest will have been paid by approximately 39% of the term, and 90% of the interest will have been paid by approximately 53% of the term, respectively. Therefore, in the case of $N = 4$ years, it might not be wise, financially, to liquidate a debt if half the term has passed since 90% of the interest will have already been paid. ∎

EXAMPLE 2.2.4 An automobile is financed by a $4,000 loan for four years at an add-on interest rate of 9%.

(a) What is the monthly payment?

(b) If the loan is liquidated after the 36th payment, what liquidation payment is necessary?

Solution

a) The simple interest on $4,000 for four years at 9% is given by Equation (2.2-1) as

$$I = B_o rt$$

$$= (4,000)(0.09)(4)$$

$$= \$1,440.$$

The total of payments is given by Equation (2.2-2) as

$$S = B_o + I$$

$$= 4,000 + 1,440$$

$$= \$5,440.$$

From Equation (2.2-3),

$$N = 12t$$

$$= (12)(4) = 48,$$

and from Equation (2.2-4),

$$P = \frac{S}{N}$$

$$= \frac{5,440}{48} = 113.33.$$

Thus, the monthly payment is $113.33.

b) After 36 payments, the number of payments remaining is

$$L = 48 - 36 = 12.$$

From Equation (2.2-7), the rebate factor is

$$R_F = \frac{L(L+1)}{N(N+1)}.$$

$$= \frac{(12)(13)}{(48)(49)} = 0.0663265,$$

or, 6.63265%. Then, the amount of interest rebate is $(0.063265)(1,440) = \$95.51$. The final liquidation payment can be determined from Equation (2.2-8)

$$P_f = 113.33(12) - 95.51$$

$$= \$1,264.45. \quad ∎$$

This is a rather hefty amount to pay in order to save less than $96. If the liquidation amount were placed in an investment at 5.25% for the remaining year, the interest would amount to $66. Then, the net cost NOT to liquidate would be $33. When the last payment is made, the liquidation payment would "be in the bank."

TRUE INTEREST RATE

The simple interest rate that is used to compute the add-on interest is not a true rate. This is because the rate is applied to the full term of the loan; however, as payments are made, the amount that is owed for the remainder of the term decreases. For the first part of the term, the amount owed is greater than half the original loan, but for the latter part of the term, the amount owed is less than half the original loan. Letting:

I be the add-on interest,
f be the number of payments per year, and
r_t be the approximate true interest rate,

$$r_t \cong \frac{2fI}{B_o(N+1)},$$ (2.2-9)

and in terms of the quoted simple interest rate, r,

$$r_t \cong 2r\left(\frac{N}{N+1}\right).$$ (2.2-10)

As N gets large, $\frac{N}{N+1}$ approaches 1 and r_t approaches $2r$. For example, if $N = 24$, $r_t \cong 1.92r$; if $N = 36$, $r_t \cong 1.95r$; and if $N = 48$, $r_t \cong 1.96r$. For add-on interest, a rule of thumb is that the true interest rate will be approximately twice the quoted rate. The Truth-in-Lending Act requires that the true interest rate, to the nearest $\frac{1}{4}$%, be supplied to the consumer. However, it may be included in the "small print" so a borrower should ask and read before signing any finance agreement. An iterative process (i.e., trial and error) is necessary in order to determine the true annual interest rate and is deferred to Chapter 4.

PROBLEM SET 2.2

For Problems 1–5, determine the monthly payments where add-on interest is used.

	Loan	Interest Rate	Term
1.	$5,000	9%	5 years
2.	$3,500	10%	4 years
3.	$6,000	8%	6 years
4.	$2,500	12%	3 years
5.	$3,000	11%	2 years

6. The list price of a used automobile is $5,995. A trade-in allowance of $2,300 is made for the present car. If the new automobile is to be financed for four years at 10% interest:

a) What will be the monthly payment?

b) What will be the total cost?

7. The list price of an automobile is $7,245. The automobile is sold at a 15% discount from the list to a person who will finance the remainder for four years at 12% interest.

a) What will be the monthly payment?

b) What will be the total cost?

In Problems 8–12, determine the rebate factor after the indicated monthly payment number.

Loan	Interest Rate	Term	Payment No.
8. $5,000	9%	5 years	35
9. $3,500	10%	4 years	30
10. $6,000	8%	6 years	65
11. $2,500	12%	3 years	30
12. $3,000	11%	2 years	5

In Problems 13–17, determine the necessary liquidation payments after the indicated payment number.

Loan	Interest Rate	Term	Payment No.
13. $5,000	9%	5 years	35
14. $3,500	10%	4 years	30
15. $6,000	8%	6 years	65
16. $2,500	12%	3 years	30
17. $3,000	11%	2 years	5

18–22 Determine the approximate true annual interest rates for Problems 1–5 respectively.

23. A color television set is priced at $550. A $50 down payment is required, and the remainder may be financed for one year with a service charge of $70.

a) What will be the monthly payment?

b) What will be the total cost?

c) What liquidation payment is necessary after eight months?

d) Instead of quoting a service charge of $70, what apparent interest rate would have been charged the seller?

e) What is the approximate true interest rate?

2.3 SHORT-TERM INVESTMENT RATE OF INTEREST

The natures of the problems we have encountered to this point have been such that the interest rate or discount rate has been known. However, suppose we know the beginning amount of an investment portfolio, cash flows into and out of the portfolio during an interest measurement period, and the ending balance at the end of the interest measurement period. The question we need answered is, what was the interest rate that was earned by the portfolio? We discuss two methods in this section: dollar-weighted and time-weighted rates of interest.

DOLLAR-WEIGHTED RATE OF INTEREST

Consider the investment activity shown in Figure 2-4 where one contribution, C, and one withdrawal, W, are the activities for an investment fund.

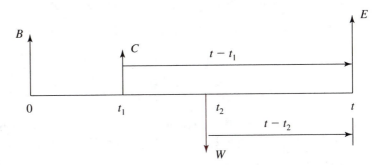

FIGURE 2-4 Dollar-Weighting, Short-Term Investment Activity

In Figure 2-4, let:

B be the beginning balance of the fund,

C be the contribution at time t_1,

W be the withdrawal at time t_2,

E be the ending balance of the fund,

t be the time from the beginning balance date to the ending balance date, and

r be the simple interest rate.

The concept of **dollar weighting** is that each activity contributes to or deducts from the ending value of the investment rate of interest. For short-term investments with maturity dates less than or equal to one year, the investment rate of interest may be determined by simple interest concepts. As such, the ending value would be determined by

$$E = B(1 + rt) + C(1 + r[t - t_1]) - W(1 + r[t - t_2]),$$

which may be solved for r as

$$r = \frac{E - (B + C - W)}{Bt + C(t - t_1) - W(t - t_2)}. \qquad (2.3\text{-}1)$$

EXAMPLE 2.3.1   An investment fund had a beginning balance of $100,000. At the end of three months, $25,000 was added to the fund, and at the end of nine months, $30,000 was withdrawn. What was the simple interest rate if at the end of 12 months the ending balance was $120,000?

Solution   For Equation (2.3-1), the values are as follows: $B = 100,000$, $C = 25,000$, $W = 30,000$, and $E = 120,000$. The value of t is 1, since the total duration is one year. Then,

$$r = \frac{120,000 - (100,000 + 25,000 - 30,000)}{100,000(1) + 25,000\left(1 - \dfrac{3}{12}\right) - 30,000\left(1 - \dfrac{9}{12}\right)},$$

which can be solved for r as 22.4% for the year. ■

Equation (2.3-1) can be generalized to allow for several contributions and withdrawals as

$$r = \frac{E - B - \sum F_i}{Bt + \sum F_i(t - t_i)}, \qquad (2.3\text{-}2)$$

where F_i represents the respective cash flow. For a contribution, the sign of F_i is positive, and for a withdrawal the sign of F_i is negative.

TIME-WEIGHTED RATE OF INTEREST

The concept of **time weighting** is to determine a simple interest rate between successive periods and link those rates over the entire term. Consider the investment activity in Figure 2-5. For precise time weighting, the investment fund balance is determined on the date of a cash flow into or out of the fund. Thus the B's, C's, and W's occur at the same time.

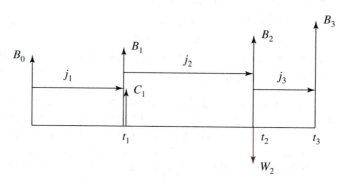

FIGURE 2-5 Time-Weighting Short-Term Investment Activity

From the fundamental definition of simple interest in Chapter 1, where the future amount was given as

$$S = P(1 + rt),$$

we have

$$B_1 = (B_0 + C_0)(1 + j_1),$$

from which,

$$1 + j_1 = \frac{B_1}{B_0 + C_0}.$$

For the second period

$$B_2 = (B_1 + C_1)(1 + j_2),$$

from which

$$1 + j_2 = \frac{B_2}{B_1 + C_1}.$$

Similarly,

$$1 + j_3 = \frac{B_3}{B_2 - W_2}.$$

Remember, in this context W_2 is subtracted since it represents a withdrawal. The simple interest rate for the three periods would be determined by

$$r = (1 + j_1)(1 + j_2)(1 + j_3) - 1.$$

EXAMPLE 2.3.2   The beginning balance of an investment fund was $100,000. At the end of three months it had increased to $105,000 at which time $25,000 was added to the fund. At the end of nine months, it had increased to $143,000 at which time $30,000 was removed. At the end of one year, the fund had a balance of $120,000. What was the time-weighted rate of interest?

Solution   From the given information,

$$1 + j_1 = \frac{105,000}{100,000 + 0} = 1.05,$$

$$1 + j_2 = \frac{143,000}{105,000 + 25,000} = 1.10,$$

and

$$1 + j_3 = \frac{120,000}{143,000 - 30,000} = 1.062.$$

Then, the time-weighted rate of interest is obtained by linking the accumulations, $(1 + j)$, as

$$r = (1.05)(1.10)(1.062) - 1 = 0.227,$$

and the rate of interest is 22.7%. ■

The foregoing may be generalized to

$$r = \left[\prod_{k=1}^{n} (1 + j_k)\right] - 1,\tag{2.3-3}$$

where the symbol $\prod$ is read as "the product of." It should be noted that time was not a factor in determining the rate because the simple interest rate, j, is determined for one period and balances are determined at the beginning and end of each period. For managers of large amounts of money, it may not be feasible to determine the balance of a fund whenever money flows into or out of the fund. Balances are determined at fixed intervals, such as monthly or quarterly. The practice is to find the dollar-weighted rate during the period and then link the rates as indicated in Equation (2.3-3). Further, it could be quite tedious to use the exact day of the period when the money flowed into or out of the fund. The standard that is used is that of the Association for Investment Management and Research, which is to assume all cash flows occur at the midpoint of the measurement period. With this standard, the simple interest for the kth period is given by

$$1 + j_k = \frac{B_k}{B_{k-1} + 0.5 F_{k-1}},\tag{2.3-4}$$

and the time-weighted rate of interest is determined by Equation (2.3-3). The caution that must be exercised in using the midpoint standard is that the cash flow should be less than approximately 5% of the balance, and the period should be no longer than three months. When this standard cannot be met, it would be wise to use the dollar-weighted rate of interest in the period as

$$1 + j_k = \frac{B_k}{B_{k-1} + F_{k-1}(t - t_k)},\tag{2.3-5}$$

where the sign of F_{k-1} is positive for contributions and negative for withdrawals at the subinterval $k - 1$.

EXAMPLE 2.3.3   The beginning balance of an investment fund is $100,000. The balance at the end of month one was $108,000. The balance at the end of month two was $116,000, and the balance at the end of month three was $128,000. $5,000 was added between months one and two, and $5,000 was added between months two and three. Using the midmonth standard, determine the time-weighted rate of interest.

Solution   From Equation (2.3-4)

$$1 + j_1 = \frac{108,000 - 100,000}{100,000 + 0} = 1.08,$$

$$1 + j_2 = \frac{116,000}{108,000 + 0.5(5,000)} = 1.05,$$

$$1 + j_3 = \frac{128,000}{116,000 + 0.5(5,000)} = 1.08,$$

and from Equation (2.3-3)
$$r = (1.08)(1.05)(1.08) - 1 = 0.2247.$$

Thus, the time-weighted interest rate is 22.47%. ∎

EXAMPLE 2.3.4 The beginning balance of an investment fund is $100,000. At the end of three months the fund had a value of $108,000 just prior to the addition of $25,000. At the end of nine months the fund had a value of $146,000 just prior to the withdrawal of $30,000. Using dollar weighting between fund values, determine the time-weighted rate of interest if the fund value at the end of 12 months is $120,000.

Solution From Equation (2.3-5),

$$j_1 = \frac{108,000 - 100,000}{100,000 + 0} = 0.08,$$

$$j_2 = \frac{146,000 - 108,000 - 25,000}{108,000 + 25,000 \left[\dfrac{9}{12} - \dfrac{3}{12} \right]} = 0.108,$$

$$j_3 = \frac{120,000 - 146,000 + 30,000}{146,000 - 30,000 \left[\dfrac{12}{12} - \dfrac{9}{12} \right]} = 0.029,$$

and from Equation (2.3-3),
$$r = (1.08)(1.108)(1.029) - 1 = 0.2313,$$

or 23.13%. This compares favorably with the result shown in Example 2.3.1. ∎

In dollar weighting, the duration from the cash flow date to the end of the measurement period can have an effect on the actual performance of a fund. However, it does give an accurate measure of the performance of the fund in that intermediate values of a fund have no effect on the overall performance. In time weighting, the beginning value of a subinterval is established on the date of a cash flow. Then the investment performance depends simply upon the value of the fund at the next cash flow date.

In general, dollar weighting provides a valid measure of the actual investment results, and time weighting provides an indicator of the fundamental investment performance. In practice, both methods usually give comparable results. Problems 7 and 8 in Problem Set 2.3 show the effect of "pure" time weighting.

PROBLEM SET 2.3

Note: If time is given in months, use 12 months for a year. If dates are given, the year "yy" is simply the year prior to "yz." The year "yz" is a 365-day year. Use Table A-1 to determine the number of the day of the year.

1. A fund had a beginning balance of $300,000. Contributions of $30,000 are made at the end of three and six months. A withdrawal of $50,000 occurs at the end of nine months. The fund value at the end of 12 months is $340,000. Determine the dollar-weighted rate of interest.

2. The balance of a fund on 12/31/yy was $200,000. Contributions of $10,000 were made on 3/31/yz, 6/30/yz, 9/30/yz, and 12/31/yz. The fund balance on 12/31/yz is $290,000. Determine the dollar-weighted rate of interest.

3. The beginning value of a fund is $500,000. Withdrawals of $50,000 are made at the end of every three months for one year. At the end of the year, the fund has a value of $350,000. Determine the dollar-weighted rate of interest.

4. The value of a fund was $500,000 on 12/31/yy. Withdrawals of $50,000 were made on 3/31/yz, 6/30/yz, 9/30/yz, and 12/31/yz. The fund has a value of $250,000 on 12/31/yz. Determine the dollar-weighted rate of interest.

5. The beginning balance of a fund is $400,000. A contribution of $30,000 and a withdrawal of $50,000 occur at the end of six months. The value of the fund at the end of one year is $475,000. Determine the dollar-weighted rate of interest.

6. The value of a fund was $100,000 on 12/31/yy. A contribution of $30,000 was on 3/31/yz, 6/30/yz, and 9/30/yz. A withdrawal of $50,000 was made on 10/31/yz. Determine the dollar-weighted rate of interest if the ending balance is $160,000 on 12/31/yz.

7. A fund has a beginning balance of $300,000. At the end of three months, the value of the fund had increased to $305,000 at which time a contribution of $30,000 is made. At the end of six months, the fund had increased in value to $340,000 at which point a contribution of $30,000 is made. At the end of nine months the fund has decreased in value to $330,000 at which time a withdrawal of $50,000 occurs. The fund value at the end of twelve months is $340,000.

    a) Determine the time-weighted rate of interest.

    b) Compare the rate of interest of part (a) with the rate of interest determined in Problem 1, above. What can you conclude about the two methods in reporting the investment rate?

8. A fund has a beginning balance of $200,000. At the end of three months, the value of the fund has increased to $205,000 at which time a contribution of $10,000 is made. At the end of six months, the fund had increased in value to $220,000 at which time a $10,000 contribution is made. At the end of nine months the fund had increased in value to $245,000 at which time a withdrawal of $50,000 occurs. The fund value at the end of 12 months is $215,000.

    a) Determine the dollar-weighted rate of interest.

    b) Determine the time-weighted rate of interest.

    c) What conclusion can you reach concerning the two methods in reporting the investment interest rate?

CHAPTER

3

Compound Interest

Suppose $1,000 is deposited in a savings account, which pays 6% simple interest. At the end of one year, the amount in the account would be $1,060. Two options are then available with respect to the interest.

1. The $60 interest may be withdrawn leaving the original principal in the account to earn another $60 interest for the second year. Repeating the process for two years, the total amount would be the original principal of $1,000 plus two interest withdrawals of $60 for a sum of $1,120.

2. All or some part of the $60 interest may be left in the account. If all of the interest is kept in the account, then the principal is increased to $1,060 for the second year. This higher principal will earn $63.60 interest for the second year giving a total amount of $1,123.60 at the end of two years. It should be readily apparent that the first $60 interest earned $3.60 interest (6% of 60) in the second year.

The process where interest earns interest is called **compound interest**, and there are two types of compound interest—discrete and continuous. This process is illustrated in Figure 3-1.

1. **Discrete compounding** is the computation of interest at *fixed intervals* during each year with the addition of that interest to successive principals at the ends of those fixed intervals.

2. **Continuous compounding** is the computation of interest on a *continuing basis* during the year and the addition of that interest to successive principals at fixed intervals during the year.

This chapter discusses both types of compounding.

Note in Figure 3-1 that each successive "Interest" is larger than the preceding "Interest." This is because the interest earned by the "new" principal includes the previous interest—the interest is compounded. The more frequently the interest is compounded, the faster the value will grow. That is, as the interval

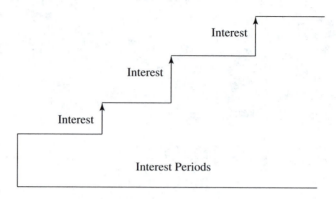

FIGURE 3-1 The Effect of Compound Interest

between the additions of interest is shortened, the "Interest" steps occur more frequently, and the growth follows an exponential line.

3.1 COMPOUNDING FREQUENCY AND PERIODIC INTEREST RATE

DISCRETE COMPOUNDING

In discrete compounding, interest is added to the previous balance at the end of an interest computation period. The fixed interval for computing interest is at the discretion of the institution that pays the interest or lends the money, which earns interest. The intervals may be fixed as shown in Table 3-1. It should be apparent that the frequency of compounding is determined by dividing the length of the year by the fixed interval. The number of times each year that interest is compounded is referred to as the **compounding frequency**.

TABLE 3-1 DISCRETE COMPOUNDING OF INTEREST

Time Interval	Periodic Interest Rate	Compounding Frequency
12 months	Annual rate	Once each year
6 months	Semi-annual rate	Twice each year
3 months	Quarterly rate	4 times each year
1 month	Monthly rate	12 times each year
1 day	Daily rate	365 times each year

During the compounding interval, the principal earns simple interest that is based on a simple interest rate, which is proportional to the corresponding annual interest rate. For example, for an annual interest rate of 6%, if the interest is compounded semi-annually, a simple interest rate of 3% is applied to each six-month period. However, at the end of the first six months, the principal will have increased by 3%, and the increased principal earns 3% for the second six months. The 3% here would be referred to as the periodic rate since it is applied for a period of time.

In general, letting:

i be the periodic interest rate (%),

r be the annual (nominal) interest rate (%), and

f be the compounding frequency per year,

the periodic rate is determined by

$$i = \frac{r}{f}. \tag{3.1-1}$$

As stated above, 6% compounded semi-annually yields a periodic rate of 3% applied for two periods each year. Then, if it were applied for three years, there would be six periods (3 years × 2 periods per year). In general, letting:

n be the total number of periods and

t be the number of years, then

$$n = ft. \tag{3.1-2}$$

EXAMPLE 3.1.1   Determine the compounding frequency, the periodic interest rate, and the number of periods for 6% compounded every four months for 10 years.

Solution   Compounding every four months gives a frequency of $\frac{12}{4}$ or $f = 3$. Then, from Equations (3.1-1) and (3.1-2),

$$i = \frac{6\%}{3} = 2\% \text{ per period,}$$

and

$$n = 3(10) = 30 \text{ periods.} \quad \blacksquare$$

CONTINUOUS COMPOUNDING

In continuous compounding, interest is computed and added to the previous balance on a continuing basis. The actual addition of the interest occurs at discrete intervals. This method will be discussed in the next section.

PROBLEM SET 3.1

For the following problems, determine:
a) The compounding frequency. **b)** The periodic interest rate. **c)** The total number of periods.

	Annual Interest Rate	Compounding Interval	Number of Years
1.	8%	Semi-annually	6
2.	8%	Quarterly	6
3.	8%	Monthly	6
4.	8%	Daily	6
5.	10%	Quarterly	7
6.	12%	Monthly	5
7.	9%	Monthly	20
8.	11%	Quarterly	10
9.	6%	Quarterly	5
10.	18%	Monthly	3
11.	10%	Semi-annually	4
12.	5%	Monthly	1
13.	5.5%	Monthly	2
14.	6.75%	Semi-annually	3
15.	7.00%	Quarterly	8
16.	7.375%	Monthly	20
17.	8.50%	Quarterly	15
18.	9.00%	Semi-annually	12.5

3.2 FUTURE AND PRESENT VALUES

FUTURE VALUE

Suppose a principal of $1,000 is invested at an annual rate of 6% for one year. Then, the sum, S, of principal and interest at the end of one year is

$$S_1 = 1,000 + 1,000(0.06)$$
$$= 1,000(1 + 0.06) = \$1,060,$$

where the subscript 1 refers to the first year. If the interest is left on deposit for a second year at the same interest rate, S_1 becomes the principal for the second year, and the sum of the new principal and interest becomes

$$S_2 = 1,060(1 + 0.06) = \$1,123.60.$$

Substituting S_1 in the form $1,000(1 + 0.06)$ gives

$$S_2 = 1,000(1 + 0.06)(1 + 0.06)$$
$$= 1,000(1 + 0.06)^2.$$

Repeating the process for a third year, with S_2 as the new principal, gives

$$S_3 = 1,123.60(1 + 0.06) = \$1,191.02.$$

The substituting of S_2, $1,000(1 + 0.06)^2$, directly gives

$$S_3 = 1,000(1 + 0.06)^2(1 + 0.06)$$
$$= 1,000(1 + 0.06)^3.$$

The subscript on S, which indicates the number of years, is the same as the exponent on $(1 + 0.06)$ in each of the equations. Then, if interest is left on deposit for n periods at $i\%$ per period, the sum of the original principal and the accumulated interest can be directly determined by

$$S_n = P(1 + i)^n. \tag{3.2-1}$$

This sum of the original principal and accumulated interest is called the **future value** of P. If i is kept constant and n is varied, Equation (3.2-1) is an exponential function, and if n is kept constant and i is varied, Equation (3.2-1) is a power function. In either case, Figure 3-2 is an illustration of the time value of money with compound interest and the conditions of Equation (3.2-1). To facilitate the calculation of compound interest, the use of either tables or a calculator with a y^x function is recommended. Calculators that have this function will also have functions for $\log(x)$, $\mathrm{Ln}(x)$, and e^x, all of which are useful in calculations of compound interest. This book is directed toward solutions using a calculator that includes a y^x function—a short table of $(1 + i)^n$ is supplied in the Appendix as Table A-2.

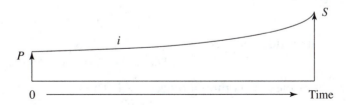

FIGURE 3-2 The Time Value of Money

EXAMPLE 3.2.1   What is the future value of $1,000 for three years at 6% compounded a) annually, b) semi-annually, c) quarterly, and d) monthly?

Solution

a) For annual compounding, $f = 1$ and, from Equations (3.1-1) and (3.1-2) i and n are determined as

$$i = \frac{r}{f} = \frac{6\%}{1} = 6\%,$$

and

$$n = ft = (1)(3) = 3.$$

From Equation (3.2-1), $S_3 = 1,000(1 + 6\%)^3$. The notation $(1 + 6\%)$ is used because i and n are the visible variables in the equation. However, in use, the i would be converted to a decimal. Then, the future value of $1,000 at 6% compounded annually for three years is

$$S_3 = 1,000(1.06)^3.$$

For many of the popular calculators, enter 1.06, depress the y^x key, enter 3, and depress the = key. The display will show 1.191016. Depress the x key, enter 1000, and depress the = key. For the Hewlett-Packard calculators, key in 1.06, depress the "Enter" key, enter 3, depress the y^x key, enter 1,000, and depress the x key. The display shows 1,191.016, which rounds to $1,191.02. This is the same value as indicated above in the year-by-year development. In Table A-2, the periodic interest rate is given across the top of the table, and the number of periods is given on each side. At the intersection of $i = 6\%$ and $n = 3$ find the value 1.19102. This represents the future amount of $1.00 at a periodic interest rate of 6% for three periods. As with the calculators, this value is multiplied by the principal (1,000 for this illustration).

b) For semi-annual compounding, $f = 2$ and, from Equations (3.1-1) and (3.1-2), we determine i and n as

$$i = \frac{r}{f} = \frac{6\%}{2} = 3\%,$$

and

$$n = ft = (2)(3) = 6 \text{ periods}.$$

Then,

$$S_6 = 1,000(1 + 3\%)^6,$$

and the future value is

$$S_6 = 1,000(1.1940523) = \$1,194.05.$$

c) For quarterly compounding, $f = 4$. Then,

$$i = \frac{6\%}{4} = 1\tfrac{1}{2}\%,$$

and

$$n = (4)(3) = 12 \text{ periods}.$$

Then,

$$S_{12} = 1,000 \left(1 + 1\tfrac{1}{2}\%\right)^{12}$$

$$= 1,000(1.19561817) = \$1,195.62.$$

d) For monthly compounding, $f = 12$ gives

$$i = \frac{6\%}{12} = \frac{1}{2}\%,$$

$$n = (12)(3) = 36 \text{ periods},$$

and the future value is
$$S_{36} = 1,000(1.1966805) = 1,196.68.$$

Comparing the results of a), b), c), and d), notice that as the compounding frequency increases, the future value *increases*.   ∎

EFFECTIVE ANNUAL INTEREST RATE

The **effective annual interest rate** is an interest rate which, when compounded annually, yields the same future value as a specified nominal annual interest rate compounded a specified number of times each year.

Let us determine the future values of $1 at 6% for one year, compounded as indicated below.

Compounding Frequency—f	Periodic Nominal Rate—i	Number of Periods—n	Future Value $(1+i)^n$
2	3%	2	1.0609000
4	$1\frac{1}{2}\%$	4	1.0613635
12	$\frac{1}{2}\%$	12	1.0616778

In each of the above, the future value is the value of $(1+i)^n$ because the principal is $1. Then, the decimal part of $(1+i)^n$ is the interest that is earned for the year. Because 1 is the base, the interest may be expressed as a percent. That interest earned on $1 at the end of one year, expressed as a percent, is defined as the effective annual interest rate, or simply the **effective interest rate**. It is also referred to as the **annual percentage rate**.

It may be seen above that the effective interest rate for 6% a) compounded semi-annually is 6.09%, b) compounded quarterly is 6.14%, and c) compounded monthly is 6.17%. Expressed differently, *as the frequency of compounding is increased, the effective interest rate increased.*

EXAMPLE 3.2.2   Find the effective interest rate for 12% compounded monthly.

Solution   For monthly compounding, $f = 12$. Therefore, $i = \frac{12\%}{12} = 1\%$ and for one year, $n = 12$. Then, the factor $(1 + 1\%)^{12} = 1.1268250$. The effective interest rate is the decimal part of this value which, when expressed as a percent, is 12.6825%.   ∎

The process that has just been described for determining the effective interest rate can be summarized by the equation

$$r_e = \left[\left(1 + \frac{r}{f} \right)^f - 1 \right] \times 100\%. \tag{3.2-2}$$

DETERMINATION OF TIME

Suppose you expect to purchase a new automobile at some point in the future and had P in a savings account. Further assume that the automobile you expected to buy will cost you S after a trade-in of your present car with $S > P$.

It becomes necessary to determine how long it will take for the $\$P$ to grow to $\$S$. For this determination, Equation (3.2-1) is solved for n by using the properties of logarithms. Then,

$$n = \frac{\text{Ln}\left(\dfrac{S}{P}\right)}{\text{Ln}(1+i)}. \tag{3.2-3}$$

Either common or natural logs may be used in the solution of Equation (3.2-3).

EXAMPLE 3.2.3 At 7% compounded quarterly, how long will it take for $4,000 to grow to $7,000?

Solution   The periodic (quarterly) interest rate is $\frac{7\%}{4}$ or 1.75%. From Equation (3.2-3),

$$n = \frac{\text{Ln}\left(\dfrac{7,000}{4,000}\right)}{\text{Ln}\left(1 + \dfrac{7}{400}\right)} = \frac{\text{Ln}(1.75)}{\text{Ln}(1.0175)}.$$

Because we are taking the quotient of logarithms, either common logs or natural logs may be used, but the same log must be used in both the numerator and denominator. Using natural logs,

$$n = \frac{0.559616}{0.017349} = 32.25 \text{ periods.}$$

Since interest is added at the end of interest conversion periods, it will take 33 periods for $4,000 to accumulate to $7,000. Actually, the accumulation will be slightly greater than $7,000 because of the continued growth for $\frac{3}{4}$ths of a period. Whenever "how long" is requested, time, not periods, is implied. The time it will take is determined from $n = ft$, or $t = \frac{n}{f}$. In this case

$$t = \frac{33}{4} = 8.25 \text{ years.} \quad \blacksquare$$

DETERMINATION OF INTEREST RATE

Equation (3.2-1) may be rearranged to enable the determination of the necessary interest rate for P to accumulate to S in a given length of time. The periodic interest rate may be determined by taking the nth root of

$$(1+i)^n = \frac{S}{P},$$

which gives

$$i = \left(\frac{S}{P}\right)^{\frac{1}{n}} - 1. \tag{3.2-4}$$

The nominal annual rate is determined by $r = fi$.

EXAMPLE 3.2.4 What nominal rate of interest, compounded quarterly, is necessary for $1,000 to accumulate to $2,000 in 10 years?

Solution   For this case, $\frac{S}{P} = \frac{2,000}{1,000} = 2$. Since the frequency of compounding is four, $n = 4(10) = 40$. Then, from Equation (3.2-4),

$$i = (2)^{\frac{1}{40}} - 1$$

$$= 1.0174797 - 1$$

$$= 0.017497 = 1.75\%.$$

For quarterly compounding, $r = 4(1.75) = 7\%$. ■

PRESENT VALUE

As with simple interest, the present value is the required principal in order to achieve a desired future value. In Equation (3.2-1)

$$S_n = P(1+i)^n.$$

S_n would be a known value, and P would be unknown. Dividing both sides by $(1+i)^n$ gives

$$P = \frac{S_n}{(1+i)^n}. \tag{3.2-5}$$

An alternative form of Equation (3.2-5) may be stated as follows:

$$P = S_n(1+i)^{-n}. \tag{3.2-6a}$$

The factor $(1+i)^{-n}$ is referred to as the present value of $1.00, and a short table is supplied in the Appendix as Table A-3. The actuarial profession defines the symbol v as

$$v = (1+i)^{-1}.$$

Then, Equation (3.2-6a) would be written as

$$P = S_n v^n \tag{3.2-6b}$$

and has the effect of using only positive exponents. Use of Equation (3.2-6b) is convenient for algebraic operations and tables. However, the form of Equation (3.2-6a) is necessary when calculators are to be used. As with Table A-2, the present value of $1.00 is located at the intersection of the periodic interest rate i and the number of periods n.

EXAMPLE 3.2.5   What is the present value of $5,000 in five years at 6% compounded a) semi-annually, b) quarterly, and c) monthly?

Solution

a) For semi-annual compounding, $f = 2$, $i = 3\%$, and $n = 10$. From Equation (3.2-6a), or Table A-3 at the intersection of $i = 3\%$ and $n = 10$, we have
$$P = 5,000(1 + 3\%)^{-10}$$

$$= 5,000(0.7441) = \$3,720.50.$$

b) For quarterly compounding, $f = 4$, $i = 1\frac{1}{2}\%$, and $n = 20$. From Equation (3.2-6a) we have

$$P = 5{,}000 \left(1 + 1\tfrac{1}{2}\%\right)^{-20}$$

$$= 5{,}000(0.7424704) = \$3{,}712.35.$$

c) For monthly compounding, $f = 12$, $i = \frac{1}{2}\%$, and $n = 60$. From Equation (3.2-6a) we have

$$P = 5{,}000 \left(1 + \tfrac{1}{2}\%\right)^{-60}$$

$$= 5{,}000(0.7413721) = \$3{,}706.86. \quad \blacksquare$$

Notice that as the compounding frequency increases, the present value *decreases*.

CONTINUOUS COMPOUNDING

The future value at compound interest is given by Equation (3.2-1) as

$$S_n = P(1 + i)^n,$$

where $i = \frac{r}{f}$ and $n = ft$. In the early 1980s, inflation was such that interest rates were in the 15%–17% range of magnitudes. In the competition of banks to attract savings, the interval for compounding interest was "reduced to a zero time span." This had the effect of compounding the interest *continuously*. However, to actually convert the "old" principal into a "new" principal continuously would have been impractical. The continuously compounded interest was "posted" at discrete intervals of time such as monthly or quarterly. The future value of a principal that earns a nominal annual interest rate, compounded continuously, is from Equation (3.2-1) as

$$S = P \left(1 + \frac{r}{f}\right)^{ft},$$

where f becomes infinite. The exponent may be multiplied and divided by r to give

$$S = P \left(1 + \frac{r}{f}\right)^{\left(\frac{f}{r}\right)rt}.$$

The factor $\left(1 + \frac{r}{f}\right)^{\frac{f}{r}}$ is of the form of $\left(1 + \frac{1}{x}\right)^x$, and as x approaches infinity this converges to the value of 2.71828183 .... and is the base of the natural logarithms. It is denoted by the letter e. Then, for continuous compounding, Equation (3.2-1) becomes

$$S_t = Pe^{rt}. \tag{3.2-7}$$

The subscript on S has been changed to t since, for continuous compounding, the "number of periods" becomes infinite, and the measure of time is years.

Many banks use continuous compounding of interest as an inducement to save with "them." Further, they may use a bank year of 360 days, but pay interest for 365 days a year. By doing so, the t in e^{rt} becomes $\left(\frac{365}{360}\right) t$. Letting $k = \frac{365}{360}$ be the bank-year correction, Equation (3.2-7) becomes

$$S_t = Pe^{krt}. \tag{3.2-8}$$

For the present value at continuous compounding of interest Equations (3.2-7) and (3.2-8) respectively become

$$P = S_t e^{-rt}, \tag{3.2-9}$$

and

$$P = S_t e^{-krt}. \tag{3.2-10}$$

EXAMPLE 3.2.6   For $1,000 at 6% compounded continuously for five years, determine a) the future value and b) the present value with and without a bank-year correction.

Solution   The value of rt is $(0.06)(5) = 0.3$.

  a) For future values

$$e^{0.3} = 1.3498588,$$

and

$$e^{k(0.3)} = 1.355495.$$

Without the bank-year correction,
$$S_5 = 1,000(1.349858) = \$1,349.86.$$

With the bank-year correction,
$$S_5 = 1,000(1.355495) = \$1,355.49.$$

  b) For the present values,

$$e^{-0.3} = 0.7408186,$$

and

$$e^{-k(0.3)} = 0.7377379.$$

Without the bank-year correction
$$P = 1,000(0.7408186) = \$740.82,$$

and with the bank-year correction,
$$P = 1,000(0.7377379) = \$737.74. \quad \blacksquare$$

In order to determine how long it will take to accumulate a specified future value at continuously compounded interest, Equation (3.2-3) transforms to

$$t = \frac{\text{Ln}\left(\dfrac{S}{P}\right)}{kr}. \tag{3.2-11}$$

EXAMPLE 3.2.7   At 7% compounded continuously, how long will it take $2,000 to grow to $4,981 using the bank-year correction factor?

Solution   Using Equation (3.2-11)

$$t = \frac{Ln\left(\dfrac{4,981}{2,000}\right)}{k(0.07)} = \frac{0.912483}{0.070972} = 12.857 \text{ years.}$$

As in the case of discrete compounding, the effective interest rate for continuous compounding is the decimal part of the future value, e^r or e^{kr}, where $t = 1$, expressed as a percent.   ■

SOME MAGIC NUMBERS: 72, 114, AND 167

If it is necessary to know the exact time it takes for a principal to grow to a future value, then Equations (3.2-3) and (3.2-11) may be used. However, it is often only necessary to know the approximate time for a specified growth to occur. When an approximation is sufficient, the following "rules" may be used. The discovery of the rule of 72 is credited to Albert Einstein who is quoted as saying that compound interest "is the greatest mathematical discovery of all time," Amat (2002). C. L. Trowbridge (1985) developed the rules of 114 and 167 in an article that is indicated in the Bibliography.

The **Rule of 72** gives the approximate number of years for a principal to double. The **Rule of 114** gives the approximate number of years that it takes a principal to triple. The **Rule of 167** gives the approximate number of years it takes for a principal to increase by a factor of five. Letting r be the nominal annual interest rate expressed as a percent, the rules for the magic numbers are as follows:

$$t_{2\,times} \approx \frac{72}{r\%} \text{ years,}$$

$$t_{3\,times} \approx \frac{114}{r\%} \text{ years,}$$

and

$$t_{5\,times} = \frac{167}{r\%} \text{ years.}$$

EXAMPLE 3.2.8   At a nominal annual interest rate of 7%, determine the approximate number of years it will take for a principal to double, triple, and increase by a factor of five.

Solution   Using the rules of 72, 114, and 167,

$$t_{2\,times} \approx \frac{72}{7} \approx 10 \text{ years,}$$

$$t_{3\,times} \approx \frac{114}{7} \approx 16 \text{ years,}$$

and

$$t_{5\,times} \approx \frac{167}{7} \approx 24 \text{ years.}$$

The rules may be expanded for multiples of two, three, and five by adding the numerators of the factors of the multiple and dividing by $r\%$. ■

EXAMPLE 3.2.9   Determine the approximate number of years for a principal to increase by a factor of six, and a factor of 18, if the interest rate is 10%.

Solution   A multiple of six is obtained by multiplying two and three. Then, an increase by a factor of six is determined by adding the numerators of the rule of 72 (for doubling) and the rule of 114 (for tripling) as follows:

$$t_{6\,\text{times}} \approx \frac{72 + 114}{10}$$

$$\approx \frac{186}{10} \approx 19 \text{ years.}$$

The factor of 18 is determined by increasing the factor of 6 three times, or

$$t_{18\,\text{times}} \approx \frac{72 + 114 + 114}{10}$$

$$\approx \frac{186 + 114}{10}$$

$$\approx \frac{300}{10} \approx 30 \text{ years.} ■$$

PROBLEM SET 3.2

Determine the future values in Problems 1–6.

	Principal	Compounding Frequency	Annual Interest Rate	Number of Years
1.	$1,000	Semi-annual	7%	5
2.	$5,000	Quarterly	7%	5
3.	$8,000	Monthly	7%	4
4.	$4,000	Semi-annual	8%	6
5.	$6,000	Quarterly	5%	10
6.	$3,000	Monthly	5%	7

7. Determine the effective interest rate for Problems 1–6.

8. Assume that the values in the Principal column of Problems 1–6 are future values. Determine the respective present values.

9. How long will it take $1,000 to grow to $2,500 at 6% compounded semi-annually?

10. How long will it take money to double at 7% compounded quarterly?

11. What is the future value of $1,000 at 6% compounded monthly for 10 years?

12. What is the future value of $1,000 at 6% compounded monthly for 20 years?

13. What is the future value of $1,000 at 6% compounded monthly for 30 years?

14. What is the present value of $1,000 at 6% compounded monthly for 10 years?

15. What is the present value of $1,000 at 6% compounded monthly for 20 years?

16. What is the present value of $1,000 at 6% compounded monthly for 30 years?

17. Solve for the future amounts in Problems 3, 4, and 5 above for continuous compounding of interest with **a)** no bank-year correction and **b)** with the bank-year correction.

18. Assume that the values in the Principal column of Problems 3, 4, and 5 are future values at continuous compounding of interest. Determine the respective present values with **a)** no bank-year correction and **b)** with the bank-year correction.

19. At 7% compounded continuously, how long will it take $1,000 to grow to $7,389.05 if the bank-year correction is not used?

20. At 8% compounded continuously, how long will it take to quadruple money if a bank-year correction factor is used?

21. A savings account pays 5% interest compounded continuously on a deposit of $1,000. After five years, the interest rate is increased to $5\frac{1}{2}$% compounded continuously. What will be the future value of the $1,000, 15 years after the initial deposit without the bank correction factor?

22. What interest rate compounded annually is necessary for $2,000 to accumulate to $16,000 in 30 years?

23. What interest rate compounded semi-annually is necessary for $12,000 to accumulate to $36,000 in 10 years?

24. What interest rate compounded quarterly is necessary for $30,000 to accumulate to $100,000 in 18 years?

25. What interest rate compounded monthly is necessary for money to double in 12 years?

Using the appropriate rules, determine the approximate number of years it will take for the following if the interest is compounded annually.

	Principal	Future Value	Interest Rate
26.	$1,000	$4,000	8%
27.	$2,000	$6,000	6%
28.	$3,000	$15,000	7%
29.	$5,000	$50,000	10%
30.	$10,000	$120,000	12%

3.3 DISCOUNT AT COMPOUND INTEREST

The amount that is repaid to the holder of a debt instrument by the issuer at the time of maturity is known as the **maturity value**. We say the instrument is sold at a **discount** because the proceeds are smaller than the maturity value. The rate of return that is earned by the investor is referred to as the **discount rate**.

To determine the compound interest, the proceeds of a discounted instrument are treated as the present value of the note on the date of the discount and the maturity value as the future value. Letting:

M be the maturity value of the instrument (S is replaced by M),

i be the discount rate,

P be the proceeds, and

n be the number of periods,

the proceeds for discrete discounting can be stated as

$$P = M(1 + i)^{-n}, \tag{3.3-1}$$

and for continuously compound discounting by

$$P = Me^{-krt}, \tag{3.3-2}$$

where k is the factor $\frac{365}{360}$, r is the nominal annual interest rate, and t is the number of years to maturity. Equation (3.3-2) may be used with or without the bank-year correction. Figure 3-3 shows a time diagram for compound discount.

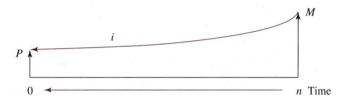

FIGURE 3-3 The Time Value of Money with Compound Discount

EXAMPLE 3.3.1   A non-interest-bearing note with a face value of $5,000 is due on July 6, yyyy. It is sold to a bank two years before the maturity date at a discount rate of 9% compounded monthly. Determine the proceeds.

Solution   Since the note is non-interest-bearing, the face value is the maturity value, and since the note is discounted two years before it is due, $t = 2$. For monthly compounding, $f = 12$, $i = \frac{3}{4}\%$, and $n = 24$.

$$P = 5,000 \left(1 + \frac{3}{4}\%\right)^{-24}$$

$$= 5,000(0.8358314) = \$4,179.16.$$

Since the bank will be paid the maturity value of $5,000 on the due date, it will have earned $820.84 interest ($5,000 − 4,179.16$). ∎

EXAMPLE 3.3.2   A promissory note for $3,000 at 10% compounded quarterly for three years is sold to a bank on the same day that the note is signed. The bank discounts the note at 12% compounded monthly. Determine the proceeds to the seller.

Solution   Since this is an interest-bearing note, the maturity value of the note must be determined before the discount can take place. For the maturity value at 10% compounded quarterly, $f = 4, I = \frac{10\%}{4} = 2\frac{1}{2}\%$, and for three years $n = 12$.

$$M = 3{,}000\left(1 + 2\frac{1}{2}\%\right)^{12}$$

$$= 3{,}000(1.3448888) = \$4{,}034.67.$$

For discounting at monthly compounding, $f = 12, i = 1\%$, and since the note is sold on the day it was initiated, the discount time is the full term, or three years. Therefore, $n = 3(12) = 36$. From Equation (3.3-1)

$$P = 4{,}034.67(1 + 1\%)^{-36}$$

$$= 4{,}034.67(0.698925) = \$2{,}819.93. \quad \blacksquare$$

The time diagram is shown below.

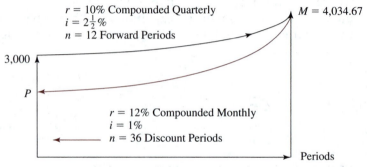

Time Diagram for Example 3.3.2

COMPARISON OF DISCOUNT METHODS

In Section 1.2, it is shown that the proceeds for simple interest discounting are

$$P = \frac{M}{1 + rt},$$

if we replace S by M.

In Section 1.3, it is shown that the proceeds for bank discount at simple interest are

$$P = M(1 - dt),$$

if we replace S by M.

In this section, it is shown that proceeds from a compound discount are

$$P = M(1 + i)^{-n},$$

or if interest is compounded continuously

$$P = Me^{-krt}.$$

In buying a discounted instrument, the objective of the investment is to maximize the return, which is the same as paying the least amount of money for the instrument. Therefore, the method that gives the *least proceeds* (i.e., the smallest present value) should be chosen.

Let us determine the proceeds on a maturity value of $1.00 for each method using $r = d = 6\%$ compounded monthly and continuously for one year.

1. Simple Discount

$$P = \frac{1}{1+r} = \frac{1}{1+0.06} = 0.9433962.$$

2. Bank Discount

$$P = (1-d) = (1-0.06) = 0.94.$$

3. Discrete, Compound Discount

$$P = \left(1 + \frac{6}{1,200}\right)^{-12} = 0.9419053.$$

4. Continuously Compounded Discount Without Bank-Year Correction

$$P = e^{-r} = e^{-0.06} = 0.941765.$$

It is obvious that bank discount at simple interest yields the least proceeds.

PROBLEM SET 3.3

Determine the proceeds for the non-interest-bearing notes in Problems 1–5. Do not use the bank-year correction factor for continuous compounding of interest.

Face Value	Discount Rate	Compounding Frequency	Time to Maturity
1. $10,000	9%	Semi-annual	2 years
2. $8,000	12%	Quarterly	1 year
3. $10,000	9%	Monthly	2 years
4. $8,000	10%	Continuously	1 year
5. $8,000	10%	Continuously	3 years

Determine the proceeds for the interest-bearing notes in Problems 6–10. Do not use the bank-year correction factor for continuous compounding of interest.

Principal	Interest Rate	Original Term	Discount Rate	Term Prior to Discount	Compounding Frequency
6. $5,000	8%	5 years	9%	2 years	Semi-annual
7. $5,000	8%	7 years	10%	3 years	Quarterly
8. $4,000	9%	4 years	12%	1 year	Monthly
9. $5,000	8%	10 years	10%	3 years	Continuous
10. $4,000	10%	4 years	10%	4 years	Continuous

11. A non-interest-bearing note for $20,000 was discounted at 9% compounded monthly two years before the maturity date. It was rediscounted a second time 1 year before the maturity date at 10% compounded continuously. Determine the proceeds from the second discount.

12. A $16,000 non-interest-bearing note was discounted at 10% compounded continuously three years before the maturity date. It was discounted a second time two years before the maturity at 12% compounded monthly. Determine the proceeds from the second discount.

13. A promissory note that is initiated today for $15,000 with interest at 8% compounded quarterly is due in two years. The holder of the note immediately sells it to the Friendly Loan Co. which discounts it at 12% compounded monthly. One year later the Friendly Loan Co. sells the note to the Loan Company Bailout Corporation which discounts it at 18% compounded monthly. Determine the proceeds to Friendly.

14. In Problem 13, determine the value of interest Friendly lost because it had to sell the note to Bailout.

4

Basic Annuities

Most every family in the United States, and other nations, will participate in an annuity program. Contributions to and distributions from pension plans, mortgage payments on a home, and life insurance premiums are some examples of annuities. The word **annuity** originally referred to annual payments, but in general, it applies to any periodic payment, usually in equal amounts. Sometimes, the periodic payment is also called the **rent** of an annuity. The period of time between successive payments is known as the **payment interval**.

There are several classifications of annuities. This chapter discusses only the simple, ordinary annuity and the simple annuity due. **Simple annuities** are annuities where the payment dates "coincide" with the dates when interest is added, interest conversion, to the principal. An **ordinary annuity** is an annuity where the payment is made at the ends of the interest conversion dates and is a term that is used by accountants. The same type of annuity is called an **annuity immediate** by actuaries in the insurance industry and an **immediate annuity** by financial planners. An **annuity due** is an annuity where payments are made at the beginnings of interest conversion dates. We will use the term **ordinary** in Chapters 4, 5, and 6 and **annuity immediate** in Chapters 9, 10, and 11.

4.1 FUTURE AMOUNT OF AN ANNUITY AND SINKING FUND

FUTURE AMOUNT OF A SIMPLE ORDINARY ANNUITY

Consider the time chart in Figure 4-1. It is an annuity of four $100 payments at 8% compounded quarterly with the payments occurring at the end of each conversion period.

Inspecting Figure 4-1, we see that the number of payments equals the number of periods. The 1st payment earned interest for three periods, the 2nd payment earned interest for two periods, the 3rd payment earned interest for one period, and the 4th payment did not earn interest because the date of the 4th payment terminated the annuity. Notice that the greatest compound amount is

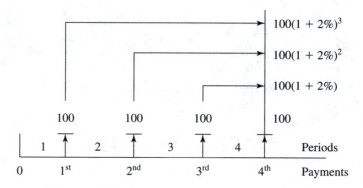

FIGURE 4-1 Time Diagram for an annuity of Four $100 Payments

for 3 $(4-1)$ periods, the next highest is for 2 $(4-2)$ periods, then for 1 $(4-3)$ period, and lastly for 0 $(4-4)$ periods. The future amount of the annuity is the sum of the four compound amounts. Then, adding upward

$$S = 100 + 100(1+2\%) + 100(1+2\%)^2 + 100(1+2\%)^3$$

$$S = 100 + 102 + 104.04 + 106.12 = 412.16.$$

To compute this value of S is relatively simple. Using the formula of $(1+i)^n$ for 100 payments, however, would be rather tedious. Obviously, a general formula is necessary and a time diagram for such is shown in Figure 4-2.

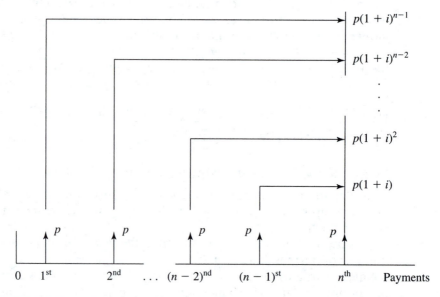

FIGURE 4-2 Time Diagram for the Sum of an Annuity: Discrete Compounding of Interest

The sum of an annuity of n payments over n periods, using the concept discussed above, is

$$S = p + p(1+i) + p(1+i)^2 + p(1+i)^3 + \cdots + p(1+i)^{n-1}. \qquad (4.1\text{-}1)$$

Equation (4.1-1) is the sum of a geometric progression with a common ratio of $(1+i)$ and reduces to

$$S = p\left[\frac{(1+i)^n - 1}{i}\right], \qquad (4.1\text{-}2)$$

where,

p is the periodic payment,

i is the periodic interest rate, $\frac{r}{f}$, and

n is the number of periods, ft, where t is the number of years.

The standard notation for the expression within the brackets is $s_{\overline{n}|}$ and is the future amount of $1. Then, in notational form,

$$S = p s_{\overline{n}|}.$$

The bracketed factor, $s_{\overline{n}|}$, is the sum of an annuity of $1 per period, and a short, illustrative table is supplied in the Appendix as Table A-4. The future amount of an annuity of $1 per period is found at the intersection of i and n. However, the primary form of solution will be to solve Equation (4.1-2). As in compound interest, essentially, only i and n are necessary for annuity computations. Using Equation (4.1-2) to determine the sum of the annuity in Figure 4-2, gives, for $i = 2\%$ and $n = 4$,

$$S = 100\left[\frac{(1.02)^4 - 1}{0.02}\right]$$

$$= 100(4.121608) = 412.16,$$

as shown above. The total of the payments is np or $4(100) = 400$. The earned interest, obviously, is $12.16.

EXAMPLE 4.1.1 Every month, on the interest conversion date, a deposit of $100 is made into a savings account that pays 5% compounded monthly. This is done faithfully for 20 years. What will be the amount in the account at the end of the annuity?

Solution For 5% compounded monthly, $i = \frac{5}{12}\%$ and $n = 240$ periods. Then, from Equation (4.1-2), where i is replaced by $\frac{5}{1,200}$, since the percent must be converted to a decimal (division by 100) and the nominal annual interest rate must be divided by the frequency of compounding (12 for monthly),

$$S = 100\left[\frac{\left(1 + \dfrac{5}{1,200}\right)^{240} - 1}{\dfrac{5}{1,200}}\right] = 100(411.033669) = 41,103.37.$$

Since $\frac{5}{1,200}$ is not a "simple" value $(0.0041666\ldots)$, it should be stored in memory of a calculator and recalled when needed to divide by the denominator. ■

SINKING FUND

A specified future amount is desired at the end of a given period of time. To reach this goal, we may want to set aside a fixed amount on a regular basis. In this sense, the **annuity** is the periodic payment necessary to achieve the specified future amount. The accumulating amount is called a **sinking fund**.

Equation (4.1-2) is the sum of an annuity for a periodic payment, p. Then, to find p, both sides of Equation (4.1-2) may be divided by $s_{\overline{n}|}$.

EXAMPLE 4.1.2   At an interest rate of 5% compounded monthly, what monthly payment is necessary to have \$10,000 in four years (for example, saving for an automobile)?

Solution   For 5% compounded monthly, $f = 12$, $i = \frac{5}{12}\%$, and $n = 48$. From Equation (4.1-2)

$$10,000 = p\left[\frac{\left(1 + \dfrac{5}{1,200}\right)^{48} - 1}{\dfrac{5}{1,200}}\right]$$

$$= p(53.01488525),$$

and solving for p gives \$188.63 per month. (A relatively easy way to save for your next car—provided the present car is paid for and is a safe vehicle to drive.) ■

PROBLEM SET 4.1

Determine the future amount for periodic payments of \$1.00 at the ends of conversion periods for Problems 1–4.

	Interest Rate	Compounding Frequency	Term
1.	6%	Monthly	1 year
2.	6%	Quarterly	1 year
3.	6%	Semiannually	1 year
4.	6%	Annually	1 year

Determine the future value for periodic payments of \$1.00 at the end of each conversion period for Problems 5–8.

	Interest Rate	Compounding Frequency	Term
5.	6%	Monthly	7 years
6.	7%	Quarterly	10 years
7.	8%	Semiannually	5 years
8.	9%	Annually	10 years

Determine the future amounts of the annuities in Problems 9–17 for payments at the ends of conversion periods.

9. A monthly payment of $100 for Problem 1.

10. A quarterly payment of $300 for Problem 2.

11. A semiannual payment of $600 for Problem 3.

12. An annual payment of $1,000 for Problem 4.

13. A monthly payment of $100 for Problem 5.

14. A quarterly payment of $300 for Problem 6.

15. A semiannual payment of $600 for Problem 7.

16. An annual payment of $1,200 for Problem 8.

17. A monthly payment of $700 for Problem 5.

18–21. Determine the necessary periodic payments in order to achieve $10,000 in Problems 5–8 respectively.

22. If a person makes a $1,000 deposit semiannually for 20 years into an account that pays $6\frac{1}{2}$% interest compounded semiannually, how much would be in the account as of the final deposit?

23. An individual makes $1,000 annual deposits, on the same day each year, into a savings account that pays $5\frac{1}{4}$% compounded annually. If this procedure is followed for 30 years, what will be the total value of the savings as of the last annual transaction?

24. As an education fund for their child, parents make annual deposits of $2,000 into a savings account that pays $5\frac{1}{4}$% interest compounded annually. If the first deposit is made one year after the child is born and the last deposit is made on the child's 18th birthday, how much will be in the account?

25. What annual deposit into an account, that will pay no less than 6% compounded annually, is needed in order to have a total of $250,000 after 40 years?

26. If the parents of the child in Problem 24 feel that by the time their child is 18, they will need $40,000, what annual deposit is necessary?

27. If a business is able to invest money at 10% compounded annually, what annual amount is necessary to have $1,000,000 at the end of 10 years?

28. If a business is able to invest money at 12% compounded quarterly, what quarterly deposit is necessary in order to have $500,000 for the replacement of equipment in 10 years?

29. You have an automobile that is paid for and should perform satisfactorily for four years. You estimate that when you buy a new automobile in four years, the net cost to you will be $7,000. If a savings account pays 5% compounded monthly,

a) What monthly deposit is required?

b) How much interest will have been earned?

30. It is now four years later than the time of Problem 29. You have the $7,000 to buy your automobile whereas your friend finances an identical car for four years where a 5% add-on interest is charged. Determine the difference between the total of the payments your friend will make for the car and the total of the payments you will have made into the sinking fund of Problem 29.

4.2 PRESENT VALUE OF AN ANNUITY AND AMORTIZATION

PRESENT VALUE OF AN ANNUITY

Consider the time diagram of Figure 4-3. It is also an annuity of four payments of $100 at 8% compounded quarterly, as is Figure 4-1, except that the annuitant is receiving the payments instead of making them.

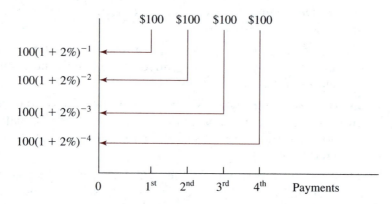

FIGURE 4-3 A Time Diagram for the Present Value of an Annuity of Four Payments

Figure 4-3 shows that in order to receive $100 at the end of the 1st period, $100(1 + 2\%)^{-1}$ must be available at time 0. Similarly, in order to receive $100 at the end of the 2nd, 3rd, and 4th periods, the present values, $100(1 + 2\%)^{-2}$, $100(1 + 2\%)^{-3}$, and $100(1 + 2\%)^{-4}$ must be available at time 0, respectively. Adding the respective present values, gives the present value of the annuity. Letting A be the present value of an annuity, upward addition gives

$$A = 100(1 + 2\%)^{-4} + 100(1 + 2\%)^{-3} + 100(1 + 2\%)^{-2} + 100(1 + 2\%)^{-1}.$$

The present value of this annuity of four payments, therefore, is

$$A = 100(0.9238) + 100(0.9423) + 100(0.9612) + 100(0.9804)$$

$$= 380.77.$$

Thus, $380.77 today will purchase four $100 payments at 3-month intervals. A general formula is necessary in order for finding the present value of an annuity of n payments. The time diagram for this, with an interest rate of $i\%$ per period, is shown in Figure 4-4.

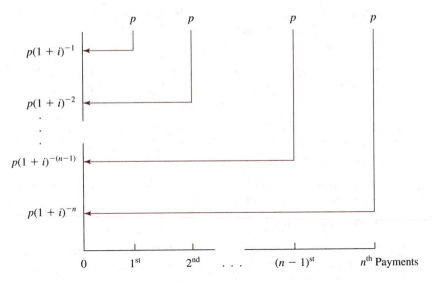

FIGURE 4-4 A Time Diagram for the Present Value of an Annuity of n Payments

Adding upward gives

$$A = p(1+i)^{-n} + p(1+i)^{-(n-1)} + \cdots + p(1+i)^{-1}, \qquad (4.2\text{-}1)$$

where,

p is the periodic payment,

i is the periodic interest rate, $\frac{r}{f}$, and

n is the number of periods, ft, where t is in years.

Equation (4.2-1) is the sum of a geometric progression with a ratio of $(1+i)^{-1}$ and reduces to

$$A = p\left[\frac{(1-(1+i)^{-n})}{i}\right]. \qquad (4.2\text{-}2)$$

The bracketed factor which has the symbol, $a_{\overline{n}|}$, is the standard notation of the Society of Actuaries and is the present value of $1 per period, and a short illustrative table is supplied in the Appendix as Table A-5. Then, in notational form,

$$A = p a_{\overline{n}|}.$$

The present value of $1 per period is found at the intersection of i and n. Using Equation (4.2-2) to solve for the present value problem of Figure 4-3, $i = 2\%$, $n = 4$, and

$$A = (100)(3.8077286987) = 380.77.$$

The $380.77 generated a total of payments of $400; therefore, the interest was $19.23.

EXAMPLE 4.2.1   An individual is planning retirement such that at the end of each quarter, $400 could be withdrawn from a bank account to pay the quarterly taxes on a house. It is desired to be able to do this for 15 years. If the bank pays interest at 5% compounded quarterly, how much must be in the account at the start of the retirement if the first tax payment is three months later?

Solution   For 5% compounded quarterly, $f = 4$, $i = \frac{5}{4}\% = 1.25\%$, and $n = 60$. Then, from Equation (4.2-2) the present value of the 60 future $400 payments is

$$A = (400)(42.0345917945) = \$16,813.84.$$

If this individual lives to make the 60 payments, a total of $24,000 will have been paid with the $16,813.84. This means that $7,186.16 was interest. Viewing it another way, the bank paid approximately 18 of the payments $\left(\frac{7,186.16}{400}\right)$ or approximately $4\frac{1}{2}$ years of the taxes. Note that this neglects the effect of income taxes on the earned interest. ∎

AMORTIZATION

The term **amortization** refers to the liquidation of a debt by periodic partial payments where the interest that is charged is based on the unpaid balance. Pension annuities and loan payments are examples of amortizing debts. Consider, for example, monthly payments for a loan at some specified annual interest rate compounded periodically (usually monthly) for a specified term. The periodic payment to the bank includes the interest for the period based on the unpaid balance and a fraction of the principal by which the balance is reduced. The sum of all payments represents a future amount to the bank. Thus, the amount of the mortgage represents the present value of that future amount. The question to be answered is; When the interest rate is given, what periodic payment is necessary to liquidate the mortgage in the specified time?

Since the amortization of loans requires the determination of the periodic payment, Equation (4.2-2) is solved for p.

EXAMPLE 4.2.2   A building is mortgaged for $500,000 at 10% compounded quarterly for 20 years. Determine the quarterly payment and the interest earned by the lending institution. A loan for the purpose of purchasing real estate has a standard term–mortgage.

Solution   For 10% compounded quarterly, $2\frac{1}{2}\%$ and $n = 80$. From Equation (4.2-2)

$$500,000 = p \left[\frac{1 - (1.025)^{-80}}{0.025} \right] = p(34.451817),$$

and solving for p gives $14,513 per quarter. The total of the payments is $(80)(14,513)$ or $1,161,040, and for a $500,000 loan, the interest earned by the lending institution is $661,040. ∎

EXAMPLE 4.2.3   Determine the periodic payment and total interest for a 20-year mortgage of $60,000 at 9% compounded monthly.

Solution For 9% compounded monthly, $i = \frac{9}{12}\% = \frac{9}{1,200}$, and $n = 12(20) = 240$. From Equation (4.2-2),

$$60,000 = p\left[\frac{1 - \left(1 + \dfrac{9}{1,200}\right)^{-240}}{\dfrac{9}{1,200}}\right] = p(111.144).$$

Solving for p gives

$$p = \frac{60,000}{111.144},$$

and the monthly payment will be $539.84. The total interest would be $240(539.84) - 60,000$ or $69,561.60. ■

SINKING FUND AND AMORTIZATION SCHEDULES

In order for a business or an individual to maintain records of the growth or depletion of funds, schedules for the accumulation in a sinking fund, and the depletion of a fund (or the amortization of a loan) are generated. Figure 4-5 shows the accumulation process for the first five years, and Table 4-1 shows the first five years and the last five years of a sinking fund schedule with an annual payment of $1,462.27 compounded at an annual rate of 5% for 40 years.

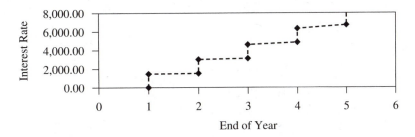

FIGURE 4-5 Accumulation of a Sinking Fund

In the Table 4-1 the numbers in the []'s refer to the column numbers. [2] + [3] means add columns [2] and [3], and 0.05[3] means multiply the value in the previous column [3] by 0.05. Note that the Balance Before Payment is the sum of the Balance After Payment from the *previous* year plus the interest for the current end of year.

Figure 4-6 shows a graph of the first five years of the amortization of a fund of $176,642.13, invested at 6% compounded annually, with annual withdrawals of $24,000 for 10 years. Table 4-2 shows the amortization schedule.

In Figure 4-6, the distance from the top to the point within each vertical line at the end of each year represents the interest part of each payment, and the remainder of each line represents the principal part of each payment. These

TABLE 4-1 SINKING FUND SCHEDULE

Annual Payments = p = $1,462.27
Annual Interest Rate = 5% Compounded Annually

End of Year	[1] Balance before Payment [2] + [3]	[2] Interest for Year 0.05[3]	[3] Balance after Payment [1] + [2] + p
1	$0.00	$0.00	$1,462.27
2	1,462.27	73.11	2,997.65
3	2,997.65	149.88	4,609.80
4	4,609.33	230.49	6,302.56
5	6,302.56	315.13	8,079.96
.	.	.	.
.	.	.	.
.	.	.	.
36	132,072.96	6,603.65	140,138.87
37	140,138.87	7,006.94	148,608.09
38	148,608.09	7,430.40	157,500.76
39	157,500.76	7,875.04	166,838.06
40	166,838.06	8,341.90	176,642.13

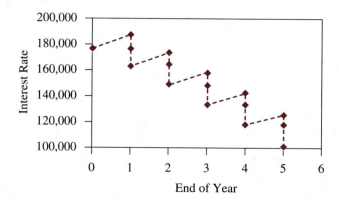

FIGURE 4-6 Amortization of a Fund

values are shown in Table 4-2, and the balance before withdrawal is the sum of the balance after withdrawal of the *previous* year plus the interest for the current year. The values in year 10 are correct to within $1.00 due to successive rounding of decimals and have been adjusted for clarity of illustration.

In Table 4-2, [4] + [3] means add the value in the previous column [4] to the current column [3], and column [3] is obtained by multiplying the previous column [4] by 0.06.

TABLE 4-2 AMORTIZATION SCHEDULE

Annual Withdrawals = p = $24,000
Annual Interest Rate = 6% Compounded Annually

End of Year	[1] Balance before Withdrawal [4] + [3]	[2] Principal Reduction p−[3]	[3] Annual Interest 0.06[4]	[4] Balance after Withdrawal [1]−[2]
0				$176,642.13
1	$187,240.65	$13,401.48	$10,598.52	163,240.65
2	173,035.08	14,205.57	9,794.43	149,035.08
3	157,977.18	15,057.90	8,942.10	133,977.18
4	142,015.81	15,961.38	8,038.62	118,015.80
5	125,096.75	16,919.06	7,080.94	101,096.74
6	107,162.55	17,934.20	6,065.80	83,162.54
7	88,152.30	19,010.25	4,989.75	64,152.30
8	68,001.44	20,150.87	3,849.13	44,001.44
9	46,641.53	21,359.92	2,640.08	22,641.27
10	24,000.00	22,641.50	1,358.50	0.00

TRUE ANNUAL INTEREST RATE

In the amortization of loans based on add-on interest or when "points" (defined below) are charged, the determination of the true annual interest rate cannot be determined explicitly by algebraic principals only. Referring to Equation (4.2-2), the true periodic interest rate (i) for a loan of $A that requires a periodic payment of $p requires the solution of the following equation for i,

$$\left[\frac{1 - (1 + i)^{-n}}{i}\right] = \frac{A}{p}. \tag{4.2-3}$$

Since i cannot be determined explicitly, the solution requires tables, a business-type calculator such as the Hewlett-Packard 12C or the Texas Instruments BA-III, or a computer. However, once i is known, the true annual interest rate is determined by

$$R_t = fi. \tag{4.2-4}$$

EXAMPLE 4.2.4   An automobile is financed for $10,000 at an add-on interest rate of 5% for 36 months. Determine the true annual interest rate.

Solution   At 5% for 36 months (or three years), the add-on interest is given by

$$I = Prt$$

$$= 10,000(0.05)(3) = 1,500,$$

and the monthly payments are determined by

$$p = \frac{10,000 + 1,500}{36} = \$319.45.$$

For the true periodic interest rate, a loan of $10,000 is being amortized by a payment of $319.45 per month for 36 months. From Equation (4.2-4),

$$\left[\frac{1-(1+i)^{-36}}{i}\right] = \frac{10,000}{319.45} = 31.3038.$$

Using a Hewlett-Packard 12C, the true monthly interest rate is determined to be 0.775975. Equation (4.2-4), gives the true annual interest rate as

$$R_t = 12(0.775975) = 0.09312,$$

which is 9.312%. ∎

MORTGAGE POINTS

Lending institutions often choose not to hold home mortgages to maturity. Rather, they sell those mortgages to secondary financial investors. However, the primary institution (i.e., the original lender) may charge a fee, which is called points. A **point** is 1% of the mortgage amount, and is an amount that must be paid by the borrower in addition to any down payment. Thus if the mortgage loan amount is $A and the number of points is N, the actual amount given to the borrower, after the deduction of points, will be

$$P = A\left(1 - \frac{N}{100}\right). \tag{4.2-5}$$

However, the periodic payment will be based on the mortgage amount A and can be determined by using Equation (4.2-2). In Example 4.2.3 it was shown that a $60,000 mortgage at 9% for 20 years requires a monthly payment of $539.84. If three points were charged for that loan, the borrower would receive, from Equation (4.2-5),

$$P = 60,000\left(1 - \frac{3}{100}\right) = \$58,200.$$

Then, from Equation (4.2-3),

$$a_{\overline{240}|} = \frac{58,200}{539.84} = 107.8097214.$$

For this example, a Hewlett-Packard 12C calculator shows i to be 0.785827310%. This transforms to a true annual interest rate of 9.430%.

The Goal Seek option under the Tools Menu of Microsoft Excel may be used to determine the true annual interest rate of a loan that is subject to points. The layout below is a portion of the Excel spreadsheet that is appropriate. The periodic payment in Cell D4 can be calculated by including a Cell F4 in which the original loan amount is entered and entering the following formula in Cell D4.

```
=ROUNDUP(F4*(E4/C4)/(1-(1+E4/C4)^(-C4*B4)),2).
```

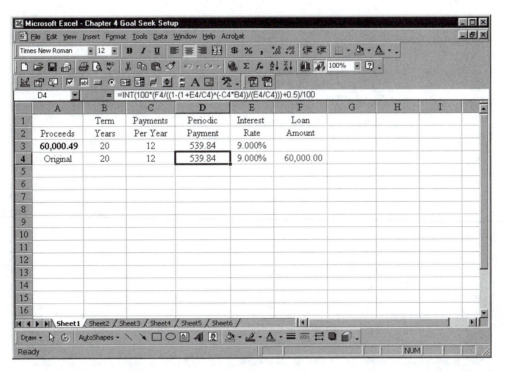

This format will round the payment up to the next higher cent, the result of which is 539.84. This value and the interest rate are entered manually into Cell D3 and E3, respectively. The periodic payment is always rounded up to the next higher monetary unit—cents in this case in order for a loan to be completely paid by the indicated number of payments—in this case 240. Thus the loan documents will specify 239 payments of the amount indicated in Cell D3 and a final smaller payment. That calculation is shown in Section 4.4. Note that the Proceeds in Cell A3 is slightly greater than 60,000 and is due to the rounding up of the periodic payment. Also note that the values of Row 3 (i.e., term, payments per year, and the original interest rate) are repeated in Row 4 in order that the calculation of the periodic payment is a separate process.

The following steps should be followed. Highlight the Proceeds Cell, A3, as shown below.

Note that the formula for A3 in the formula bar is for the present value of an annuity. Depress the Tools command button on the menu bar and then, the Goal Seek command button on the submenus. The following dialog box and the spreadsheet below will appear.

In the Goal Seek dialog box, since Cell A3 is highlighted, the Set Cell entry is already entered. The proceeds of the loan are entered in the To Value box, and the Cell location that includes the unknown interest rate is entered in the By Changing Cell box. Depress the OK button to proceed with the Goal Seek. A rapid set of iterations to convergence, if convergence is possible, is shown.

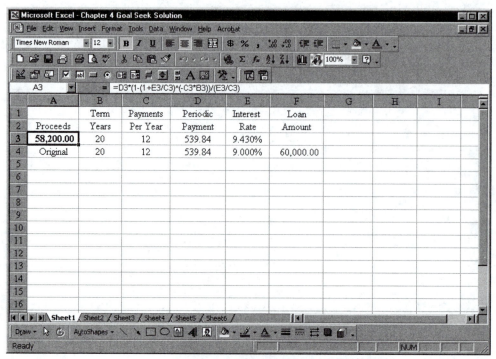

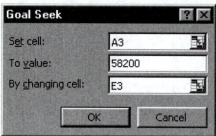

The true annual interest rate will appear in both a second dialog box and in Cell E3 of the spreadsheet below the dialog box. Depress OK to exit Goal Seek.

Whenever a solution to an equation cannot be determined explicitly by algebraic methods, the Goal Seek option affords an efficient way to obtain a solution by trial and error.

PROBLEM SET 4.2

For Problems 1–8, solve Equation (4.2-2) using a value of p = 1. These are the present values of annuities of $1 per period.

	Interest Rate	Compounding Frequency	Term
1.	6%	Monthly	5 years
2.	7%	Quarterly	6 years

	Interest Rate	Compounding Frequency	Term
3.	8%	Semiannually	7 years
4.	9%	Annually	10 years
5.	6%	Monthly	15 years
6.	7%	Quarterly	20 years
7.	8%	Semiannually	25 years
8.	9%	Annually	30 years

Determine the present values in Problems 9–12 for:

9. A monthly payment of $100 for Problem 1.

10. A quarterly payment of $300 for Problem 2.

11. A semiannual payment of $600 for Problem 3.

12. An annual payment of $1,200 for Problem 4.

Determine the periodic payments in Problems 13–16 for:

13. A present value of $10,000 for Problem 5.

14. A present value of $5,000 for Problem 6.

15. A present value of $10,000 for Problem 7.

16. A present value of $1,000,000 for Problem 8.

17. A pension of $25,000 per year is desired for 10 years after retirement. The interest rate during the preretirement years never falls below 6% compounded annually, and the interest rate during the postretirement years never falls below 5% compounded annually. Determine the necessary annual deposit that must be made from age 30 to age 65, assuming that the first deposit is made at age 30.

18. Repeat Problem 17 for 15 retirement years.

19. Your parents estimated that by the time you enter college, yearly costs would amount to $15,000. Starting one year after you were born, and for 17 years thereafter, they made annual deposits into a savings account that paid 5% compounded annually. The money will be withdrawn once each year for four years. What were the annual deposits?

20. If payments are made at the end of each year, an Individual Retirement Account (IRA) funded with $2,000 per year will accumulate to $27,632.90 in 10 years at an interest rate of 7% compounded annually. Construct a Sinking Fund Schedule.

21. If payments are made at the end of each year, an Individual Retirement Account (IRA) funded with $2,000 per year will accumulate to $11,501.48 in five years at an interest rate of 7% compounded annually. Construct a Sinking Fund Schedule.

22. If payments are made at the end of each year, an Individual Retirement Account (IRA) funded with $2,000 per year will accumulate to $28,973.12 in

10 years at an interest rate of 8% compounded annually. Construct a Sinking Fund Schedule.

23. If payments are made at the end of each year, an Individual Retirement Account (IRA) funded with $2,000 per year will accumulate to $11,733.20 in five years at an interest rate of 8% compounded annually. Construct a Sinking Fund Schedule.

24. Assuming that the accumulated funds in Problem 20 must be withdrawn over a period of five years, determine the annual withdrawal amount if the interest rate is assumed to remain constant at 6% and construct an Amortization Schedule with the first withdrawal occurring in one year.

25. Assuming that the accumulated funds in Problem 21 must be withdrawn over a period of five years, determine the annual withdrawal amount if the interest rate is assumed to remain constant at 7% and construct an Amortization Schedule with the first withdrawal occurring in one year.

26. Assuming that the accumulated funds in Problem 22 must be withdrawn over a period of five years, determine the annual withdrawal amount if the interest rate is assumed to remain constant at 8% and construct an Amortization Schedule with the first withdrawal occurring in one year.

27. Assuming that the accumulated funds in Problem 23 must be withdrawn over a period of five years, determine the annual withdrawal amount if the interest rate is assumed to remain constant at 8% and construct an Amortization Schedule with the first withdrawal occurring in one year.

28. A lending institution is charging 8% interest with four points on a mortgage of $200,000. Determine the true annual interest for a 20-year loan with quarterly payments.

29. A lending institution is charging 10% interest with two points on a loan of $500,000. Determine the true annual interest rate for a 25-year loan with semiannual payments.

30. Using 4% add-on interest for an automobile loan of $10,000 for 48 months determine the true annual interest rate.

4.3 ANNUITY DUE

Recall that a simple, ordinary annuity is an annuity where payment dates and interest conversion dates coincide, and payments occur at the end of interest conversion periods. These two conditions are not always practical. There may be situations where payments are made at the beginning of interest conversions such as for lease payments, insurance premiums, and pension payments. When the payments of an annuity are made at the beginning of interest conversion periods, the annuity is called an **annuity due**. The symbolic notations for annuities due are S(due) for the future amount and A(due) for the present value. Other notations for the elements of annuities due are the same as for ordinary annuities and are summarized as follows.

p is the periodic payment,

n is the number of payments,

i is the interest rate per period,

r is the nominal annual interest rate,

f is the number of interest conversions per year, and

t is the number of years.

FUTURE AMOUNT

As indicated above, the payment of an annuity due is made at the beginning of the period while that of an ordinary annuity is made at the end of the period. By converting an annuity due payment, p, from the beginning of the period to a value, $p(1 + i)$, at the end of the same period, the process becomes the same as an ordinary annuity, and we can use Equation (4.1-2) to calculate S(due). The time diagram, showing the conversion, for the future amount of an annuity due is shown in Figure 4-7.

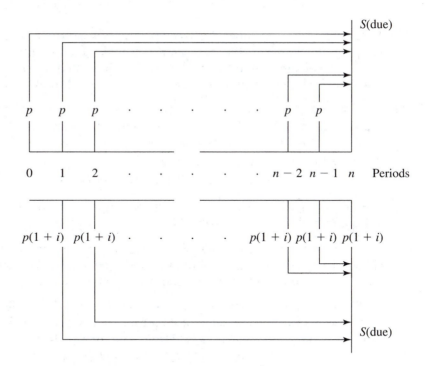

FIGURE 4-7 Time Diagram for the Future Amount of an Annuity Due

Using Equation (4.1-2) and the conversion of p to $p(1 + i)$ gives

$$S(\text{due}) = p(1 + i)\left[\frac{(1 + i)^n - 1}{i}\right].$$

The equation can be used more efficiently on a calculator by rewriting it in the form

$$S(\text{due}) = p\left[\frac{(1+i)^n - 1}{i}\right](1+i). \qquad (4.3\text{-}1)$$

Since we use S as the notation for the future amount of an ordinary annuity, the future amount of an annuity due is simply

$$S(\text{due}) = S(1+i). \qquad (4.3\text{-}2)$$

The standard notation that is used by actuaries for future amount for an annuity due would be expressed as

$$\ddot{S} = p\ddot{s}_{\overline{n}|}, \qquad (4.3\text{-}3)$$

where,

$$\ddot{s} = \left[\frac{(1+i)^n - 1}{i}\right](1+i). \qquad (4.3\text{-}4)$$

EXAMPLE 4.3.1   On December 31, you open a bank account with a $100 deposit. Through November of the next year, you make $100 monthly deposits on the last day of each month. If the bank pays 5% interest compounded monthly, how much will be in the account on December 31, of the following year?

Solution   Since 1 deposit is made in the first year and 11 in the second year, the number of payments is 12. Therefore, $n = 12$. The periodic interest rate is $\frac{5}{12}$%; therefore $i = \frac{5}{12}$%. Then, the term in the brackets of Equation (4.3-1) = 12.27885549 and $(1+i) = 1.00416667$. From Equation (4.3-2)
$$\ddot{S} = 100(12.27885549)(1.00416667) = \$1{,}233.$$

In order to establish a sinking fund with an annuity due, either Equation (4.3-1) or Equation (4.3-2) may be solved for p. ■

EXAMPLE 4.3.2   It is desired to have $5,000 (toward the purchase of a new car) in a savings account at the end of three years. If the bank pays interest at the rate of 5% compounded monthly, what monthly payment, beginning now, is required?

Solution   Since the monthly payments are to begin "now," there will be 36 payments in three years with the accrued amount remaining on deposit for one additional month. Then $n = 36$ and $i = \frac{5}{12}$%. Using Equation (4.3-1),
$$5{,}000 = p(38.75333555)(1.004166667),$$

and

$$p = \frac{5{,}000}{38.91480778} = 128.48579.$$

This value would be rounded up to the next higher penny, and monthly deposits that must be $128.49. ■

PRESENT VALUE

The present value of the annuity due is the amount that would be necessary at the time of the first payment of an ordinary annuity as shown in Figure 4-8.

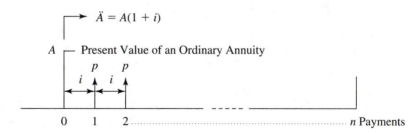

FIGURE 4-8 Time Diagram for the Present Value of an Annuity Due

Figure 4-7 shows that

$$A(\text{due}) = A(1 + i),\tag{4.3-5}$$

or

$$A(\text{due}) = p\left[\frac{1 - (1 + i)^{-n}}{i}\right](1 + i).\tag{4.3-6}$$

The standard notation that is used by actuaries for the present value for annuities due are as follows.

$$\ddot{A} = p\ddot{a}_{\overline{n}|},\tag{4.3-7}$$

where,

$$\ddot{a}_{\overline{n}|} = \left[\frac{1 - (1 + i)^{-n}}{i}\right](1 + i).\tag{4.3-8}$$

EXAMPLE 4.3.3   A retiree draws $1,000 per month and will do this for 10 years. If the account earns 6% interest compounded monthly, how much was in the account on the day of the first withdrawal?

Solution   This is an annuity due because the first payment occurs at the beginning of the payment period. Then, for 6% compounded monthly, $i = \frac{1}{2}\%$, and for 10 years, $n = 120$. The term in the brackets of Equation (4.3-6) = 90.07345332. Then,
$$\ddot{A} = 1{,}000(90.073455332)(1.005) = 90{,}523.82.$$

Thus, $90,523.82 at the start of the annuity will generate $120,000 over the 10-year term of the annuity. ■

PROBLEM SET 4.3

Find the a) amount and b) the present value of the annuities due in Problems 1–5.

	Payment	Payment Interval	Term Years	Interest Conversion Rate
1.	$200	1 month	4	5%, monthly
2.	$500	1 quarter	5	6%, quarterly
3.	$1,000	6 months	10	7%, semiannually
4.	$2,000	1 year	10	8%, annually
5.	$5,000	6 months	4	10%, semiannually

6. You open a savings account with an initial deposit of $50. On the same day of each month thereafter you make $50 deposits. If the bank pays interest at 5% compounded monthly, how much will be in the account just before the 49th deposit?

7. On June 1, you must send a check for $7,500 to your college for next semester's tuition. How much must be in the account at the time of the 1st withdrawal in order to have prepared for the remaining seven semiannual tuition payments? Assume the account pays interest at $5\frac{1}{2}$% compounded semiannually and that tuition will not change.

8. Beginning on your 25th birthday and every birthday thereafter, you deposit $1,000 into a retirement account until your 64th birthday. If the account pays interest at 7% compounded annually, how much will be available on your 65th birthday?

9. You have been assigned to a foreign office of your company and will remain there for one year. Your apartment rents for $800 per month in advance, and you wish to retain the lease because of periodic visits to the home office. Instead of making monthly payments, you negotiate with the landlord for a single payment at the start of the year. If a 10% interest rate compounded monthly is agreed upon, what single amount, in advance, will satisfy the lease?

10. Repeat Problem 8 for annual deposits of $2,000 and interest rate 8%.

4.4 PRACTICAL APPLICATIONS

It is often necessary to determine the number of periods that will be required to accumulate a specified future amount, given the equal periodic payments. Or, given the equal periodic payments, it is often necessary to determine the number of payments that will deplete a present value. When a loan of n payments is negotiated, the loan documents indicate $n - 1$ equal payments and a smaller n^{th} payment. This is because the payment that is calculated is rounded up to the next higher monetary unit; therefore, each equal payment is slightly greater than necessary and serves to reduce the principal a slightly greater amount. Thus, in order to indicate what the final payment would be on the loan documents, the balance after $n - 1$ payments must be calculated. That balance is multiplied by $(1 + i)$ in order to determine the final payment. More generally, the balance after any number of payments can be determined.

THE NUMBER OF PAYMENTS TO ACHIEVE A SPECIFIED FUTURE AMOUNT

Given the payment amount, the number of payments to achieve a specified future amount may be determined by solving for n in Equation (4.1-2).

$$n = \frac{Ln\left(1 + \dfrac{Si}{p}\right)}{Ln(1 + i)}. \tag{4.4-1}$$

where Ln represents natural logarithms. Equation (4.4-1) is utilized for simple, ordinary annuities. For annuities due, p is replaced by $p(1 + i)$ as follows:

$$n = \frac{Ln\left(1 + \dfrac{Si}{p(1 + i)}\right)}{Ln(1 + i)}. \tag{4.4-2}$$

Typically, the results of Equation (4.4-1) and Equation (4.4-2) include a fraction of a period. For compound interest, the interest is added at the end of the period; therefore, the future amount at the end of the next to the last period would be accumulated for one additional period. This could result in an accumulation, which is greater than the desired amount, and a final fractional payment would not be necessary.

EXAMPLE 4.4.1 How many payments of $2,000 at the end of each year are necessary to achieve a future amount of $300,000 if the interest rate remains constant at 6% compounded annually?

Solution Since the frequency of compounding is one, $i = 6\%$. Then, using Equation (4.4-1)

$$n = \frac{Ln\left(1 + \dfrac{(300,000)(0.06)}{2,000}\right)}{Ln(1 + 0.06)} = 39.52.$$

Since the frequency of compounding is one, the number of years is $\frac{39.52}{1}$ or 39.52 years. The "exact" value of 39.52 may be interpreted as 39 annual payments of $2,000, and a possible final 40th payment at the end of year 40. (Although the accumulating fund will reach the goal of $300,000 at approximately midyear between years 39 and 40, for annual compounding, the interest for the year 39.52 would be added at the end of year 40.) In order to determine the amount of the final payment, we must first calculate the future amount of 39 annual payments of $2,000, which is $290,116.92. The accumulation of this amount for one year at 6% would give $307,523.93, which is greater than the specified goal of $300,000. Therefore, 39 annual payments of $2,000 at the end of each year for 39 years will amount to $307,523.93 at the end of year 40. ∎

THE NUMBER OF PAYMENTS TO DEPLETE A SPECIFIED PRESENT VALUE

Given the payment amount, the number of payments to deplete a specified present value may be determined by solving for n in Equation (4.2-2).

$$n = -\frac{\text{Ln}\left(1 - \frac{Ai}{p}\right)}{\text{Ln}(1 + i)}, \tag{4.4-3}$$

Equation (4.4-3) is for simple ordinary annuities. For annuities due, p is replaced by $p(1 + i)$ as follows.

$$n = -\frac{\text{Ln}\left(1 - \frac{Ai}{p(1 + i)}\right)}{\text{Ln}(1 + i)}. \tag{4.4-4}$$

Typically, the results of Equation (4.2-3) and Equation (4.2-4) include a fraction of a payment that would be rounded up to next whole number. This may be interpreted as the whole number part of $n - 1$ payments of p, plus a final smaller payment at the end of the nth period.

EXAMPLE 4.4.2   How many payments of $25,000 at the end of each year are necessary to deplete a fund of $300,000 if the interest rate remains constant at 6% compounded annually?

Solution   Since the frequency of compounding is one, $i = 6\%$. Then, using Equation (4.2-2)

$$n = -\frac{\text{Ln}\left(1 - \frac{(300,000)(0.06)}{25,000}\right)}{\text{Ln}(1 + 0.06)},$$

which reduces to 21.85 and rounds up to 22 periods. Then, at a constant 6% annual interest rate, 21 annual withdrawals of $25,000 plus a 22nd payment of an amount less than $25,000 are available from a present value of $300,000. ■

BALANCE OF A FUND DEPLETION AFTER A SPECIFIED NUMBER OF WITHDRAWALS

The balance of a fund after m withdrawals have been made at the end of interest conversion periods can be determined retrospectively by

$$B_m = A(1 + i)^m - p\left[\frac{(1 + i)^m - 1}{i}\right]. \tag{4.4-5}$$

In the depletion of a fund, if payments are to begin "now," at the beginning of interest conversion periods, the balance after m payments is given by

$$B_m = (A - p)(1 + i)^{m-1} - p\left[\frac{(1 + i)^{m-1} - 1}{i}\right]. \tag{4.4-6}$$

If there are n total payments and $m = n - 1$, the final amount available is determined by accumulating the mth balance for one interest conversion period as follows:

$$p_n = B_{n-1}(1 + i). \tag{4.4-7}$$

EXAMPLE 4.4.3 What will be the final amount available at the end of year 22 for Example 4.4.2?

Solution Equation (4.4-5) becomes

$$B_{21} = 300,000(1.06)^{21} - 25,000 \left[\frac{(1.06)^{21} - 1}{0.06} \right],$$

and the balance after 21 withdrawals of $25,000 becomes

$$B_{21} = 1,019,869.08 - 999,818.17 = 20,050.91.$$

The 22nd and final amount available is determined by Equation (4.4-7) as

$$p_{22} = 20,050.91(1.06) = \$21,253.97. \quad \blacksquare$$

EXAMPLE 4.4.4 How many withdrawals of $25,000 at the beginning of each year are necessary to deplete a fund of $300,000 if the interest rate remains constant at 6% compounded annually, and what is the amount of the final withdrawal?

Solution Equation (4.4-4) becomes

$$n = -\frac{\text{Ln}\left(1 - \frac{(300,000)(0.06)}{(25,000)(1.06)}\right)}{\text{Ln}(1.06)} = 19.51,$$

which rounds up to 20 payments. The balance of the fund after 19 payments is determined by Equation (4.4-6) as

$$B_{19} = (300,000 - 25,000)(1.06)^{18} - 25,000 \left[\frac{(1.06)^{18} - 1}{0.06} \right].$$

The balance after 19 withdrawals then becomes

$$B_{19} = 784,943.27 - 772,641.31 = \$12,301.95.$$

The 20th and final, payment amount is determined by Equation (4.4-7) as

$$P_{20} = 12,301.95(1.06) = \$13,040.07. \quad \blacksquare$$

The amortization of a loan is equivalent to the depletion of a fund in that the lending institution has a "fund" with the borrower. The borrower's payments to the lender constitute the lender's "withdrawal" from the "fund."

THE PRINCIPAL AND INTEREST COMPONENTS OF AN AMORTIZATION OF A FUND

The periodic payment that depletes a fund by amortization can be separated into the component parts for any payment. The following will be exact if the periodic payments are equal. However, the usual practice is to increase the payment amount to the next higher monetary unit that is being used; therefore, the following will give very close approximations. This knowledge enables us to determine the payment number when the principal part of a payment exceeds the interest. Letting:

P be the principal part of a payment,

I be the interest part of a payment,

r be the nominal annual interest rate,

f be the frequency of compounding,

p be the periodic payment,

n be the total number of payments, and

m be the payment number in question,

the principal part of an amortization is given by

$$P = p(1+i)^{-(n-m+1)}, \qquad\qquad (4.4\text{-}8)$$

where, $i = \frac{r}{f}$. Since the total payment is comprised of principal and interest, the interest component would be

$$I = p - P. \qquad\qquad (4.4\text{-}9)$$

EXAMPLE 4.4.5　A fund of \$176,642.13 is being depleted by amortization with 10 annual payments of \$24,000 at a 6% annual interest rate. Determine the interest and principal components of the 6th payment.

Solution　The principal component can be determined from Equation (4.4-8) as
$$P = 24{,}000(1+6\%)^{-(10-6+1)} = 24{,}000(1.06)^{-5}$$
$$= 24{,}000(0.7472582) = \$17{,}934.20.$$

The interest component becomes, from Equation (4.4-9)
$$I = 24{,}000 - 17{,}934.20 = \$6{,}065.80. \quad\blacksquare$$

These values confirm the values in Table 4-2. The interest component can be determined directly by

$$I = p[1 - (1+i)^{-(n-m+1)}]. \qquad\qquad (4.4\text{-}10)$$

　　The payment number when the principal component of an amortization payment equals the interest component is the point in time when reduction of the balance accelerates. Letting k be the desired fraction of the total payment, $P = kp$. Then,

$$kp = p(1+i)^{-(n-m+1)}.$$

The p cancels, therefore, the result depends upon n and i only. Solving for m gives the desired payment number as

$$m = n + 1 + \frac{\text{Ln}(k)}{\text{Ln}(1+i)}. \tag{4.4-11}$$

The value of m will usually include a decimal part, and as such the required number of payments will be the value of m increased to the next whole number. The principal part of a payment will exceed the interest part from this value of m forward.

EXAMPLE 4.4.6   Determine the payment number in Example 4.4.5 when the interest component is $\frac{1}{4}$ the payment.

Solution   If the interest component is $\frac{1}{4}$ the payment, the principal component will be $\frac{3}{4}$ the payment. From Example 4.4.5, $n = 10$, and $i = 6\% = 0.06$. Then,

$$m = 10 + 1 + \frac{\text{Ln}\left(\frac{3}{4}\right)}{\text{Ln}(1.06)} = 11 - 4.9 \approx 6. \quad \blacksquare$$

This may be verified in Example 4.4.5, since the interest component of the 6th payment is \$6,065.80, which is approximately $\frac{1}{4}$ the payment.

The value of k would be $\frac{1}{2}$ when the principal and interest components are equal.

PROBLEM SET 4.4

Determine the number of whole payments for Problems 1–5 if payments are considered to be at the end of interest conversion periods.

	Future Amount	Periodic Payment	Interest Rate	Compounding Frequency
1.	\$3,750	\$300	10%	1
2.	\$4,300	\$250	8%	2
3.	\$6,000	\$250	10%	4
4.	\$10,000	\$100	6%	12
5.	\$50,000	\$5,000	7%	1

6. Find the amount of the final payment for Problem 1 if a fractional payment was indicated.

7. Find the amount of the final payment for Problem 2 if a fractional payment was indicated.

8. Find the amount of the final payment for Problem 3 if a fractional payment was indicated.

9. Find the amount of the final payment for Problem 4 if a fractional payment was indicated.

10. Find the amount of the final payment for Problem 5 if a fractional payment was indicated.

Determine the number of withdrawals for Problems 11–14 if the withdrawals are considered to be at the beginning of interest conversion periods.

	Present Value	Periodic Payment	Interest Rate	Compounding Frequency
11.	$3,750	$500	10%	1
12.	$4,300	$250	8%	2
13.	$6,000	$250	10%	4
14.	$10,000	$100	6%	12
15.	$50,000	$5,000	7%	1

16. Find the amount of the final withdrawal for Problem 11.

17. Find the amount of the final withdrawal for Problem 12.

18. Find the amount of the final withdrawal for Problem 13.

19. Find the amount of the final withdrawal for Problem 14.

20. Find the amount of the final withdrawal for Problem 15.

21–30. Repeat Problems 1–10 for payments at the beginnings of interest conversion periods.

31–40. Repeat Problems 11–20 for withdrawals at the ends of interest conversion periods.

41. Your parents borrow $150,000 in order to purchase a home. If the interest rate is 7% compounded monthly and the term of the loan (mortgage) is 30 years,

    a) What will be the monthly payment?

    b) What will be the amount of the final payment?

    c) At what payment number will the principal component exceed the interest component of the payment?

    d) What will be the principal component at this payment number?

42. Your parents borrow $200,000 in order to purchase a home. If the interest rate is 7% compounded monthly and the term of the loan (mortgage) is 30 years,

    a) What will be the monthly payment?

    b) What will be the amount of the final payment?

    c) At what payment number will the principal component exceed the interest component of the payment?

    d) What will be the principal component at this payment number?

43. Your parents borrow $150,000 in order to purchase a home. If the interest rate is 6% compounded monthly and the term of the loan (mortgage) is 30 years,

    a) What will be the monthly payment?

    b) What will be the amount of the final payment?

    c) At what payment number will the principal component exceed the interest component of the payment

    d) What will be the principal component at this payment number?

44. A corporation borrows $1,000,000 for capital improvements. If the interest rate is 9% compounded quarterly and the term of the loan is 30 years,

    a) What will be the quarterly payment?

    b) What will be the amount of the final payment?

    c) At what payment number will the principal component exceed the interest component of the payment?

    d) What will be the principal component at this payment number?

45. A corporation borrows $1,000,000 for capital improvements. If the interest rate is 6% compounded quarterly and the term of the loan is 30 years,

    a) What will be the quarterly payment?

    b) What will be the amount of the final payment?

    c) At what payment number will the principal component exceed the interest component of the payment?

    d) What will be the principal component at this payment number?

CHAPTER

5

Other Annuities

Simple annuities were discussed in Chapter 4. Recall that a simple, ordinary annuity is an annuity where payment dates and interest conversion dates coincide. There are annuities where:

A. The start of the annuity is deferred to some specified future date.

B. Payment dates and interest conversion dates do not coincide.

C. Payments are made for an indefinite number of periods.

D. Payment amounts increase/decrease by a constant rate each year.

Simple annuities and categories A, B, and D above are known as **annuities certain** because the number of payments is specified. Category C is known as either **annuity in perpetuity** or a **life annuity** because the payments never cease or they terminate upon the death of the recipient. This chapter discusses the annuities certain; life annuities are discussed in Chapter 9.

5.1 DEFERRED ANNUITIES

When the payments of an annuity begin at a future date, the annuity is called a **deferred annuity**. The symbolic notations are S_d for the future amount and A_d for the present value.

FUTURE AMOUNT

Since there is no accumulation of payments and interest during the deferred term, the future amount of a deferred annuity may be determined in the same manner as the future amount of an ordinary annuity.

PRESENT VALUE

For deferred annuities, the **period of deferment** is defined as the number of interest conversion periods before the beginning of an ordinary annuity. The beginning of an ordinary annuity is one period before the first payment. A time diagram for the present value of a deferred annuity is shown in Figure 5-1.

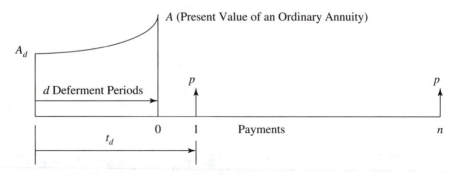

FIGURE 5-1 Time Diagram for the Present Value of a Deferred Annuity

Letting:

d be the number of periods of deferment,
t_d be the number of years to the first payment,
f be the compounding frequency per year, and
A_d be the present value of a deferred annuity.

$$d = ft_d - 1. \tag{5.1-1}$$

The present value of the ordinary annuity, which is deferred, is given by

$$A = A_d(1 + i)^d, \tag{5.1-2}$$

which gives the present value of the deferred annuity as

$$A_d = A(1 + i)^{-d}. \tag{5.1-3}$$

Since A is the present value of an ordinary annuity of n periods, the present value of a deferred annuity may be calculated by

$$A_d = p\left[\frac{1 - (1 + i)^{-n}}{i}\right](1 + i)^{-d}. \tag{5.1-4}$$

EXAMPLE 5.1.1 You wish to purchase a long-term security that will mature in 18 years in order to be able to withdraw $5,000 per year for four subsequent years to pay college tuition for your child. If the interest rate on the security is at 6% compounded annually, how much must you deposit now?

Solution   Since the first payment of the annuity does not occur for 18 years, it is a deferred annuity. Because the interest is to be compounded annually, $f = 1$, and from Equation (5.1-1) the number of periods of deferment

$$d = (1)(18) - 1 = 17 \text{ periods.}$$

The number of periods in the annuity is four because the compounding frequency and annual withdrawals occur once each year. Thus, $n = 4$, $i = \frac{r}{f} = \frac{6\%}{1} = 0.06$, and using Equation (5.1-4)

$$A_d = 5,000 \left[\frac{1 - (1.06)^{-4}}{0.06} \right] (1.06)^{-17}$$

$$= 5,000(3.465106)(0.371364) = 6,434,08.$$

Thus, $6,434.08 invested now in a security paying 6% interest compounded annually will allow a total of $20,000 to be withdrawn at the rate of $5,000 per year starting 18 years from now. ∎

PROBLEM SET 5.1

Find the present values for the following deferred annuities.

	Payment	Number of Payments	Deferred Periods	Annual Interest Rate
1.	$200	15	10	5%, monthly
2.	$500	40	50	6%, quarterly
3.	$1,000	20	60	7%, semiannually
4.	$2,000	10	15	8%, annually
5.	$5,000	8	18	10%, semiannually

6. Fifteen years ago a rich uncle deposited $20,000 in a savings account that paid interest at the rate of 5% compounded monthly. Starting now, you wish to make monthly withdrawals in an amount such that the account will be depleted in seven years. How much can you withdraw each month?

7. Repeat Problem 6 if the account pays 7% compounded monthly during the withdrawal years.

8. For the conditions of Problem 6, for how many months can a withdrawal of $1,000 per month be made?

9. The U.S. Government lends $100,000,000 to another country. The loan is to be repaid over a period of 25 years, beginning 10 years from now, at an interest rate of 10% compounded annually. What will be the annual repayment of the loan?

10. Repeat Problem 9 if the beginning of repayments is to be deferred 20 years.

5.2 GENERAL (COMPLEX) ANNUITIES

Chapter 4 and Section 5.1 of this chapter discussed simple annuities where payment dates coincide with interest conversion dates for discrete compounding of interest. However, timing of payments to coincide with interest conversion may not always be practical. When payment dates do not coincide with interest conversion dates, the annuity becomes a **general annuity** (also called a **complex annuity**). Included in the general annuity are the ordinary, due, and deferred forms of payments with two possible modes of payments.

1. Payments may be less frequent than interest conversion.
2. Payments may be more frequent than interest conversion.

The methods of this section will allow either situation to be treated in the same manner.

GENERAL (COMPLEX) ORDINARY ANNUITIES

For *simple* annuities, the number of payments and the number of interest conversion periods are equal and the letter n was used rather "carelessly" to imply either number. *General* annuities require more rigor. The symbolic notation is as follows for general annuities:

t = the number of years actual payments are made,

r = the nominal annual interest rate,

f' = the number of actual payments per year,

f = the number of interest conversions per year, and

n = the number of actual payments in the term of the annuity.

In order to determine the future amount or present value of a general (complex) annuity, an equivalent interest rate between payments must be determined such that the annuity can be treated as a simple annuity. In order to accomplish this, we must determine an interest rate between payments that is equivalent to the interest rate between conversion periods. That is, the actual payment dates and the equivalent interest conversion dates are forced to coincide. If we let i, defined as $\frac{r}{f}$, be the interest rate between interest conversions, and j be the interest rate between payments that would accumulate to i at each conversion, the equivalency is shown in Figure 5-2.

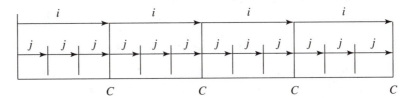

FIGURE 5-2 Equivalency between Interest Rates, j, per Payment Period and Interest Rates, i, per Conversion Period

This equivalency is given by

$$(1 + j)^{f'} = (1 + i)^{f}, \tag{5.2-1}$$

from which

$$1 + j = (1 + i)^{\frac{f}{f'}}, \qquad (5.2\text{-}2)$$

and from which

$$j = (1 + i)^{\frac{f}{f'}} - 1. \qquad (5.2\text{-}3)$$

The number of payments is determined by

$$n = f't. \qquad (5.2\text{-}4)$$

EXAMPLE 5.2.1 For an interest rate of 8% compounded quarterly, what is the equivalent interest rate per payment period for 12 payments per year?

Solution For $r = 8\%$ compounded quarterly, $f = 4$, $i = \frac{8}{400} = 0.02$, and for monthly payments, $f' = 12$. Then, from the Equation (5.2-1), the definition for j,

$$(1 + j)^{12} = (1.02)^4,$$

and taking the 12th root of both sides,

$$1 + j = (1.02)^{\frac{4}{12}} = (1.02)^{\frac{1}{3}} = 1.00662.$$

Then, $j = 0.00662$ or 0.662%. ■

Both the working formulas and the standard notation of the Society of Actuaries for the general annuities will be shown in the following.

FUTURE AMOUNT OF A GENERAL ANNUITY

The future amount of a general annuity that has an interest rate of $i\%$ per conversion period is the future amount of a simple annuity at $j\%$ per payment period and is given by

$$S = p \left[\frac{(1 + j)^n - 1}{j} \right]. \qquad (5.2\text{-}5a)$$

The actuarial notation for the future amount of $1.00 per period is $s_{\overline{n}|j}$, and Equation (5.2-5a) may be referenced as

$$S = p s_{\overline{n}|j}. \qquad (5.2\text{-}5b)$$

PRESENT VALUE OF A GENERAL ANNUITY

Using the definitions of j and n (number of payments), the present value of a simple ordinary annuity in terms of j is given by

$$A = p \left[\frac{1 - (1 + j)^{-n}}{j} \right]. \qquad (5.2\text{-}6a)$$

The actuarial notation for the present value of $1.00 per period is $a_{\overline{n}|j}$, and Equation (5.2-6a) may be referenced as

$$A = p a_{\overline{n}|j}. \qquad (5.2\text{-}6b)$$

EXAMPLE 5.2.2   Find a) the future amount and b) the present value of an annuity of $1,000 payable at the end of each quarter for five years if the interest rate is 8% compounded monthly.

Solution   In this situation, payments are less frequent than interest conversions. For monthly compounding, $f = 12$ and for quarterly payments, $f' = 4$. Then,

$$i = \frac{r}{f} = \frac{8}{1,200} = 0.006667.$$

The number of payments is obtained from Equation (5.2-4) as

$$n = f't = 4(5) = 20.$$

From the definition of j, Equation (5.2-1),

$$(1 + j)^4 = (1.006667)^{12},$$

from which

$$1 + j = (1.006667)^{\frac{12}{4}} = 1.02013,$$

and from which j is determined as 0.02013 (2.013% per payment period).

a) For the future amount, Equation (5.2-5a) gives

$$S = 1,000 \left[\frac{(1.02013)^{20} - 1}{0.02013} \right]$$

$$= 1,000(24.32885) = \$24,328.85.$$

b) For the present value, Equation(5.2-6a) gives

$$A = 1,000 \left[\frac{1 - (1.02013)^{-20}}{0.02013} \right]$$

$$= 1,000(16.33094) = \$16,330.94. \quad \blacksquare$$

EXAMPLE 5.2.3   Find a) the future amount and b) the present value of an annuity of $1,000 per month for 10 years if the interest rate is 8% compounded quarterly.

Solution   In this situation, payments are more frequent than interest conversions. For quarterly compounding $f = 4$, and for monthly payments, $f' = 12$. Then, $i = \frac{8}{400} = 0.02$, and

$$(1 + j)^{12} = (1.02)^4,$$

from which

$$1 + j = (1.02)^{\frac{4}{12}} = 1.00662,$$

and from which $j = 0.00662$ or 0.662% per payment period.
The number of payments is determined from its definition as

$$n = f't = 12(10) = 120.$$

a) For the future amount, substituting into Equation (5.2-5a) gives

$$S = 1,000 \left[\frac{(1.00662)^{120} - 1}{0.00662} \right]$$

$$= 1,000(182.37561) = \$182,375.61.$$

b) For the present value, substituting into Equation (5.2-6a) gives

$$A = 1,000 \left[\frac{1 - (1.00662)^{-120}}{0.00662} \right]$$

$$= 1,000(82.62285) = \$82,622.85. \quad \blacksquare$$

GENERAL (COMPLEX) ANNUITY DUE

In Chapter 4, Section 4.3, the annuity due was shown to be an ordinary annuity with the sum and present value allowed to earn interest for one additional period. Since the payment and interest conversion dates coincided, the period was the same for both the payment and conversion. However, the word *annuity* refers to payments, and the period that is being referenced is the payment period. Therefore, the annuity due is the same as an ordinary annuity with the sum and present value allowed to earn interest for one additional payment period. Thus,

$$S(\text{due}) = S(1 + j), \tag{5.2-7}$$

which becomes

$$S(\text{due}) = p \left[\frac{(1 + j)^n - 1}{j} \right] (1 + j). \tag{5.2-8a}$$

The standard notation for the future value of $1.00 per payment period is $\ddot{s}_{\overline{n}|j}$, and Equation (5.2-8a) may be referenced as

$$\ddot{S} = p\ddot{s}_{\overline{n}|j}. \tag{5.2-8b}$$

For the present value,

$$\ddot{A} = A(1 + j), \tag{5.2-9}$$

which becomes

$$\ddot{A} = p \left[\frac{1 - (1 + j)^{-n}}{j} \right] (1 + j). \tag{5.2-10a}$$

The actuarial notation for the present value of $1.00 per period is $\ddot{a}_{\overline{n}|j}$, and Equation (5.2-10a) may be referenced as

$$\ddot{A} = p\ddot{a}_{\overline{n}|j}. \tag{5.2-10b}$$

EXAMPLE 5.2.4   Find a) the future amount and b) the present value of an annuity of $1,000 payable at the beginning of each quarter for five years if the interest rate is 8% compounded monthly.

Solution   In this situation, payments are less frequent than interest conversions. Since the payments are made at the beginning of each quarter, the annuity is an annuity due. For monthly compounding, $f = 12$ and for quarterly payments, $f' = 4$. From its definition $i = \frac{8}{1,200} = 0.006667$. From Equation (5.2-1),

$$(1 + j)^4 = (1.006667)^{12}.$$

The solution gives $1 + j = 1.02013$, and $j = 0.02013$. The number of payments in the annuity is

$$n = f't = 4(5) = 20.$$

a) For the future amount, substitution into Equation (5.2-8) gives

$$\ddot{S} = 1,000 \left[\frac{(1.02013)^{20} - 1}{0.02013} \right] (1.02013)$$

$$= \$24{,}818.59.$$

b) For the present value, substitution into Equation (5.2-10) gives

$$\ddot{A} = 1,000 \left[\frac{1 - (1.02013)^{-20}}{0.02013} \right] (1.02013)$$

$$= \$16{,}659.68. \quad \blacksquare$$

GENERAL (COMPLEX) DEFERRED ANNUITY

The future amount of a general deferred annuity is the same as the future amount of an ordinary annuity because no payments are made during the term of deferment. The present value of a general deferred annuity is the value at the beginning of the deferment period. For a general deferred annuity the period of deferment is the number of payment periods before the beginning of the annuity. In order to consider the annuity as an ordinary annuity, the beginning of the annuity is one payment period prior to the first payment. Then, the number of payment periods to the beginning of the annuity is given by

$$d = f't_d - 1, \tag{5.2-11}$$

and

$$A_d = p \left[\frac{1 - (1 + j)^{-n}}{j} \right] (1 + j)^{-d}. \tag{5.2-12}$$

EXAMPLE 5.2.5   What is the present value of an annuity that will pay \$5,000 every six months for eight years, beginning 10 years from now, if the interest rate is constant at 8% compounded quarterly?

Solution   In this situation, payments will be less frequent than interest conversions. For quarterly compounding $f = 4$, and for semiannual payments $f' = 2$. Then $i = \frac{8}{400} = 0.02$, and from Equation (5.2-1),

$$(1 + j)^2 = (1.02)^4.$$

The solution gives $1 + j = 1.0404$, and $j = 0.0404$. The number of payments is determined by

$$n = f't = 2(8) = 16.$$

The number of years (t_d) to the first payment is 10 years and from Equation (5.2-11) the number of deferred payment (and interest conversion at j) periods is determined by

$$d = f't_d - 1 = 2(10) - 1 = 19.$$

Substitution into Equation (5.2-12) gives the present value as $27,371.23. This is left for the student to verify. ■

PROBLEM SET 5.2

In Problems 1–6 find a) the future amount and b) the present value of the annuities where the payments are at the end of interest conversion periods.

	Payment	Payment Interval	Term	Annual Interest Rate Compounded
1.	$ 100	1 month	5 years	6%, quarterly
2.	$ 500	1 quarter	8 years	6%, monthly
3.	$ 500	1 quarter	10 years	6%, semiannually
4.	$1,000	6 months	20 years	8%, quarterly
5.	$5,000	1 year	15 years	10%, semiannually
6.	$5,000	6 months	15 years	12%, annually

7.–12. Determine **a)** the future amount and **b)** the present value of the annuities of Problems 1–6 if the payments are made at the beginning of the interest conversion periods.

In Problems 13–18 determine the present values of Problems 1–6 as deferred annuities.

13. Number of years to first payment $= 5$.

14. Number of years to first payment $= 8$.

15. Number of years to first payment $= 15$.

16. Number of years to first payment $= 20$.

17. Number of years to first payment $= 25$.

18. Number of years to first payment $= 30$.

19. A company establishes a sinking fund for the replacement of equipment. It makes its first payment of $2,000 one year later and annually thereafter for 10 years. If the interest rate for the fund is 9% compounded semiannually, what will be the value of the fund after the last deposit is made?

20. Repeat Problem 19 if payments are $500 at the end of each quarter.

21. In order to save for a college education for their children, a couple opens a bank account with $100. The interest rate is 5% compounded quarterly. At the end of every six months, they deposit $1,000 into the account. How much will be in the account after 15 years? Don't forget that the opening deposit also earns interest.

22. Repeat Problem 21 if the payments are $200 at the end of each month. Don't forget that the initial $100 deposit also earns interest.

23. The company of Problem 19 wishes to accumulate $100,000 during the 10-year period. **a)** What annual payment is required? **b)** What quarterly payment is necessary?

24. The couple of Problem 21 believes they will need $100,000 for the education of two children. **a)** What semiannual deposit is necessary to achieve this goal? **b)** What monthly deposit is necessary?

25. A potential retiree desires to receive $5,000 every month for 25 years during retirement. If the interest rate is 9% compounded annually, what amount must be available one month before the first payment?

26. Repeat Problem 25 for an interest rate of 9% compounded quarterly.

27. One method that may be used to decide between alternatives is the present worth on total investment. This present worth may be defined as the difference between the present value of the profits that an alternative may earn and the first cost of the alternative. That is present worth $= A - F$, where F is the first cost. If the life of two machines is 10 years each and if machine A costs $20,000 with an expected annual profit of $10,000, and machine B costs $30,000 with an expected annual profit of $12,000, which alternative should be selected if the interest rate is 12% compounded quarterly?

28. Repeat Problem 27 for expected monthly profits of $1,000 from machine A, and $1,100 from machine B.

29. A rich uncle establishes a trust fund for you on the day you were born, such that when you reach age 25 and on every birthday thereafter for 20 years, you will receive $5,000. If the interest rate is a constant 6% compounded semiannually, how much money is deposited on the day of your birth?

30. Repeat Problem 29 for monthly payments of $500.

5.3 ANNUITY IN PERPETUITY

Annuities in perpetuity are annuities where the term of the annuity is indefinite. The payments continue forever, and therefore, the only meaningful determination is the present value of the annuity. True perpetuities exist for nonprofit organizations, such as colleges, that establish endowment funds to be used to grant scholarships forever, or in perpetuity. Most states in the United States also allow the establishment of private trusts from which payments to children and their issue are controlled in perpetuity. However, in a few states, perpetuity is defined as the life of the "youngest in being" plus 21 years (approximately 100 years) from the time a perpetual trust is established. This means that life expectancies of children born after that time do not contribute to the duration of the trust.

The payment for such an annuity is the interest earned on the principal, and if a specified periodic payment is to continue forever, that principal (present value) cannot be reduced below the amount required in order to generate the necessary interest. There are three possibilities for annuities in perpetuity.

1. Simple perpetuities (ordinary and due)
2. General perpetuities (ordinary and due)
3. Deferred perpetuities

Each of the three types of perpetuities is discussed below, and the symbol that denotes the present value of perpetuity is A_∞.

SIMPLE ORDINARY PERPETUITY AND PERPETUITY DUE

Between interest conversion periods, money earns interest at the interest rate per conversion period. Then, for one period, the interest (payment) is

$$p = A_\infty i,$$

and the present value of a simple ordinary perpetuity is

$$A_\infty = \frac{p}{i}. \qquad (5.3\text{-}1)$$

The time diagram for a simple ordinary perpetuity is shown in Figure 5-3.

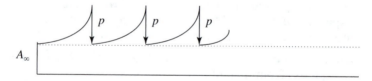

FIGURE 5-3 Time Diagram for a Simple Ordinary Perpetuity

EXAMPLE 5.3.1   What is the present value of a simple perpetuity of $1,000 at the end of each quarter if the interest rate is 10% compounded quarterly?

Solution   Since $r = 10\%$, and $f = 4$ (quarterly compounding), $i = \frac{10}{400} = 0.025$. Then, from Equation (5.3-1), the present value is

$$A_\infty = \frac{1,000}{0.025} = 40,000.$$

Thus, $40,000 will earn $1,000 interest in three months, and $1,000 may be paid to the annuitant (recipient of payment) without reducing the principal. ■

For the simple perpetuity due, one payment is added to the present value of a simple perpetuity. Thus,

$$\ddot{A}_\infty = \frac{p}{i} + p. \qquad (5.3\text{-}2)$$

GENERAL (COMPLEX) ORDINARY PERPETUITY AND PERPETUITY DUE

The general ordinary perpetuity is an annuity where the payment's frequencies are different than the interest conversion frequencies, and the payments are made at the end of several interest conversion periods. The procedure is to determine a "fictitious" interest rate, j, between payments that will convert the complex perpetuity into a simple perpetuity. Then, the present value of the converted perpetuity is given by Equation (5.3-1) as

$$A_\infty = \frac{p}{j}, \tag{5.3-3}$$

where j is determined from Equation (5.2-3) to be

$$j = (1+i)^{\frac{f}{f'}} - 1.$$

EXAMPLE 5.3.2   Find the present value of a perpetuity that pays $5,000 every six months if the interest rate is 12% compounded monthly.

Solution   For monthly compounding of interest, $i = \frac{12}{1,200} = 0.01$. Since $f' = 2$, and $f = 12$,

$$j = (1.01)^{\frac{12}{2}} - 1 = 0.06152,$$

and substituting into Equation (5.3-3) gives the present value of the general perpetuity as

$$A_\infty = \frac{5,000}{0.06152} = \$81,274.38.$$

For the general perpetuity due, one payment is added to the ordinary perpetuity.

$$\ddot{A}_\infty = \frac{p}{j} + p. \quad \blacksquare \tag{5.3-4}$$

DEFERRED PERPETUITY

If a perpetuity is not to begin for a specified number of years, the present value of the perpetuity is the discounted present value of a simple perpetuity. Then,

$$A_{d\infty} = A_\infty(1 + j)^{-d}. \tag{5.3-5}$$

EXAMPLE 5.3.3   It is desired to establish a perpetuity of $1,000 payable at the end of every six months in perpetuity. The first payment is to be made in 10 years. If the interest rate is 10% compounded quarterly, what is the present value?

Solution   The term of deferment (t_d) is 10 years. Therefore, the number of payment periods in the deferment is

$$d = f't_d - 1 = 2(10) - 1 = 19,$$

and $i = \frac{10}{400} = 0.025$. From Equation (5.2-3),

$$j = (1.025)^{\frac{4}{2}} - 1 = 0.050625.$$

Then, from Equation (5.3-5),
$$A_{d(\infty)} = \frac{1,000}{0.050625}(1.050625)^{-19}$$
$$= 19,753.086(0.391285) = 7,729.08.$$

If \$7,729.08 is invested at 10% compounded quarterly, \$1,000 semiannual payments can begin in 10 years and continue forever. ■

PROBLEM SET 5.3

Find the present value of the following perpetuities.

	Payment	Interest Rate Compounded
1.	\$500 at the end of each quarter	8%, quarterly
2.	\$500 at the end of each month	12%, monthly
3.	\$1,000 at the end of each six months	10%, semiannually
4.	\$5,000 at the end of each year	10%, annually
5.	\$1,000 at the beginning of each quarter	12%, quarterly
6.	\$500 at the beginning of each month	9%, monthly
7.	\$2,500 at the beginning of each six months	8%, semiannually
8.	\$5,000 at the beginning of each year	12%, annually
9.	\$500 at the end of each quarter	8%, monthly
10.	\$1,000 at the end of each six months	12%, quarterly
11.	\$5,000 at the end of each year	7%, semiannually
12.	\$500 at the beginning of each quarter	9%, monthly
13.	\$1,000 at the beginning of each six months	12%, monthly
14.	\$5,000 at the beginning of each year	8%, quarterly
15.	\$5,000 at the beginning of each year	10%, semiannually

16. A wealthy alumnus of a college wishes to establish a \$5,000 per year scholarship at the college. If the interest rate is 9% compounded annually, how much must be donated to the college if the first scholarship is to be awarded one year later?

17. Repeat Problem 16 if the first scholarship is to be awarded at the time of the donation.

18. Repeat Problem 16 if the first scholarship is to be awarded five years after the donation.

19. Repeat Problems 16, 17, and 18, for quarterly compounding of interest.

20. It is estimated that a warehouse will bring a net (after all expenses) rental profit of \$3,000 per month "forever." If interest is available at 10% compounded annually, what should be the minimum selling price of the warehouse? Rent is paid at the beginning of each month.

21. Repeat Problem 20 for interest at 10% compounded semiannually.

22. A trust is obligated to pay a claim of $2,000 per month forever, beginning now. If interest is computed at $\frac{3}{4}$% per month, what amount is in the trust now?

23. Repeat Problem 20 for interest at 8% compounded quarterly.

5.4 GEOMETRICALLY VARYING ANNUITIES

In the discussion of annuities in Chapter 4, and in Sections 5.1–5.3, the periodic payment for each type of annuity has been assumed to be constant. In many practical annuities, however the periodic (monthly, quarterly, or semiannual) amounts are constant for each year, but the annual amount is increased by a rate factor such as inflation, e.g., federal social security payments to retired or disabled workers, state pensions, some private pensions, and housing rents. This section discusses this type of annuity.

PRESENT VALUE OF A GEOMETRICALLY VARYING ORDINARY ANNUITY

Figure 5-4 is a time diagram for this type of annuity.

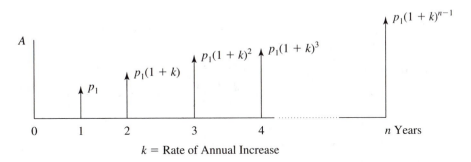

k = Rate of Annual Increase

FIGURE 5-4 Time Diagram for a Geometrically Increasing Ordinary Annuity

The present value for this sequence of payments is the sum of the individual present values,

$$A = \frac{p_1}{(1+i)} + \frac{p_1(1+k)}{(1+i)^2} + \frac{p_1(1+k)^2}{(1+i)^3} + \cdots + \frac{p_1(1+k)^{(n-1)}}{(1+i)^n}, \qquad (5.4\text{-}1)$$

where p_1 is the initial annual amount, and k is the growth rate of the annual payment. Equation (5.4-1) is the sum of a geometric progression. If we indicate the first payment by the letter p_1, then

$$A = p_1 \left[\frac{1 - \left(\frac{1+k}{1+i}\right)^n}{i - k} \right].$$ (5.4-2)

The sign of k would be negative for geometrically decreasing annuities. For the special case where $k = i$

$$A = \frac{np_1}{1+i}.$$ (5.4-3)

EXAMPLE 5.4.1   What is the present value of 25 annual withdrawals from a fund if the first withdrawal of $18,000 is to occur one year from now and is to increase at a rate of 3% per year thereafter? What will be the final withdrawal amount? Assume the interest rate never falls below 5% compounded annually.

Solution   It is given that $p = 18,000$, $k = 0.03$, $f = 1$, $i = \frac{5}{100} = 0.05$, and $n = 25$. Substituting into Equation (5.4-2) gives

$$A = 18,000 \left[\frac{1 - 0.618298}{0.05 - 0.03} \right] = \$343,531.80.$$

The final withdrawal amount, which is shown in Figure 5-4, is given by

$$p_n = 18,000(1.03)^{24} = \$36,590.29. \quad \blacksquare$$

EXAMPLE 5.4.2   What would be the present value for Example 5.4.1 if the interest rate is assumed to be 3%?

Solution   Substituting into Equation (5.4-3),

$$A = \frac{(25)(18,000)}{1.03} = \$436,893.20. \quad \blacksquare$$

PRESENT VALUE OF A GEOMETRICALLY VARYING ANNUITY DUE

As with annuities with fixed periodic payments, the present value of the annuity due is simply the present value of the ordinary annuity times $(1 + i)$. Thus,

$$\ddot{A} = A(1 + i).$$ (5.4-4)

The present value for the annuity due of Example 5.4.1 if payments are at the beginning of the year would be

$$\ddot{A} = 343,531.80(1.05) = \$360,708.39,$$

and the present value for the annuity due of Example 5.4.2 would be

$$\ddot{A} = 436,893.20(1.03) = \$450,000.$$

FUTURE AMOUNT OF A GEOMETRICALLY VARYING ORDINARY ANNUITY

Since the present value of any annuity is a single quantity of money, the future value may be determined by multiplying the present value by $(1 + i)^n$. Then, from Equation (5.4-2), the future amount of a geometrically varying annuity becomes

$$S = p_1 \left[\frac{(1 + i)^n - (1 + k)^n}{i - k} \right]. \qquad (5.4\text{-}5)$$

For the special case where $k = i$, Equation (5.4-5) reduces to

$$S = n p_1 (1 + i)^{n-1} \qquad (5.4\text{-}6)$$

EXAMPLE 5.4.3   At the end of each year for 40 years, a sum of money is transferred to a tax-deferred investment that never returns less than 7% compounded annually. If the initial annual amount is \$1,000 and is increased by 3% each year thereafter, how much will be in the account at the end of 40 years? What will be the final annual amount to be invested? What will be the accumulation if the investment rate is never less than 3%?

Solution   For Equation (5.4-4), $f = 1$, $i = \frac{r}{f} = \frac{7}{100} = 0.07$, $k = 0.03$, $n = 40$, and $p = 1,000$. Then,

$$S = 1,000 \left[\frac{(1.07)^{40} - (1.03)^{40}}{0.07 - 0.03} \right],$$

which reduces to \$292,810.50. The final annual investment amount will be

$$p_{40} = 1,000(1.03)^{39} = \$3,167.03$$

For an investment rate of 3%, a rate equal to the growth rate, Equation (5.4-6) gives

$$S = (40)(1,000)(1.03)^{39} = \$126,681.08. \quad \blacksquare$$

FUTURE AMOUNT OF A GEOMETRICALLY VARYING ANNUITY DUE

The future amount of a uniformly varying annuity due is determined by multiplying the future amount of the uniformly varying ordinary annuity by $(1 + i)$. Then

$$\ddot{S} = S(1 + i). \qquad (5.4\text{-}7)$$

The respective future amounts in Example 5.4.3 become \$313,307.24 and \$130,481,51.

FINAL PERIODIC AMOUNT OF A GEOMETRICALLY VARYING ANNUITY

It is most important to determine the final periodic amount when embarking on an investment strategy that includes geometrically increasing periodic amounts (withdrawals or investments). The formula for the final amount is shown in

Figure 5-4 and has been used in the examples above. It is given here formally as

$$p_n = p_1(1 + k)^{n-1}, \qquad (5.4-8)$$

where p_n = the final annual amount, p_1 = the initial annual amount, and k = the annual rate of increase or decrease.

PROBLEM SET 5.4

1. What is the present value of **a)** an ordinary annuity and **b)** an annuity due of 15 annual withdrawals with an initial withdrawal of $10,000 if the interest rate is 5% compounded annually and the annual amounts are to be increased by 2% each year? What will be the final withdrawal?

2. What is the present value of **a)** an ordinary annuity and **b)** an annuity due of 16 annual withdrawals with an initial withdrawal of $10,000 if the interest rate is 5% compounded annually and the annual amounts are to be increased by 3% each year? What will be the final withdrawal?

3. What is the present value of **a)** an ordinary annuity and **b)** an annuity due of 17 annual withdrawals with an initial withdrawal of $10,000 if the interest rate is 6% compounded annually and the annual amounts are to be increased by 2% each year? What will be the final withdrawal?

4. What is the present value of **a)** an ordinary annuity and **b)** an annuity due of 18 annual withdrawals with an initial withdrawal of $10,000 if the interest rate is 3% compounded annually and the annual amounts are to be increased by 3% each year? What will be the final withdrawal?

5. What is the present value of **a)** an ordinary annuity and **b)** an annuity due of 19 annual withdrawals with an initial withdrawal of $10,000 if the interest rate is 7% compounded annually and the annual amount is to be increased by 2% each year? What will be the final withdrawal?

6. What is the present value of **a)** an ordinary annuity and **b)** an annuity due of 20 annual withdrawals with an initial withdrawal of $10,000 if the interest rate is 3% compounded annually and the annual amount is to be increased by 3% each year? What will be the final withdrawal?

7. What is the future amount of **a)** an ordinary annuity and **b)** an annuity due of 45 annual investments with an initial investment of $2,000 if the interest rate is 5% compounded annually and the annual amounts are to be increased by 2% each year? What will be the final investment amount?

8. What is the future amount of **a)** an ordinary annuity and **b)** an annuity due of 40 annual investments with an initial investment of $2,000 if the interest rate is 3% compounded annually and the annual amounts are to be increased by 3% each year? What will be the final investment amount?

9. What is the future amount of **a)** an ordinary annuity and **b)** an annuity due of 35 annual investments with an initial investment of $2,000 if the interest rate is 6% compounded annually and the annual amounts are to be increased by 3% each year? What will be the final investment amount?

10. What is the future amount of **a)** an ordinary annuity and **b)** an annuity due of 30 annual investments with an initial investment of $2,000 if the interest rate is 3% compounded annually and the annual amounts are to be increased by 3% each year? What will be the final investment amount?

11. What is the future amount of **a)** an ordinary annuity and **b)** an annuity due of 25 annual investments with an initial investment of $2,000 if the interest rate is 7% compounded annually and the annual amount is to be increased by 2% each year? What will be the final investment amount?

12. What is the future amount of **a)** an ordinary annuity and **b)** an annuity due of 20 annual investments with an initial investment of $2,000 if the interest rate is 7% compounded annually and the annual amount is to be increased by 3% each year? What will be the final investment amount?

13. In the depletion of an individual retirement fund by a uniformly varying amount, the philosophy is to increase the annual amount by the rate of inflation. Another aspect, however, is that as one ages, the "quality of life" could diminish, and, perhaps, lesser amounts would be necessary in the future. Suppose one wished to be able to make 15 withdrawals from a retirement fund beginning with $50,000 and decrease the amount by 3% each year. If the interest rate is 7%, what would be the necessary present value of the fund if withdrawals occurred at the beginning of the year? What would be the final withdrawal?

14. The 15-year period in Problem 13 represents the life expectancy of a 65-year-old male. The life expectancy for a 65-year-old female is 19 years. Repeat Problem 13 for a 65-year-old female.

15. Given that the present value of a retirement fund is known at the time of retirement, if one stipulates an initial annual amount to be increased by $k\%$ each year, it becomes necessary to determine the term of such an annuity. Solve Equation (5.4-2) for n.

6

Bond Valuation

One of the methods that governments and corporations use to borrow money for operations or capital improvements is to issue bonds. Purchasers of bonds become lenders of money, and as such expect a periodic interest payment from the parties, who issue the bonds, and the repayment of the principal at some future, specified date. The interest that is paid to bond purchasers is normally at a rate that is comparable or greater than normal rates paid by banks and is, most certainly, less than the interest rates for borrowing money from banks.

Commonly used terms for bonds are as follows:

1. **Face** or **par value**. The amount of the bond. It is the value stated on the bond. Bonds are usually issued in denominations of $1,000 or multiples of $1,000.

2. **Redemption value**. The amount of money that the bond is worth when surrendered to the issuing party. Bonds are usually redeemed at maturity dates for face value, but if the issuer reserves the right to redeem the bonds prior to the maturity date they may set redemption values above the face value to make them more attractive. The redemption date of the bonds prior to maturity is referred to as the **call date**.

3. **Coupon rate**. The interest rate stated on the bond. It is this rate that determines the periodic interest payment. Although physical coupons are no longer common on bonds, some government bonds still have coupons attached to them with the amount of interest indicated. On the date specified, the coupon may be clipped and redeemed for cash. Otherwise, the coupon interest is paid automatically.

4. **Yield rate**. The actual interest rate earned by the investor. Since bonds may be purchased in a bond market, in essentially the same way as in the stock market, the price of the bond may be more or less than the face value of the bond. If the bond is purchased at the face value, the yield rate and bond rate are the same. If the bond is purchased at a value other than the face value, the yield rate will differ from the bond rate.

5. **Interest dates**. Normally, bond interest is paid semiannually, and the dates are normally indicated on the bond or on the coupons.
6. **Purchase price**. Since the bond coupon rate is fixed at issue, the purchase price of a bond, on any date after issue, is a value based on the yield rate demanded by purchasers on that date. As indicated in number 4 above, the price may be different from the face value. If the bond is purchased between interest dates, the price of the bond will be the price based on the yield rate plus accrued interest from the previous interest date.

Corporate bonds may be purchased through brokerage firms and government bonds may be purchased through brokerage firms, banks, or directly from a Federal Reserve Bank. In order to approach a brokerage firm or bank for the purchase of a bond, the purchaser should have some information concerning the bond. Such information is available on the financial pages of many newspapers, and Table 6-1 has been extracted from the December 31, 2001 edition of *Barron's*.

TABLE 6-1 BOND TABLE ENTRIES

52 Weeks High	52 Weeks Low	Name and Coupon	Cur Yld	Sales $1,000	← Weekly → High	← Weekly → Low	← Weekly → Last	Net Change
$110\frac{3}{4}$	65	UtdAir 10.67s04	13.6	673	80	77	$78\frac{1}{4}$	$+1\frac{1}{4}$

The price that is indicated for the purchase of a bond is a percent of face value. Thus, a reading of $110\frac{3}{4}$ means that the price of the bond is 110.75% of the face value or $1,110.75 for a $1,000 bond.

Referring to Table 6-1 the bond headings have the following meanings:

- The 52-week column shows the highest and lowest prices that were paid during the previous year.
- The corporation identification. For example, UtdAir identifies United Airlines. If one has an interest in bonds, he or she will soon learn these abbreviations.
- There is a group of numbers that indicate the annual interest rate and the maturity year. Thus, 10.67s04 indicates an annual coupon rate of 10.67% and a maturity in 2004. The s is just a separator and has no meaning.
- There may be a column headed "Cur Yld" and signifies the interest rate determined by dividing the annual coupon interest by the "Last" price $\left(\frac{10.67}{78.25}\right)$.
- The "Sales" column indicates the number (673) of $1,000 bonds sold during the period (week).
- The "Weekly" shows the highest, lowest, and last prices that were paid during the week.
- A "Net Chg" column indicates the difference between "Last" offers between the previous week and the date of the table $(+1\frac{1}{4}\%)$.

Bonds are rated in the following manner where the rating signifies the risk associated with a particular corporation. Financial houses such as the Standard & Poor's Corporation and Moody's Investment Services establish bond ratings such as shown in Table 6-2. Reilly (1989) gives a more detailed description of the various grades.

TABLE 6-2 BOND RATINGS

	Standard & Poors's	Moody's
High grades	AAA and AA	Aaa and Aa
Medium grades	A and BBB	A and Baa
Speculative bonds	BB and B	Ba and B
Bonds that are in danger of, or are in default	CCC, CC, C, and D	Caa, Ca, and C

There are other rating systems, such as Duff and Phelps, and Fitch, that have similar grading patterns.

6.1 PRICE OF A BOND ON AN INTEREST PAYMENT DATE

The two factors that determine the price of a bond are:

1. The present value of the face value of the bond.
2. The present value of the remaining periodic interest payments.

The following are notations used in the calculations involving bonds, but it is not necessary to memorize each term. Their use will be apparent in the appropriate equations, and they are included here to be a single source of reference, if needed, within the chapter.

P is the price of the bonds.

R is the annual yield rate.

B is the annual bond interest rate.

b is the periodic bond interest payment.

f is the number of interest payments per year.

i is the periodic yield rate.

n is the number of remaining interest periods.

t is the number of remaining years for maturity.

F is the face value (par value) of the bond.

bF is the periodic coupon interest.

M is the redemption or maturity value of the bond (usually, but not always, the face value).

For the calculations,

$$i = \frac{R}{f}, \tag{6.1-1}$$

$$b = \frac{B}{f}, \tag{6.1-2}$$

and

$$n = ft. \tag{6.1-3}$$

The price of the bond on an interest payment date is the sum of the two present values—the present value of the redemption amount plus the present value of the annuity of the coupon interest. The price may be determined by

$$P = M(1+i)^{-n} + bF\left[\frac{1-(1+i)^{-n}}{i}\right], \tag{6.1-4}$$

or using actuarial notation,

$$P = Mv^n + bFa_{\overline{n}|}. \tag{6.1-5}$$

EXAMPLE 6.1.1 A $1,000 bond has seven years remaining until maturity at face value. The bond earns interest at a rate of 8% annually, and the interest is paid semiannually. If the bond is purchased to yield 10% annually, what will be the price on the interest date that is exactly seven years prior to maturity?

Solution For semiannual interest payments, $f = 2$. Then, the periodic yield rate is $i = \frac{10}{200} = 0.05$, the periodic bond rate is $b = \frac{8}{200} = 0.04$, and the periodic interest is $I = 1,000(0.04) = \$40$. The number of remaining interest periods is $n = 2(7) = 14$. Substituting into Equation (6.1-4) gives

$$P = 1,000(1.05)^{-14} + 40\left[\frac{1-(1.05)^{-14}}{0.05}\right].$$

Performing the calculations gives
$$P = 1,000(0.505068) + 40(9.898641),$$

which becomes $505.07 + \$395.95 = \$901.02$. ∎

EXAMPLE 6.1.2 What will be the price of the bond in Example 6.1.1 if it will be redeemed at a) 98% par value and b) 105% par value?

Solution Since the periodic interest is based on the face (par) value, $I = \$40$ as shown in Example 6.1.1,

a) For redemption at 98% par value,
$$M = (0.98)(1,000) = 980,$$

and from Equation (6.1-5)
$$P = (980)(0.505068) + (40)(9.898641)$$
$$= \$890.92.$$

b) For redemption at 105% par value,
$$M = (1.05)(1,000) = 1,050,$$

and from Equation (6.1-5)
$$P = 1,050(0.505068) + 40(9.898641)$$

$$= \$926.27. \quad \blacksquare$$

BOND YIELD RATE

Because i is the interest rate per coupon payment period, the yield to maturity cannot be determined explicitly from Equation (6.1-5) and a trial and error solution is necessary. The Goal Seek tool of Excel is one method of determining exact bond yield.

EXAMPLE 6.1.3   Determine the annual yield rate for a bond with face value of $10,000, which is purchased 9 years before maturity for $9,300. The bond rate is 10%.

Solution   A replica of an Excel spreadsheet is shown below. A temporary value such as the bond rate should be entered into Cell C2 in order that division by zero is not attempted by the formula in Cell A2.

	A	B	C	D	E
1	Price	Face Value	Yield	Bond Rate	Term (Years)
2	10,000.00	10,000.00	10.000%	10.000%	9

Cell A2 would contain the following formula.

`=B2*(1+C2/2)^(-2*E2)+(D2/2)*B2*(1-(1+C2/2)(-2*E2))/(C2/2).`

If the temporary value in Cell C2 is the same as the bond rate, the Price will be the same as the face value as shown in the spreadsheet above. Highlight Cell A2 and depress the Goal Seek command button of the Tools Menu. As shown in the Goal Seek dialog box below, enter 9300 for "To value", C2 for "By changing cell", and depress the OK button. A rapid set of iterations in another dialog box will occur, and upon convergence, the true annual interest rate will appear in both the second dialog box and in Cell C2.

Click on OK to exit the Goal Seek option.

Goal Seek	? X
Set cell:	A2
To value:	9300
By changing cell:	C2
	OK Cancel

The final value in Cell C2 will be 11.257% and is the yield to maturity. More rigorously, two additional columns should be added in which the maturity and purchase dates are entered. Then Cell E2 would be the number of years obtained by subtracting the dates and dividing by 365. ■

APPROXIMATE YIELD RATE

The determination of the yield to maturity when the price of a bond is known requires a business calculator or a computer program. However, an approximate yield rate may be determined analytically. Kellison (1991), Shao and Shao (1991), and Cissell, Cissell, and Flaspohler (1990) show a procedure that is called the *Bond Salesman's Method*. However, the following method also gives an approximation based on the current yield and uses a simple arithmetic approach. If it is assumed that the current yield is approximately equal to an annual interest rate, which is compounded semiannually, an approximate yield rate can be determined. The validity of the assumption is as follows. Letting r be the periodic simple interest rate,

$$r = \frac{bF}{P}.$$

The book value, V, of the bond at the end of one year would be the price plus the simple interest compounded, or

$$V = P(1 + r)^2.$$

For an annual yield rate, R, the value of the bond at the end of one year would be

$$V = P(1 + R).$$

Equating the two expressions for V and dividing through by P gives

$$1 + R = (1 + r)^2.$$

Solving for R gives,

$$R = (1 + r)^2 - 1,$$

which may be reduced to

$$R = r(2 + r). \tag{6.1-6}$$

EXAMPLE 6.1.4   Determine an approximate annual yield rate for a bond with face value of $10,000, which is purchased nine years before maturity for $9,300. The semiannual interest payment is $500.

Solution $I = 500$ and $P = 9,300$. Then,

$$r = \frac{500}{9,300} = 0.0538.$$

From Equation (6.1-6),

$$R = 0.0538(2 + 0.0538) = 0.110493,$$

or 11.05%. This approximation compares favorably with the 11.257% obtained by using the Goal Seek tool of Excel. ■

When a bond is to be purchased at a price that exceeds the face (or redemption) value, the average periodic excess must be subtracted from bF. Thus if a $10,000, 10% bond with 10 interest periods remaining has a price of $10,600, $60 per interest period $\frac{600}{10}$ must be subtracted from bF. For this case, $bF = 500 - 60 = \$440$ and $r = 0.0415$. Then, from Equation (6.1-6),

$$R = 0.0415(2 + 0.0415) \times 100 = 8.47\%.$$

The use of Equation (6.1-6) will give an annual yield rate, which is in the "ballpark" of the true rate. It is appropriate for quick approximations. This approximation compares favorably with exact yield of 8.502% obtained by the Goal Seek tool of Excel.

In a typical situation, the number of remaining years until the maturity of a bond must be determined because the known information is the purchase date and the maturity date. Letting

Y_0 be the purchase year,

Y_1 be the next year after the purchase year, and

Y_2 be the maturity year,

$$t = Y_2 - Y_1 + \text{fraction of } Y_0 \tag{6.1-7}$$

EXAMPLE 6.1.5   A bond pays interest every February 1 and August 1. If a bond was purchased on August 1, 1997, and matures on February 1, 2012, how many years are there until maturity?

Solution   The maturity year, Y_2, is 2012, the purchase year $Y_0 = 1997$, and the next year, Y_1, is 1998. Since the bond is purchased on August 1, $\frac{1}{2}$ of the purchase year, Y_0, remains. Then,

$$t = 2012 - 1998 + \tfrac{1}{2}(\text{of } 1997) = 14.5 \text{ years.} ■$$

A time diagram for this is shown below.

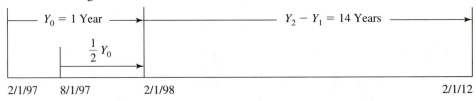

PROBLEM SET 6.1

Find the price of the bond in each of the following. All interest payments are semiannual.

	Face Value	Redemption Value	Annual Bond Rate	Remaining Years to Maturity	Annual Yield
1.	$10,000	par	7%	6	8%
2.	$5,000	par	5%	12	6%
3.	$1,000	par	6%	10	5%
4.	$1,000	par	8%	15	6%
5.	$10,000	95%	10%	5	12%
6.	$5,000	97%	10%	8	8%
7.	$1,000	102%	7%	$5\frac{1}{2}$	8%
8.	$5,000	105%	10%	$11\frac{1}{2}$	9%
9.	$10,000	par	9%	$12\frac{1}{2}$	9%
10.	$1,000	par	6%	20	7%

11. A $10,000, 8% bond matures on June 30, 1998. Interest is payable semiannually on June 30 and December 31. Determine the price of the bond on June 30, 1983 in order to give the investor a 9% annual yield.

12. Repeat Problem 11 if the bond was purchased on December 31, 1986.

13. A $5,000, 9% bond has 10 years remaining until maturity. Interest is paid semiannually. If the price is $4,600, what will be the annual yield rate if the bond is held to maturity?

14. A $10,000, 10% bond has a maturity date of January 1, 2005. It is purchased on January 1, 1988 for $9,800. What will be the annual yield rate if the bond is held to maturity?

15. Suppose the price of the bond in Problem 11 is $10,200. Determine the annual yield rate if the bond is held to maturity?

16. Determine an approximate annual yield for the bond of Problem 13.

17. Determine an approximate annual yield for the bond of Problem 14.

18. Determine an approximate annual yield for the bond of Problem 15.

6.2 PRICE OF A BOND BETWEEN INTEREST PAYMENT DATES

It would be coincidental to be able to purchase a bond exactly on an interest payment date. At the least, a few days are necessary to complete a buy transaction between buyer and seller. Thus, the interest payment date would have to be known and the transaction to buy begun with the exact lead time necessary for culmination on the interest payment date. To plan for a purchase on an interest payment date is really unnecessary. Since the buyer will receive the interest payment on the bond due from the last interest payment date prior to the purchase, the buyer simply pays interest to the seller for the time between the last interest date and the purchase date. A time diagram for this situation is shown in Figure 6-1. The actual price P is called the *flat* price.

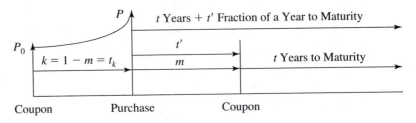

FIGURE 6-1 Purchase Date between Interest Payment Dates

P_0 is the price of the bond on the interest payment date prior to the purchase date, and the price of the bond on the prior interest payment date is determined by

$$P_0 = M(1+i)^{-(n+1)} + bF \left[\frac{1 - (1+i)^{-(n+1)}}{i} \right], \qquad (6.2\text{-}1)$$

where $n = ft$ as indicted in Equation (6.1-4). The price of the bond on the purchase date would be the accumulated value of P_0 at the yield rate for the fraction of a period, k, from the prior interest payment date to the purchase date. Thus,

$$P = P_0(1+i)^k. \qquad (6.2\text{-}2)$$

Equation (6.2-2) gives the theoretically correct flat price, but a practical method, Kellison (1991) would be using the first two terms of the binomial expansion of $(1+i)^k$ or

$$P = P_0(1 + ki). \qquad (6.2\text{-}3)$$

EXAMPLE 6.2.1 A $1,000 bond has seven years and two months remaining until maturity. The bond rate is 8%, and the interest is paid semiannually. If the yield rate is 10% annually, what will be the price of the bond?

Solution For semiannual interest payments, $f = 2$ and

$$i = \frac{10\%}{2} = 5\%,$$

$$b = \frac{8\%}{2} = 4\%,$$

and

$$bF = 0.04(1{,}000) = 40.$$

The whole number of years remaining from the purchase date to maturity is seven. Therefore,

$$n = ft = 2(7) = 14.$$

Then, from Equation (6.2-1), $n + 1 = 15$, and

$$P_0 = 1,000(1 + 5\%)^{-15} + 40 \left[\frac{1 - (1 + 5\%)^{-15}}{5\%} \right]$$

$$= 1,000(0.481017) + 40(10.379658)$$

$$= \$896.21.$$

The time remaining from the purchase date to the next interest payment date is three months, and since a period is comprised of six months,

$$m = \frac{t'}{6} = \frac{2}{6} = \frac{1}{3} \text{ period.}$$

The time from the prior interest payment date to the purchase date is determined by

$$k = 1 - m = 1 - \frac{1}{3} = \frac{2}{3} \text{ period.}$$

Then from Equation (6.2-2),

$$P = 896.21(1 + 5\%)^{\frac{2}{3}} = \$925.84.$$

The practical method of Equation (6.2-3) gives

$$P = 896.21 \left(1 + \tfrac{2}{3}(0.05) \right) = \$926.08. \quad \blacksquare$$

Naturally, the seller would prefer pricing using the practical method since the price is greater than that using the theoretical method. However, we will use the theoretical approach in this text.

The practical purchase of a bond is accomplished by determining what price sellers of bonds are asking (also what price investors are willing to pay). This market price can be determined from the financial pages of newspapers or from brokerage houses. The price that is cited is called the quoted price. Now, if the quoted price occurs between interest dates, the seller will be entitled to the bond interest due between the previous interest date and the purchase date. This quoted price would normally be the price of the bond as determined by the calculations shown in Example 6.2.1 less the accrued bond interest. Therefore, the price of the bond would be the quoted price plus the accrued bond interest. Letting Q be the quoted price of a bond,

$$P = Q + BF t_k, \tag{6.2-4}$$

where t_k is the value of k expressed as a fraction of the year from the prior interest payment date to the purchase date. This price, which is the price given by Equation (6.2-2) or Equation (6.2-3) is referred to as the flat price. Thus the quoted price of a bond would be

$$Q = P - BF t_k. \tag{6.2-5}$$

EXAMPLE 6.2.2   What would be the quoted price of the bond in Example 6.2.1?

Solution   In Example 6.2.1, the flat price is determined to be $925.84. Since $t' = $ two months and the period is six months, $t_k = 6 - 2$ months after the prior interest payment date at which time the seller of the bond is entitled to that fraction of the bond coupon interest. Then, from Equation (6.2-5), the quoted price would be

$$Q = 925.84 - (0.08)(1,000)\left(\frac{4}{12}\right)$$

$$= 925.84 - 26.67 = 899.17.$$

The quoted price is usually expressed a percent of face value, therefore Q would be divided by 1,000 and would be stated as 89.917%.   ■

EXAMPLE 6.2.3   The bond pages of a newspaper on July 2, 1997 gave the following information for the bond of a corporation. The coupon rate is $9\frac{3}{8}\%$ with a yield $= 9.4\%$. The quote is $99\frac{7}{8}$. Suppose bond interest is paid every March 30 and September 30. What would have been the price of a $1,000 bond on July 2?

Solution   Using Table A-1 of the Appendix, the number of days from the previous interest date (3/30) to July 2 is determined to be 94. For the accrued bond interest, the fraction of a year from the previous coupon date (3/30) to the purchase date is $t_k = \frac{94}{365}$, and therefore, the annual coupon rate would be used in Equation (6.2-3).

$$P = 998.75 + (0.09375)(1,000)\left(\frac{94}{365}\right)$$

$$= 998.75 + 24.14 = \$1,022.89.   ■$$

EXAMPLE 6.2.4   Determine the quoted price of Example 6.2.3 if the bond matures on 3/30/05.

Solution   From the given information, $R = 9.4\%$ and $B = 9\frac{3}{8}\%$. For semiannual payments, $f = 2$ and

$$i = \frac{9.4\%}{2} = 4.7\%,$$

$$b = \frac{9.375\%}{2} = 4.5875\%,$$

and

$$I = 1,000(0.046875) = 46.88.$$

The number of years from the next interest payment date, 9/30/97, is determined as follows. The time line is

94 Days

3/30/97	7/2/97	9/30/97	3/30/05

From Equation (6.1-7),

$$t = Y_2 - Y_1 + \text{fraction of } Y_0.$$

Now, $Y_0 = 1997$, $Y_1 = 1998$, and $Y_2 = 2005$. The fraction $Y_0 = 6$ months. Then,
$$t = 2005 - 1998 + \frac{6}{12} = 7.5 \text{ years}$$
to the maturity date of 3/30/05. The number of whole periods to maturity is
$$n = ft = 2(7.5) = 15.$$
Using Equation (6.2-1) for $n + 1 = 16$ gives
$$P_0 = 1,000(0.479571) + 46.88(11.072953)$$
$$= 479.57 + 519.10 = \$998.67.$$
Since there are 2 periods per year, the number of days from the prior interest payment date gives
$$k = 2\left(\frac{94}{365}\right) = 0.515 \text{ periods.}$$
The flat price would be determined using Equation (6.2-2) as
$$P = 998.67(1.047)^{0.515} = \$1,022.57.$$
The accrued bond interest (from Example 6.2.3) gives
$$Q = 1,022.57 - 24.14 = \$998.43.$$
The quote on this bond should be 99.843%.   ■

PROBLEM SET 6.2

Find the price of the bonds in Problems 1–12. All interest payments are semi-annual.

	Face Value	Annual Bond Rate	Remaining Years to Maturity	Annual Yield
1.	$10,000	7%	6 years and 2 months	8%
2.	$5,000	5%	12 years and 4 months	6%
3.	$1,000	6%	8 years and 3 months	7%
4.	$10,000	10%	5 years and 1 month	12%
5.	$5,000	9%	15 years and 3 months	10%
6.	$1,000	8%	9 years and 5 months	9%

	Face Value	Quoted Price	Annual Bond Rate	Time from Prior Interest Payment
7.	$10,000	92%	9%	2 months
8.	$5,000	88%	7%	4 months
9.	$1,000	91%	8%	3 months
10.	$10,000	98%	10%	25 days
11.	$5,000	95%	10%	54 days
12.	$1,000	99%	11%	68 days

13. A $10,000, 8% bond matures on June 30, 1998. Interest is payable semiannually on June 30 and December 31. In order for an investor to achieve a 9% annual yield, what must be the price of the bond on September 1, 1990? Use the number of days for the fraction of a year.

14. A $5,000, 7% bond matures on August 15, 2005. Interest is payable semiannually on February 15 and August 15. For an 8% annual yield, what will be the price of the bond on June 30, 1990? Use the number of days for the fraction of a year.

15. What will be the price of the bond in Problem 13 if it is quoted at 94 on September 1, 1990?

16. What will be the price of the bond in Problem 15 if it is quoted at 96 on June 30, 1990?

6.3 ZERO COUPON BONDS

A concept whereby the bond coupon interest payments were withheld was developed in the 1980s as a way for investors to avoid paying income tax on interest income. The Internal Revenue Service (IRS) established regulations whereby the interest income would be "imputed" based on the yield of the bond. **Imputed** means that an assumption is made that a bond returned a taxable amount of interest when, actually, no interest was received. Naturally, such bonds "no longer interest investors" except for tax-sheltered private investments, such as for private pension funds including IRAs.[1]

The price of a zero coupon bond is simply the present value of the redemption amount, and that aspect is the attractiveness of those bonds. A relatively large redemption amount can be purchased for a relative small investment with a known, fixed yield for the duration of the bond. The price to purchase zero coupon bonds is given by

$$P = F(1+i)^{-n}, \qquad (6.3\text{-}1)$$

where n is the number of periods to the maturity date and is determined by

$$n = f\left(t + \frac{m}{y}\right), \qquad (6.3\text{-}2)$$

where y is compatible with m. It would be 12 if m is in months, and either 360, 365, or 366, as appropriate, if m is in days.

EXAMPLE 6.3.1   Zero coupon bonds with a maturity value of $20,000 were purchased on 6/15/yy to yield 9% compounded semiannually. The maturity date of the bonds is 8/15/(yy+10). What was the price based on ordinary interest?

---

[1] IRA is an acronym for Individual Retirement Arrangement as indicated in the Internal Revenue Publication 590. However, common usage refers to the acronym as Individual Retirement Account.

Solution   The duration of the bonds is 10 years + 61 days, and the number of periods is given by

$$n = 2\left(10 + \frac{61}{360}\right) = 20.339.$$

For $i = \frac{9}{200} = 0.045$,

$$P = 20,000(1.045)^{-20.339}$$

$$= 20,000(0.408502) = \$8,170.04.$$

The actual price may be higher due to fees charged by banks and brokers to obtain them for the investor. ■

PROBLEM SET 6.3

Find the price of the zero coupon bond in each of the following. All yields are compounded semiannually, and redemption is at par unless specified otherwise.

	Face Value	Remaining Time to Maturity	Annual Yield
1.	$10,000	6 years	8%
2.	$5,000	12 years	6%
3.	$1,000	10 years	5%
4.	$1,000	15 years	6%
5.	$10,000	5 years	12%
6.	$5,000	8 years	8%
7.	$1,000	5 years	8%
8.	$5,000	11 years	9%
9.	$10,000	30 years	9%
10.	$1,000	20 years	7%
11.	$10,000	6 years and 2 months	8%
12.	$5,000	12 years and 4 months	6%
13.	$1,000	8 years and 3 months	7%
14.	$10,000	5 years and 1 month	12%
15.	$5,000	15 years and 3 months	10%
16.	$1,000	9 years and 5 months	9%

6.4 DURATION AND VOLATILITY

It is apparent from the equations for the price of a bond that as the interest rate and time to maturity increase, the price of the bond decreases. That is, the price of a bond varies inversely as the changes in the interest rate and term. Bonds represent relatively safe investments with interest rates that reflect the sum of the real rate premium, i.e., liquidity premium and inflation. The coupon cash flow during the term and face amount at maturity will be available with some reasonable assurance. Investors are willing to buy the bonds of a company or a government for longer terms if the interest rates are relatively high because their money will not be available for shorter-term investments. That is, long-term bonds return a higher interest rate at the sacrifice of liquidity (the immediate availability of money). Conversely, if an investor wants relative liquidity, shorter-term bonds could be purchased at the sacrifice of a higher interest rate. Thus, beginning with 3-month Treasury bills and ending with 30-year bonds of the U.S. Treasury, curves of yields to maturity, i.e., interest rates, have shapes similar to those shown in Figure 6-2. The U.S. Treasury has stopped issuing 30-year bonds as of 2002; however, those bonds that were issued in the past will have yield rates until final maturity.

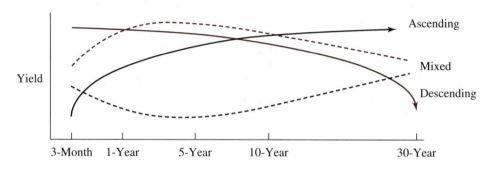

FIGURE 6-2 Yield Curves

In the United States, and perhaps throughout the industrialized world, bonds issued by the U.S. Treasury Department are considered to be risk-free. However, risk-free refers to the guarantee of the periodic payments of coupons and face value at maturity. Between the purchase and maturity dates, bond prices are volatile, much the same as stocks. After a bond has been bought, if it must be sold during a period of rising interest rates, the investor could sustain a loss. Similarly, if interest rates fall, the bonds could be sold at a capital gain. Even if held to maturity, the bond investor faces another form of risk—reinvestment rate risk. If a bond matures at a point in time when interest rates have fallen significantly, as shown by the lower curve in Figure 6-2, the bond investor is faced with the dilemma of purchasing new bonds at lower interest rates, which translates to a higher price.

The term structure of interest rates in mid-July, 1999 is shown in Table 6-3, and follows the ascending curve that is shown in Figure 6-2.

In order to measure reinvestment rate risk, an investor may use a measure that represents the average time to maturity of a bond, or a portfolio of bonds may be used. One such measure is called duration. The concept of equated time

TABLE 6-3 TERM STRUCTURE OF INTEREST RATES	
	Interest Rate
3-Month T-Bills	4.58%
1-Year T-Bills	4.73%
5-Year Notes	5.62%
10-Year Notes	5.73%
30-Year Bonds	5.91%

was introduced in Section 2.1 and is reproduced here as

$$\bar{t} = \frac{\sum Ft}{\sum F}.$$

DURATION

Duration is the equated time based on the present values of the coupon flow and the maturity value of the bond. Letting:

$\bar{d}$ represent the duration of a bond, and

F represent the coupon interest, then

$$\bar{d} = \frac{bF \sum_{t=1}^{n} t(1+i)^{-t} + nM(1+i)^{-n}}{bF \sum_{t=1}^{n} (1+i)^{-t} + M(1+i)^{-n}}. \qquad (6.4\text{-}1)$$

In Equation (6.4-1), M would be the redemption value of the bond.

EXAMPLE 6.4.1   Determine the duration of two $1,000 bonds with five years to maturity. Bond A pays an annual coupon of 10%, and bond B pays an annual coupon of 5%. Use an interest rate of 7%.

Solution   The duration for bond A is determined as follows:

$$\bar{d} = \frac{100[(1)(1.07)^{-1} + (2)(1.07)^{-2} + (3)(1.07)^{-3} + (4)(1.07)^{-4} + (5)(1.07)^{-5}]}{100[(1.07)^{-1} + (1.07)^{-2} + (1.07)^{-3} + (1.07)^{-4} + (1.07)^{-5}]} ,$$
$$+5(1.07)^{-5}(1,000)$$
$$+(1.07)^{-5}(1,000)$$

from which

$$\bar{d} = \frac{4,739.62}{1,123.006} = 4.22.$$

For bond B, the $100 payments would be replaced by $50 payments, and the duration would calculate to be 4.52. Thus, bond A could be considered to have an average payment date of 4.22 years, and bond B could have an average payment date of 4.52 years.   ∎

The use of Equation (6.4-1) could be quite laborious unless a spreadsheet is used. The following may be used for an algebraic solution. Recalling the notation for the present value of $1.00 per period, let

$$a_{\overline{n}|} = \frac{1 - (1 + i)^{-n}}{i},$$

where i is yield to maturity, and n is the number of coupons to maturity. Then, the duration of a bond may be calculated as

$$\overline{d} = \left[\frac{bF \left(\dfrac{a_{\overline{n}|}(1 + i) - n(1 + i)^{-n}}{i} \right) + nM(1 + i)^{-n}}{bFa_{\overline{n}|} + M(1 + i)^{-n}} \right]. \qquad (6.4\text{-}2)$$

The expression within the parentheses is the present value of an arithmetically increasing annuity with a common difference of one and has the actuarial notation $(Ia)_{\overline{n}|}$, Kellison (1991). The units of $\overline{d}$ are periods. For semiannual payments, the duration in years would be half the duration in periods. Note that if the redemption value $M = F$—the face value—then F may be factored and canceled. Duration is a factor that determines the type of fixed income investments a money manager may consider. The reinvestment rate risk may be serious if "current" high yielding bonds are to mature during a low-yield period of time. An example of this is that bond yields in the late 1970s and early 1980s were in the 18% to 20% range; currently, they are in the 3% to 7% range. Therefore, money managers may invest in short-duration bonds with the hopes that yields will be higher when the bonds mature.

VOLATILITY

Volatility is a measure of how rapidly the present value of a series of payments changes as the rate of interest changes. It is defined as the negative of the rate of change in the present value divided by the present value at the point in time when the interest rate changes. The derivation of a formula for volatility requires calculus, which is beyond the level of this book. The calculus and subsequent algebra give the formula for volatility as

$$\overline{v} = \frac{\overline{d}}{1 + i}. \qquad (6.4\text{-}3)$$

Volatility may be considered to be a modified duration, Kellison (1991).

PROBLEM SET 6.4

Determine a) the duration, and b) the volatility for Problems 1–10 for semiannual coupon payments.

	Face Value	Redemption Value	Annual Bond Rate	Years to Maturity	Yield to Maturity
1.	$10,000	par	7%	5	8%
2.	5,000	par	5%	10	6%
3.	1,000	par	6%	10	5%
4.	1,000	par	8%	15	6%
5.	10,000	par	10%	5	12%
6.	5,000	par	10%	8	8%
7.	1,000	par	7%	5	8%
8.	5,000	par	10%	11	9%
9.	10,000	par	9%	12	9%
10.	1,000	par	6%	20	7%

7

Elements of Linear Programming

In many business and financial situations, the equal sign is too strong a constraint to place on a variable. We may state that the future value of an investment at compound interest must *equal* $S. Because the accumulated amount of $S at compound interest will normally occur between interest conversion periods, we must wait until the end of the final conversion period for the addition of the final interest. Therefore, we really mean that the future value of that investment must be equal to or greater than $S (at least $S). We may be willing to spend $C for an asset, but if we can acquire it for less we would do so. Therefore, we would stipulate that we will not spend more than $C for that asset. Linear programming is a method for finding the optimum combination of independent variables subject to constraints that are represented by a system of inequalities in which the "right-side" constants are upper or lower limits on the sum of the terms of the inequalities. Of course, the "right-side" constraint may be a fixed value such as the total amount to be invested.

7.1 THE LINEAR PROGRAMMING SYSTEM

A **linear program** is comprised of an **objective function** that is one equation in two or more unknowns and a system of two or more **constraint inequalities** equal to the number of unknowns as illustrated in Example 7.1.1.

EXAMPLE 7.1.1 Solve the following system for the combination of independent variables, x_1 and x_2, that will give the optimized value of $F(x_1, x_2)$.

$$\text{Maximize:} \quad F(x_1, x_2) = 6x_1 + 7x_2,$$
$$\text{Subject to:} \quad 2x_1 + 3x_2 \leq 12, \text{ and}$$
$$2x_1 + x_2 \leq 8.$$

Solution   The graphic solution is obtained by first plotting the system of constraints. Note that the arrows on the axes point in the direction of the inequality.

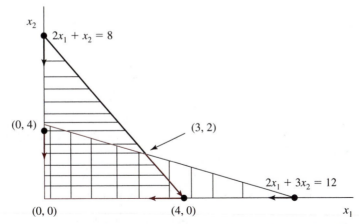

The area with horizontal lines represents the inequality with a limit of eight, and the area with vertical lines represent the inequality with a limit of 12. The region of feasible solutions is the area where both the vertical and horizontal lines intersect, and the theory of linear programming states that the optimum solution will be at one of the vertices of this region. Thus, we test the points of the vertex in the feasible solution area and the point that gives the maximum value of the objective function will be the solution.

$$F(0, 0) = 6(0) + 7(0) = 0,$$

$$F(0, 4) = 6(0) + 7(4) = 28,$$

$$F(3, 2) = 6(3) + 7(2) = 32, \longleftarrow$$

$$F(4, 0) = 6(4) + 7(0) = 24.$$

It can be seen that the combination of three units of x_1 and two units of x_2 gives the maximum value of 32. ■

The foregoing may be formalized as follows. Let:

F represent the objective function that would be optimized as a maximum or a minimum,

x_i represent the independent variable, with $i = 1, 2, \ldots, n$,

f_i represent coefficients in the objective function,

$a, b, \ldots, m$ represent coefficients in the constraint inequalities, and

L_j represent the constraint limit,

the system may be written as follows for maximization.

$$\text{Maximize:} \quad F = \sum_{i=1}^{n} f_i x_i$$

$$\text{Subject to:} \quad \sum_{i=1}^{n} a_i x_i \leq L_1,$$

$$\sum_{i=1}^{n} b_i x_i \leq L_2$$

$$\vdots$$

$$\sum_{i=1}^{n} m_i x_i \leq L_m. \tag{7.1-1}$$

An illustration of minimization for the system of Example 7.1.1 is shown in Example 7.1.2.

EXAMPLE 7.1.2 Solve the following system for the combination of independent variables, x_1 and x_2, that will give the optimized value of $F(x_1, x_2)$.

$$\text{Minimize:} \quad F(x_1, x_2) = 6x_1 + 7x_2,$$
$$\text{Subject to:} \quad 2x_1 + 3x_2 \geq 12, \text{ and}$$
$$2x_1 + x_2 \geq 8.$$

Solution The graphic solution is obtained by first plotting the system of constraints. This is the same system as that of Example 7.1.1 except that the region of feasible solutions lies outside inequalities as indicated by the arrows and the hatched area. The polygon is closed at (∞, ∞).

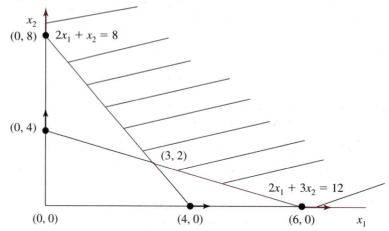

The optimum solution occurs at one of the corners of the region of feasible solutions. Thus, substituting the three ordered pairs of the corners, we get

$$F(0, 8) = 6(0) + 7(8) = 56,$$

$$F(3, 2) = 6(3) + 7(2) = 32, \longleftarrow$$

and

$$F(6, 0) = 6(6) + 7(0) = 36.$$

It can be seen that the combination of three units of x_1 and two units of x_2 give the minimum value of 32. ■

The minimization process can be formalized as follows.

$$\text{Minimize:} \quad F = \sum_{i=1}^{n} f_i x_i$$

$$\text{Subject to:} \quad \sum_{i=1}^{n} a_i x_i \geq L_1,$$

$$\sum_{i=1}^{n} b_i x_i \geq L_2, \tag{7.1-2}$$

$$\vdots$$

$$\sum_{i=1}^{n} m_i x_i \geq L_m.$$

For both types of systems, a non-negativity constraint is usually stipulated in that only positive values of the independent variable are allowed. This makes practical sense since the independent variable usually represents quantities and as such cannot be negative. The non-negativity constraints will be assumed throughout this chapter. Also, in order to ensure a closed polygon for maximization, there must be at least one $\leq$ inequality, and for minimization, there must be at least one $\geq$ inequality.

EXAMPLE 7.1.3 A person wishes to invest \$150,000 to purchase stocks of companies A and B. The price of the stock of Company A is \$125 per share, and the price of the stock of Company B is \$25 per share after purchase commissions. Company A pays an annual dividend of \$4.00 per share, and Company B pays an annual dividend of \$1.00 per share. Not more than \$90,000 may be invested in any one stock. It is desired to maximize the dividend revenue.

Solution Let D be the total dividends, x_1 be the number of shares of Company A, and x_2 be the number of shares of Company B. In order to establish the system of constraints it is often convenient to create a table of the values.

	Company A	Company B	Direction	Limit
Constraints	x_1	x_2		L
Maximum Amount	125	25	$=$	150,000
Company A	125		$\leq$	90,000
Company B		25	$\leq$	90,000

Then, the system may be represented as follows:

$$\text{Maximize:} \quad D(x_1, x_2) = 4x_1 + x_2,$$
$$\text{Subject to:} \quad 125x_1 + 25x_2 = 150,000,$$
$$125x_1 \qquad \leq \ 90,000,$$
$$25x_2 \leq \ 90,000,$$

and the non-negativity constraints are also imposed. The graph for the region of feasible solutions is shown below.

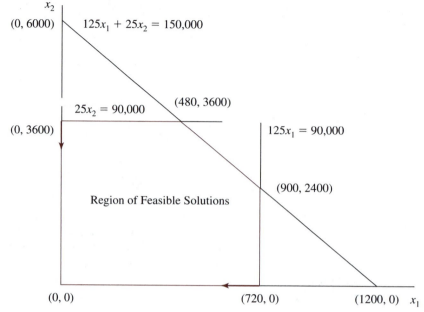

The optimization is obtained by substituting the corners of the region of feasible solutions into the objective function as follows:

$$D(0, 0) = 4(0) + 1(0) = 0,$$

$$D(0, 3{,}600) = 4(0) + 1(3{,}600) = 3{,}600,$$

$$D(480, 3{,}600) = 4(480) + 1(3{,}600) = 5{,}520, \longleftarrow$$

$$D(720, 2{,}400) = 4(720) + 1(2{,}400) = 5{,}280,$$

and

$$D(720, 0) = 4(720) + 1(0) = 2{,}880.$$

The combination of 480 shares of Company A at a cost of $60,000 and 3,600 shares of Company B at a cost of $90,000 gives a maximum dividend of $5,520, subject to the constraints on the investments.

The possibility of multiple solutions exists. This will occur if the slope of a constraint inequality is the same as the slope of the objective function and will be apparent if the corresponding coefficients are in the same ratio. Suppose that in Example 7.1.3, the price of each share of stock of Company A is $100. Then, the system to be solved would be as follows:

Maximize: $D(x_1, x_2) = 4x_1 + x_2,$

Subject to: $100x_1 + 25x_2 = 150{,}000,$

$$100x_1 \qquad\quad \leq\ 90{,}000,$$

$$25x_2 \leq\ 90{,}000.$$

Note that the ratio of the coefficients of x_1 and x_2 is four to one in the objective function and the $150,000 total investment constraint. The graph for the region of feasible solutions would be as follows.

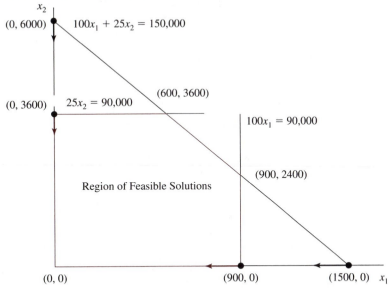

Substituting the corners of the region of feasible solutions into the objective function gives:

$$D(0, 0) = 4(0) + 1(0) = 0,$$

$$D(0, 3{,}600) = 4(0) + 1(3{,}600) = 3{,}600,$$

$$D(600, 3{,}600) = 4(600) + 1(3{,}600) = 6{,}000, \longleftarrow$$

$$D(900, 2{,}400) = 4(900) + 1(2{,}400) = 6{,}000, \longleftarrow$$

and

$$D(720, 0) = 4(720) + 1(0) = 2{,}280.$$

The solution shows that 600 shares of Company A at a cost of $60,000 and 3,600 shares of Company B at a cost of $90,000 are optimal. Alternatively, 900 shares of Company A at a cost of $90,000 and 2,400 shares of Company B at a cost of $60,000 are also optimal. Both options will give a maximized dividend of $6,000. Also any ordered pair of coordinates that lies on the line $100x_1 + 25x_2 = 150{,}000$ will generate a $6,000 dividend. ■

PROBLEM SET 7.1

1. Maximize: $F(x_1, x_2) = 9x_1 + 8x_2,$

Subject to: $4x_1 + 5x_2 \leq 20,$

$3x_1 + 2x_2 \leq 19,$

$x_1, x_2 \geq 0.$

2. Maximize: $F(x_1, x_2) = 5x_1 + 8x_2,$
   Subject to: $x_1 + x_2 \leq 13,$
                $x_1 + 2x_2 \leq 22,$
                $2x_1 + x_2 \leq 20,$
                $x_2 \leq 8,$
                $x_1, x_2 \geq 0.$

3. Maximize: $F(x_1, x_2) = 3x_1 + 2x_2,$
   Subject to: $x_1 + 3x_2 \leq 24,$
                $2x_1 + x_2 \leq 18,$
                $x_1 + x_2 \leq 10,$
                $x_1 \leq 7,$
                $x_2 \leq 7,$
                $x_1, x_2 \geq 0.$

4. Minimize: $F(x_1, x_2) = 1.5x_1 + 2.5x_2,$
   Subject to: $x_1 + 2x_2 \geq 16,$
                $5x_1 + x_2 \geq 20,$
                $x_1 + x_2 \geq 12,$
                $x_1, x_2 \geq 0.$

5. Minimize: $F(x_1, x_2) = 4x_1 + x_2,$
   Subject to: $x_1 + x_2 \leq 4,$
                $3x_1 + x_2 \geq 6,$
                $x_1, x_2 \geq 0.$

6. Minimize: $F(x_1, x_2) = 77x_1 + 102x_2,$
   Subject to: $x_1 + x_2 \leq 4,000,$
                $x_1 + x_2 \geq 2,000,$
                $x_1 \leq 3,000,$
                $x_1 \geq 1,000,$
                $x_1, x_2 \geq 0.$

7. An investor wishes to invest no more than $2,000,000 between U.S. Treasury bonds that have a 6% interest rate and corporate bonds that have a 9% interest rate. The stipulation is that the amount of money in corporate bonds must be no more than 50% of the total amount of money. The investor wishes to maximize the interest income from these investments. Determine the amount of money that should be invested in each of the securities.

8. An investor wishes to invest no more than $2,000,000 between U.S. Treasury bonds that have a 6% interest rate and corporate bonds that have a 9% interest rate. The stipulation is that the amount of money in corporate bonds must be no more than 50% of the amount of money in Treasury bonds. The investor wishes

to maximize the interest income from these investments. Determine the amount of money that should be invested in each of the securities.

9. To produce a plastic part, two operations are performed on the raw material. One worker must shape the plastic, while another worker must put threaded holes in the piece. The first worker's operation requires at least two hours while the second worker's operation requires at least three hours. The production day is no more than eight hours. If the first worker is paid $10.00 per hour and the second worker is paid $12.50 per hour, determine the number of hours each worker must spend on each part that is produced in order to minimize the cost.

10. Two products are to be manufactured and sold. Product A returns a profit of $500 per unit and product B returns a profit of $400 per unit. The total number required each day is at least 16 but not more than 50 can be produced. Also, the daily demand for Product A is no more than 20 units, and the daily demand for Product B is no more than 36 units. Determine the number of units of each that will maximize the daily profit.

7.2 THE LINEAR PROGRAMMING SOLVER

The linear programming examples in Section 7.1 were limited to two unknowns, and as such, the corners of the regions of feasible solutions could be determined and graphed. However, when the number of unknowns is greater than two, graphing the region of feasible solutions is impractical, if not impossible in a two-dimension diagram. For solving multidimension problems, an algebraic technique has been developed, where inequalities are converted to equalities by adding "slack" variables to create a system of equalities, and the system of equalities is solved. A routine of the algebra has been developed and is referred to as the Simplex Method. That routine has been programmed into Microsoft Excel, and is called the "Solver." In the Excel spreadsheet, the system of inequalities is reduced to one row for the objective function, and two columns for the system constraints. The dialog box for the "Solver" is obtained from the "Tools" menu on the standard tool bar. When the "Solver" menu item of the "Tools" menu is clicked, the dialog box that is shown in Figure 7-1 appears. The "Target Cell" would be the cell on an Excel spreadsheet that would contain the formula for the objective function, e.g., Cell A1. The "Equal To" line would have one of the items clicked ("Min" shown here), reflecting the nature of the objective function. Cells B1, C1, D1, etc., would represent the unknowns, x_1, x_2, x_3, etc. Any initial value may be inserted, but beginning with "0" sets the initial values x_1, x_2, x_3, etc., to zero. The entry in the "By Changing Cells" refers to spreadsheet cells whose trial values are allowed to change for optimizing the objective function. For a linear programming problem with three independent variables, we put B1:D1 (the ending cell in row 1). Using the "Add" command button would complete the "Subject to the Constraints" box. On the spreadsheet, the system of constraints would occupy only two columns, one for the constraint formula and one for the constraint limit. As an example, given a

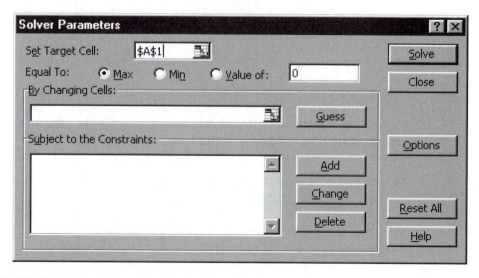

FIGURE 7-1 The Dialog Box for the Excel Solver

constraint $2x_1 + 3x_2 \leq 500$, the constraint formula $2x_1 + 3x_2$ could be written in Cell B2 as 2*B1+3*C1, and the limit value 500 could be written in Cell C2. When the "Add" command button is depressed, the dialog box in Figure 7-2 will appear, wherein B2 would be inserted under "Cell Reference" and C2 would be inserted under "Constraint." Clicking on the down arrowhead, will display a submenu for the direction of the constraint. Depressing the "Add" command button will transfer the constraint to the Solver dialog box, Figure 7-1. This process must be repeated for each constraint in the system. After the final constraint is added, the "Add Constraint" dialog box is closed by clicking on the OK button or the X in the upper right corner.

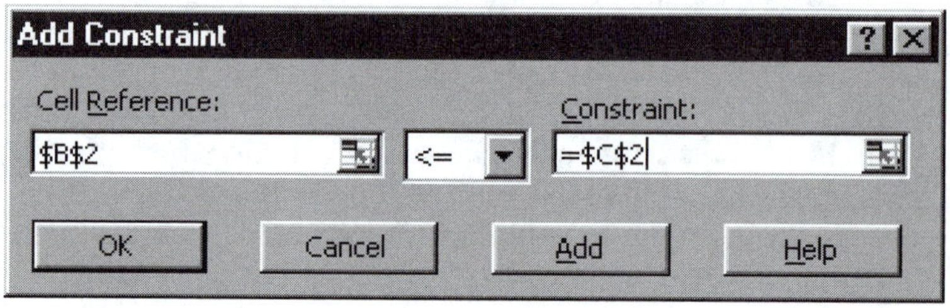

FIGURE 7-2 The Add Constraint Dialog Box

After the white areas of Figure 7-1 are completed with the parameters of the system, the "Solve" command button is clicked to generate a solution. Prior to that, however, one of the "Options" needs to be selected to ensure a linear solution. The dialog box for the "Options" command button is shown in Figure 7-3. The "Assume Linear Model" and "Assume Non-Negative" boxes should be checked, after which "OK" may be clicked. The latter check removes

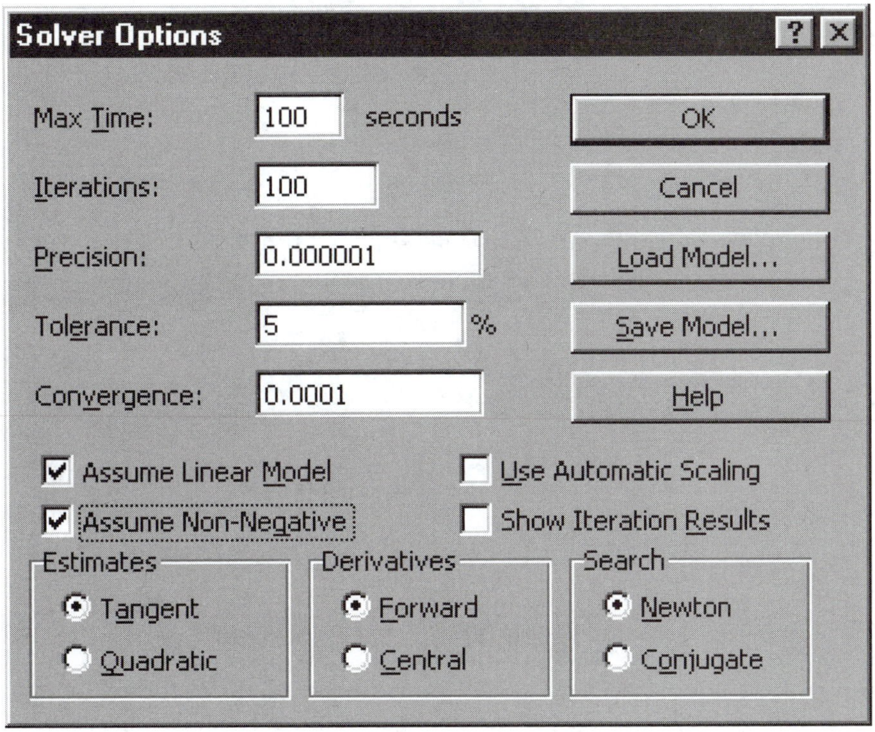

FIGURE 7-3 The Solver Options Dialog Box

the need to indicate the nonnegativity constraints in the constraint box of the Solver dialog box of Figure 7-1. Leave the default numbers and selections as shown. The choice of these numbers and selections, which determines the method and the technical details for solving a linear program problem, is beyond the scope and needs of this book. When the "Solver" command button is depressed, the iterations of the Simplex Method takes place, and the solution will be displayed in cells A1, B1, C1, D1, etc. To illustrate the use, Example 7.1.1 will be solved using the Solver.

EXAMPLE 7.2.1   Solve the following system for the combination of independent variables, x_1 and x_2, that will give the optimized value of $F(x_1, x_2)$.

$$\text{Maximize:} \quad F(x_1, x_2) = 6x_1 + 7x_2,$$
$$\text{Subject to:} \quad 2x_1 + 3x_2 \leq 12, \text{ and}$$
$$2x_1 + x_2 \leq 8.$$

Solution   The Excel spreadsheet could be set up as shown below where Column B is for x_1, and Column C is for x_2. The constraint equations are shown in rows 2 and 3.

	A	B	C
1	=6*B1+7*C1	1	1
2		=2*B1+3*C1	12
3		=2*B1+ C1	8

The "Solver Parameters" dialog and Results boxes for this setup are shown below.

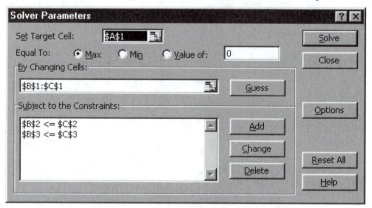

Upon depressing the "Solve" command button, the following replacements of the original cells will occur. The "Solver Results" dialog box will indicate an optimal solution, if any, has been found. For this example, the maximum will be 32 when three units of x_1 and two units of x_2 are produced as shown in the table below. As indicated, the solution may be retained or the original values may be restored. The Reports that are available to be printed will be discussed in Section 7.3. Proficiency with the Solver will be achieved with practice. ■

	A	B	C
1	32	3	2
2		12	12
3		8	8

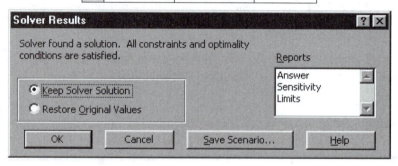

EXAMPLE 7.2.2 The following system is to be solved for the combination of independent variables, x_1, x_2, and x_3.

$$\text{Maximize:} \quad F(x_1, x_2, x_3) = 3x_1 + 5x_2 + 2x_3,$$
$$\text{Subject to:} \quad x_1 + x_2 + 2x_3 \leq 150,$$
$$2x_1 \qquad\qquad \leq 30,$$
$$x_2 \qquad \leq 40, \text{ and}$$
$$x_3 \leq 60.$$

Solution The spreadsheet could be set up as shown below where Column B is for x_1, Column C is for x_2, and Column D is for x_3. The equations are displayed in the formula bar of Excel, and to see them in the cells, depress the Ctrl and $\sim$ keys simultaneously. The constraint equations are shown in rows 2, 3, and 4 and columns B and C. The completed "Solver" dialog box is shown below.

	A	B	C	D
1	=3* B1+5*C1+2*D1	1	1	1
2		=B1+C1+D1	150	
3		=B1	30	
4		=C1	40	
5		=D1	60	

Upon depressing the "Solve" command button, the cells in the spreadsheet are replaced as shown below the Solver dialog box. The maximum value is 410 when $x_1 = 30$ units, $x_2 = 40$ units, and $x_3 = 60$ units. Note that constraints given in rows 3, 4, and 5 are fully satisfied, but only 130 units of the total 150 units in Row 2 are used. This indicates that there are 20 units of "slack" in the constraint. ■

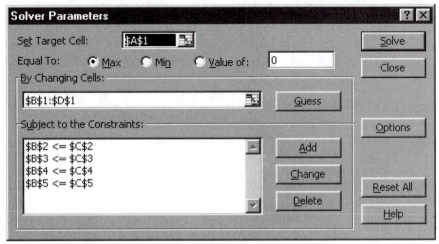

	A	B	C	D
1	410	30	40	60
2		130	150	
3		30	30	
4		40	40	
5		60	60	

PROBLEM SET 7.2

Use the Solver Tool for the following problems. Save the solutions on diskette for use in Problem Set 7.3.

1. Maximize: $F(x_1, x_2, x_3) = 5x_1 + 6x_2 + x_3,$

Subject to: $2x_1 + 3x_2 + 5x_3 \leq 30,$

$6x_1 + 2x_2 + 3x_3 \leq 14,$

$x_1 + 3x_2 + 4x_3 \leq 24,$

$x_1, x_2, x_3 \quad \geq 0.$

2. Maximize: $F(x_1, x_2, x_3) = 20x_1 + 42x_2 + 56x_3,$

Subject to: $2x_1 + 3x_2 + x_3 \leq 6,$

$4x_1 + 2x_2 + 4x_3 \leq 12,$

$4x_1 + 2x_2 + x_3 \leq 8,$

$x_1, x_2, x_3 \quad \geq 0.$

3. Maximize: $F(x_1, x_2, x_3, x_4) = 80x_1 + 10x_2 + 16x_3 + 12x_4,$

Subject to: $x_1 + x_2 + x_3 + x_4 \leq 40,$

$2x_1 + x_2 + 4x_3 + x_4 \leq 90,$

$x_1, x_2, x_3, x_4 \quad \geq 0.$

4. Maximize: $F(x_1, x_2, x_3) = 5x_1 + 9x_2 + 2x_3,$

Subject to: $2x_1 + x_2 + 2x_3 \leq 80,$

$x_1 + x_2 + x_3 \leq 100,$

$x_1 \qquad\qquad \leq 30,$

$x_2 \qquad \leq 50,$

$x_1, x_2, x_3 \quad \geq 0.$

5. An investor wishes to invest $2,000,000 among stocks, bonds, and money market funds. Since 1982, stocks have had a return of about 18% per year, bonds have had a return of about 12% per year, and money market funds have had a return of about 6% per year. The investor has limited the investment among stocks and bonds to at most $1,500,000 and no more than $1,500,000 to be invested among stocks and money market funds. If the returns of each market "next" year are the same as the history of each category, how much should be invested in each category in order to maximize the total projected dollar return?

6. For Problem 5, add the constraint that the maximum investment in stocks is to be $600,000.

7. For Problem 5, add the constraint that the maximum investment in bonds is to be $600,000.

8. For Problem 5, add the constraint that the maximum investment in money market funds is to be $600,000.

9. Suppose the commission paid to a broker is 3% of the total cost of stocks and 1% of the total cost of bonds. Also, suppose that the management fee for the money market funds is 0.5%. The investor wishes to invest $2,000,000 among stocks, bonds, and money market funds. The minimum investment in stocks is to be $750,000, and the minimum investment in bonds is to be $600,000. In order to maintain some immediate liquidity, the investor stipulates that $250,000 is to be invested in money market funds. How much should be invested in stocks and bonds to minimize the commission costs?

10. For Problem 9, add the constraint that no more than $1,000,000 is to be invested in stocks.

7.3 SENSITIVITY AND LIMIT ANALYSES

THE PRIMAL SYSTEM

The linear programming solution is valid only as long as there is no change in any of the coefficients in the original setup. That solution is called the primal solution. However, if the coefficients in either the objective function or in any of the constraints change there can be an effect on the optimal solution mix. For a complete analysis of a system, it is valuable to know the range of the coefficients in the objective function that will leave the solution mix unchanged. Those coefficients could be profit margins, unit prices, or unit costs, and any of them could change. Also, each constraint has an effect on the optimal solution mix, and the availability of the constraints may change. For example, if a constraint is the number of hours a production machine will be available, the number of units in the solution mix will depend on that time availability. If that machine must be diverted to some other use, the contribution of that machine toward the optimal solution will be affected. Knowledge of the range of the available amounts for each constraint is also valuable information in order to plan for contingencies. This determination of the range of coefficients in the objective function and the range of constraints is called a sensitivity analysis, and the Excel Solver provides this function as well as an optimal solution.

Suppose a company manufactures two products, A and B, and each need to be processed on two machines, Machines 1 and 2. Each unit of Product A generates a profit of $6.00, and each unit of Product B generates a profit of $7.00. Product A requires two hours per unit on Machine 1 and three hours per unit on Machine 2. Each unit of Product B requires two hours per unit on Machine 1 and one hour per unit on Machine 2. Machine 1 is available for 12 hours each day, and Machine 2 is available for eight hours each day. The company desires to maximize profit. This system is that of Example 7.1.1. Letting:

P be the profit,

x_1 be the number of units of Product A, and

x_2 be the number of units of Product B,

the system for the primal solution becomes

$$\text{Maximize:} \quad P = 6x_1 + 7x_2$$

$$\text{Subject to:} \quad 2x_1 + 3x_2 \leq 12 \quad \text{Machine 1}$$

$$2x_1 + x_2 \leq 8. \quad \text{Machine 2}$$

The formula setup (using the Ctrl $\sim$ keys in combination) in an Excel spreadsheet is shown in Figure 7-4.

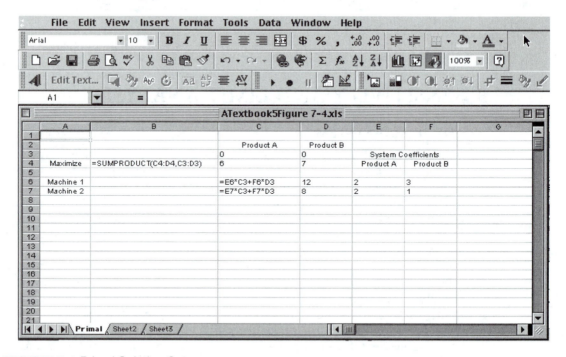

FIGURE 7-4 Primal Solution Setup

The Solver Parameters setup, as discussed in Section 7.2, is shown in Figure 7-5. Be sure to depress the Options Command button in order to check the "Assume

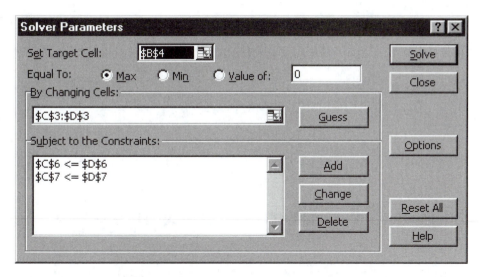

FIGURE 7-5 The Solver Setup

Linear Model" and "Assume Non-Negative" boxes as shown in Figure 7-3. After completing the setup, depressing the "Solve" command button gives the Solver solution and is shown in Figure 7-6. As in Example 7.1.1, the solution mix is that a daily production of three units of Product A and two units of Product B will give a maximum profit of $32. Notice, in Figure 7-6 the dialog box with the heading, "Reports" and along the bottom edge of the screen will be sheets named "Answer Report 1," "Sensitivity Report 1," and "Limits Report 1." After the Solver solution, holding a depressed Ctrl key and left-clicking on "Answer," "Sensitivity," and "Limits" followed by clicking on OK will create the Answer Report, the Sensitivity Report, and the Limits Report as Worksheets.

Answer Report 1 is shown in Figure 7-7. The heading, which identifies the name that has been given to the Excel Worksheet, is generated automatically. Normally the Report Created line gives the date and time of the creation; they have been omitted in this text. Both the Target and Adjustable cells show the original and final values. The reason both are equal is that the Solver was implemented a second time after the optimal solution was determined with the starting point of the second run being the ending point of the first run. The Answer Report shows the cell values associated with the constraints. The two columns **Status** and **Slack** indicate whether or not a constraint has been fully utilized. If the equality part of the weak ≤ inequality is the controlling factor, the constraint is indicated to be **binding** and the slack in the constraint will be zero. Otherwise, the constraint is indicated to be **not binding** and the amount of slack can vary from zero to the limit of the constraint.

The Answer Report is an Excel Worksheet and Excel attempts to identify the columns and rows from the source Worksheet. Note in Figure 7-7 that the names of the Adjustable Cells are Product A and Product B, and that the names of the constraint rows are Machine 1 and Machine 2. The product names appear

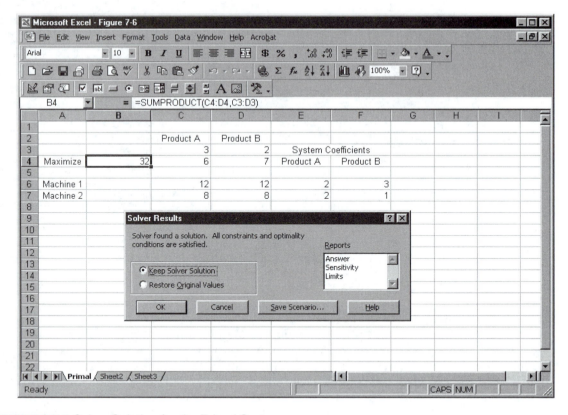

FIGURE 7-6 Solver Solution for the Primal System

in rows 12 and 13 in Figure 7-7. However, Cell C19 will indicate Machine 2 Product A, and Cell C20 will indicate Machine 1 Product A. To remove the term, Product A, cells C19 and C20 were highlighted and the words were erased in the Formula Bar.

Click the Worksheet Sensitivity Report 1 tab and Figure 7-8 will appear. In the Adjustable Cells section, the Allowable Increase and Allowable Decrease columns indicate the range over which the Objective Coefficient may be varied. If the objective coefficient six is increased by eight units to 14, and *all other coefficients remain the original values*, the maximized objective will change, but the combination of three units of Product A and two units of Product B will remain the optimum solution mix. Similarly, if the objective coefficient is decreased by 1.33, the maximized objective will change, but the solution mix will remain the same. The emphasis is that *all other coefficients remain the original values*. The Sensitivity Report shows that the Allowable Increase in the profit margin of Product B is $2.00 and the allowable decrease is $4.00.

The Reduced Cost for each variable is the *objective function contribution per unit of the variable minus the cost of the resources that are used by the variable priced at the Shadow Prices* in the Constraints Section. The primal system is

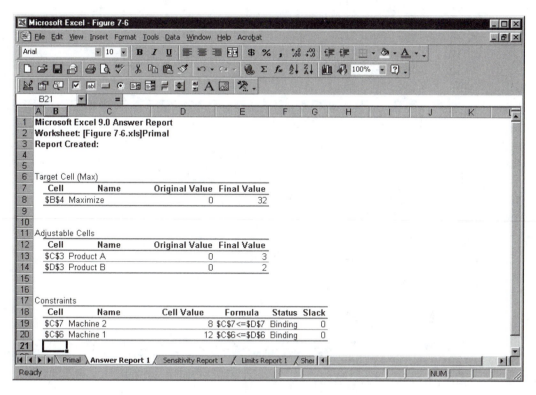

FIGURE 7-7 The Answer Report

reproduced below for convenience in this discussion.

$$\text{Maximize:} \quad P = 6x_1 + 7x_2$$
$$\text{Subject to:} \quad 2x_1 + 3x_2 \leq 12 \quad \text{Machine 1}$$
$$2x_1 + \ x_2 \leq \ 8. \quad \text{Machine 2}$$

The resources that are used are as follows. Product A requires two hours per unit on Machine 1 for a resource cost of $4.00 (2 hours per unit times the $2.00 per hour shadow price). It also requires two hours per unit on Machine 2 for a Resource Cost of $2.00 (2 hours per unit times $1.00 per hour shadow price). This generates a total resource cost of $6.00. Since Product A contributes $6.00 per unit profit, the Reduced Cost is zero ($6.00 profit per unit minus $6.00 resource cost per unit). A similar analysis will show that the Reduced Cost for Product B = $7 - [3(2) + 1(1)]$ and equals zero. The Reduced Cost will be zero whenever the variables in the solution mix are positive.

The Shadow Prices are the period contributions to the objective function that are made by each constraint and may be thought of as the "Cost of the Constraint." As shown in Figure 7-8, the Shadow Price for Machine 1 is $2.00 per hour of machine time, and the Shadow Price for Machine 2 is $1.00 per hour of machine time. The Final Value for Machine 1 is 12 and for Machine 2 is

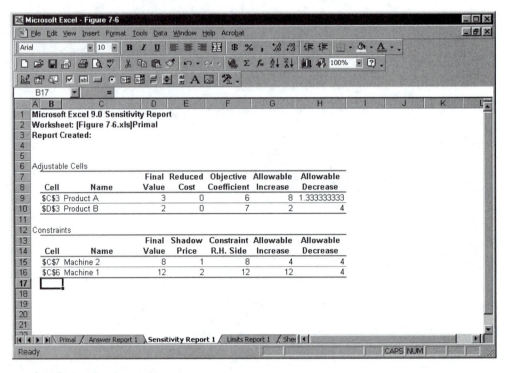

FIGURE 7-8 The Sensitivity Report

eight, and the Shadow Prices point to the maximum profit as $2(12) + 1(8) = \$32$. The availability of the machines may change due to necessary other uses. The columns in the Constraints Section labeled Allowable Increase and Allowable Decrease show the amount of change in either direction that can be tolerated without the corresponding variable leaving the solution mix. For example, with *all other values left at their original levels*, increasing the time Machine 1 is to be used by 12 hours—from 12 hours to 24 hours—will remove Product A from the solution mix. This is shown in Figure 7-9 where Cell C3 is 0. Excel may show this value in a form such as n.nnnE-11, which is n.nnn $\times 10^{-11}$. This may be considered to be zero.

One concern should be addressed when constraints are set at the maximum Allowable Increase and Allowable Decrease. Alternative optimal solutions may be present, but the Solver will find only one solution. The presence of alternative optimal solutions may be noted by looking at the Sensitivity Report. The Allowable Increase or Allowable Decrease of one or more constraints will be zero if alternative solutions are present. Alternative solutions will also be indicated if the coefficients in the constraint inequalities are equal to or are the same multiple of the corresponding coefficients in the objective function. For example, if the coefficients in the constraint for Machine 1 were six for x_1 and seven for x_2, or the same multiple of six and seven in the objective function, as

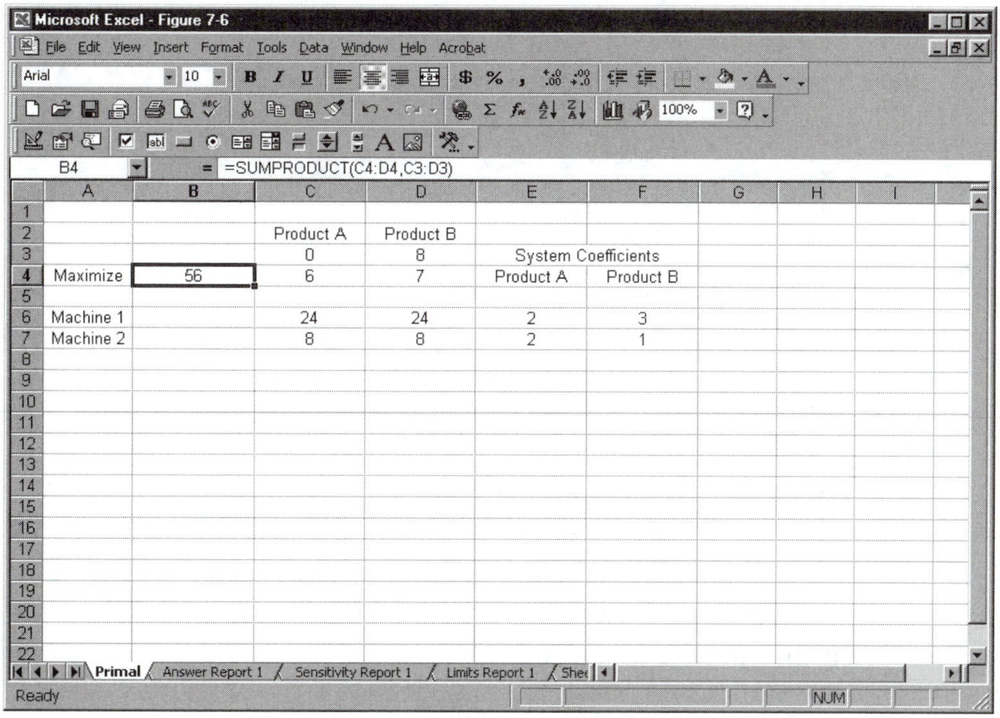

FIGURE 7-9 Loss of a Product from a Solution Mix

shown below, multiple optimal solutions will be present.

$$\text{Maximize:} \quad P = 6x_1 + 7x_2$$

$$\text{Subject to:} \quad 12x_1 + 14x_2 \leq 12 \quad \text{Machine 1}$$

$$2x_1 + x_2 \leq 8. \quad \text{Machine 2}$$

The final Solver Report is the Limit Report and is shown in Figure 7-10.

The Limits Report shows the limits that each of the objective function coefficients may reach, *while the other coefficients remain at the original values.* The items in this report are straightforward. It is obvious that the lower limit of each objective function is zero, and the upper limit is the originally indicated coefficient.

The foregoing analysis completes a study of a primal system. Accompanying the primal analysis is an analysis that is called the *dual*.

THE DUAL SYSTEM

The dual system and solution is a special optimization that is the opposite of the optimization of the primal system. If each constraint has a "cost" associated with it and the objective is to maximize profit, the objective of the dual would be to minimize the "cost" of the constraint. Conversely, if the objective were to minimize an objective function, the dual would be to maximize the use of each

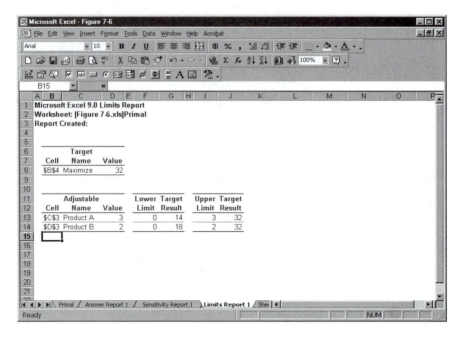

FIGURE 7-10 The Limits Report

constraint. Letting:

C represent the objective function of the dual system to be optimized,

y_j represent the dual variable, with $j = 1, 2, \ldots n,$

L_j represent coefficients in the objective function of the dual and are the constraint limits in the primal,

$a, b, c \ldots$ represent coefficients in the constraint inequalities, and

f_i represent the coefficient in the primal objective function,

the dual system may be written as follows for a minimization.

$$\text{Minimize:} \quad C = \sum_{j=1}^{m} L_j y_j$$

$$\text{Subject to:} \quad \sum_{j=1}^{m} a_i y_j \geq f_1$$

$$\sum_{j=1}^{m} b_i y_j \geq f_2 \qquad\qquad (7.3\text{-}1)$$

$$\vdots$$

$$\sum_{j=1}^{m} m_i y_j \geq f_m$$

Rewriting the primal for the production system above,

$$\text{Maximize:} \quad P = 6x_1 + 7x_2$$

$$\text{Subject to:} \quad 2x_1 + 3x_2 \le 12 \quad \text{Machine 1}$$

$$2x_1 + x_2 \le 8. \quad \text{Machine 2}$$

the L's of the dual will be 12 and eight respectively, and the f's, as the limits for the constraints, will be six and seven respectively. Then the dual system is written as

$$\text{Minimize:} \quad C = 12y_1 + 8y_2$$

$$\text{Subject to:} \quad 2y_1 + 2y_2 \ge 6 \quad \text{Profit A}$$

$$3y_1 + y_2 \ge 7. \quad \text{Profit B}$$

The Answer and Sensitivity Reports are shown in Figures 7-11 and 7-12 respectively.

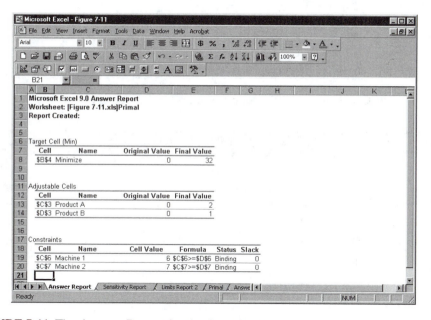

FIGURE 7-11 The Answer Report for the Dual System

Note in Figure 7-11 that minimum "cost" is equal to the maximum profit that was indicated in the primal solution. Also note that the Final Values for the dual system are the same as the shadow prices of the primal system. In Figure 7-12, note that the Shadow Prices for the dual are the same as the solution mix for the primal.

The foregoing elements of linear programming have been intended as an introduction to the concept and are not intended to be exhaustive. Advanced topics, such as integer and nonlinear programming are beyond the scope of this book.

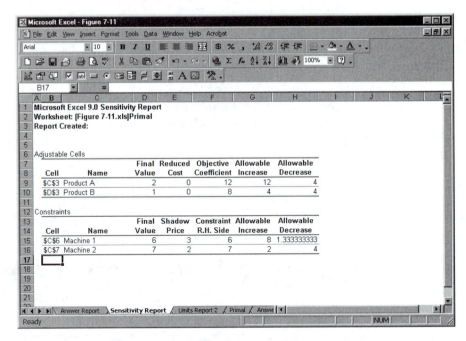

FIGURE 7-12 The Sensitivity Report for the Dual System

PROBLEM SET 7.3

Use the Solver tool for the following problems. Generate the Answer, Sensitivity, and Limits Reports. Problems 1–10 are the same as problems in Section 7.2, and prior solutions may be retrieved if they have been saved on diskette.

1. Maximize: $F(x_1, x_2, x_3) = 5x_1 + 6x_2 + x_3$,

   Subject to: $2x_1 + 3x_2 + 5x_3 \le 30$,

   $6x_1 + 2x_2 + 3x_3 \le 14$,

   $x_1 + 3x_2 + 4x_3 \le 24$,

   $x_1, x_2, x_3 \quad \ge \ 0$.

2. Maximize: $F(x_1, x_2, x_3) = 20x_1 + 42x_2 + 56x_3$,

   Subject to: $2x_1 + 3x_2 + \ x_3 \le \ 6$,

   $4x_1 + 2x_2 + 4x_3 \le 12$,

   $4x_1 + 2x_2 + \ x_3 \le \ 8$,

   $x_1, x_2, x_3 \quad \ge \ 0$.

3. Maximize: $F(x_1, x_2, x_3, x_4) = 80x_1 + 10x_2 + 16x_3 + 12x_4$,

   Subject to: $x_1 + \ x_2 + \ x_3 + x_4 \le 40$,

   $2x_1 + \ x_2 + 4x_3 + x_4 \le 90$,

   $x_1, x_2, x_3, x_4 \quad \ge \ 0$.

4. Maximize: $F(x_1, x_2, x_3) = 5x_1 + 9x_2 + 2x_3$,

Subject to: $2x_1 + x_2 + 2x_3 \leq 80$,

$$x_1 + x_2 + x_3 \leq 100,$$

$$x_1 \leq 30,$$

$$x_2 \leq 50,$$

$$x_1, x_2, x_3 \geq 0.$$

5. An investor wishes to invest $2,000,000 among stocks, bonds, and money market funds. Since 1982, stocks have had a return of about 18% per year, bonds have had a return of about 12% per year, and money market funds have had a return of about 6% per year. The investor has limited the investment among stocks and bonds to at most $1,500,000, and no more than $1,500,000 to be invested among stocks and money market funds. If the returns of each market "next" year are the same as the history of each category, how much should be invested in each category in order to maximize the total projected dollar return?

6. For Problem 5, add the constraint that the maximum investment in stocks is to be $600,000.

7. For Problem 5, add the constraint that the maximum investment in bonds is to be $600,000.

8. For Problem 5, add the constraint that the maximum investment in money market funds is to be $600,000.

9. Suppose the commission paid to a broker is 3% of the total cost of stocks and 1% of the total cost of bonds. Also, suppose that the management fee for the money market funds is 0.5%. The investor wishes to invest $2,000,000 among stocks, bonds, and money market funds. The minimum investment in stocks is to be $750,000, and the minimum investment in bonds is to be $600,000. In order to maintain some immediate liquidity, the investor stipulates that $250,000 is to be invested in money market funds. How much should be invested in stocks and bonds to minimize the commission costs?

10. For Problem 9, add the constraint that no more than $1,000,000 is to be invested in stocks.

11. A company produces four products: A, B, C, and D. Each unit of A requires two hours of machine time, one hour of assembly, and $10 of administrative costs. Each unit of B requires one hour of machine time, three hours of assembly, and $5 of administrative costs. Each unit of C requires 2.5 hours of machine time, 2.5 hours of assembly, and $2 of administrative costs. Each unit of D requires 5 hours of machine time, no assembly, and $12 of administrative costs. Each month, 120 hours of machine time are available, 160 hours of assembly time are available, and administrative costs cannot exceed $1,000. The per unit profits that each product return are $40 from the sales of A, $24 from the sales of B, $36 from the sales of C, and $23 from the sales of D. Due to shipping constraints, not more than 20 units of Product A, and not more than 16 units of Product B can be made during the month. Due to a contractual requirement at least 10 units each of products C and D must be produced. Present a complete analysis for the maximization of profit.

12. Set up and present a complete analysis for the minimization of the Dual System for Problem 11. The text indicated that the Shadow Prices applied to the constraints in a Primal Solution point to the optimum solution. In this solution of the Dual of Problem 11, the Minimum "Cost" of the constraints is different than the Maximum Profit in the primal. Explain why the Solver gives an apparent discrepancy in the solution.

13. A department store has 110,000 square feet to be allocated for Women's Wear, Men's Wear, and Cosmetics. Current planning is that no more than 50% of the area be for Women's Wear, no more than 30% for Men's Wear, and the up to 25% of the area be for Cosmetics. It is expected that Women's Wear will generate a profit of $60 per square foot, Men's Wear will generate a profit of $40 per square foot, and Cosmetics will generate a profit of $50 per square foot.

a) Determine the optimal area allocation.

b) Determine the expected maximum profit.

c) Determine the maximum profit margin that Women's Wear can generate without changing the area allocation.

d) Determine the maximum area that can be allocated to Women's Wear without it being removed from the solution.

14. A loan company makes the four types of loans shown in the table below.

Type of Loan	Annual Interest Rate
Unsecured	18%
Automobile	5%
First Home Mortgage	8%
Second Home Mortgage	10%

The follow limitations have been placed on the amount of each type of loan. Unsecured loans cannot exceed 10% of the total amount of the loans. The amount of unsecured and automobile loans cannot exceed 20% of the loans. Automobile loans must be at least 20% of the combined unsecured and automobile loans. First mortgages must be at least 40% of the mortgages and at least 20% of the total amount of the loans. Second mortgages may not exceed 25% of the total amount of the loans. With $1.5 million dollars to lend, the company wishes to maximize the annual interest from the loans. Present a complete analysis for this maximization.

7.4 CAPITAL RATIONING[1]

Financial management has two important aspects: capital budgeting and working capital management. The former refers to the management of long-term uses of corporate funds, while the latter to management's uses of short-term corporate funds. Here, long-term is defined as a period over one year. Short-term is defined as one year or less. The application of linear programming in capital budgeting is the subject of this section. The discussion of an application in working capital management is postponed until the next section.

Long-term funds are used for buying land, buildings, equipment, and other assets that may be used and kept by a corporation for long periods of time. Sometimes, the assets under consideration are grouped into projects. For instance, a company may be interested in building a new plant overseas. The construction of a plant involves the acquisition of land, building, equipment, materials, and labor, and the decision maker may want to compare alternative locations and sizes of this overseas expansion. These assets or projects are usually expensive, and their acquisition signifies a long-term commitment on the part of the investor. Therefore, this type of investment must be carefully evaluated. Many criteria can be used for evaluating and selecting the assets or projects. The most acceptable criterion is the ranking of projects on the basis of net present value, which is the difference between present value of cash inflows and present value of cash outflows.

Before investing in any asset or project, a decision maker must consider the constraints on long-term capital spending. Legal constraints, production constraints, marketing constraints, financial constraints, to name a few, are usually taken into consideration before the decision is finalized. The topic of evaluating and selecting long-term assets subject to budgetary constraint is known as capital rationing. The following example will help to clarify the above concepts.

EXAMPLE 7.4.1   Computronics, a computer manufacturer, is in the process of evaluating various expansion projects to cope with the growing demand for its computers and related products. These projects, including the expansion of existing facilities in the United States and the construction of new facilities overseas, are coded as projects x_1, x_2, x_3, x_4, and x_5. Currently, the Board of Directors of Computronics has allocated a total budget of 175 million dollars for all these projects. The available funds for the subsequent two years are expected to be \$5 million annually. After that, no additional funds will be allocated for these projects. The present values of initial cash outlay as well as after-tax annual cash flows are given below. All projects are to be terminated at the end of the 4th year. The board of Directors instructs the CEO to make a recommendation on the project selection based on the net present value method.

---

[1] I am indebted to Dr. Hsi Li, Professor of Finance, Bryant College for contributing this section to the textbook.

**Present Values of Projected Annual Cash Flows
in Millions of Dollars**

Project	0	1	2	3	4
			Year		
x_1	−90	30	40	−10	60
x_2	−85	30	−20	30	40
x_3	−72	−15	45	42	38
x_4	−70	28	−10	46	55
x_5	−68	−20	−35	74	75

Solution   Since the board of directors requires all projects to be evaluated on the net present value basis, the objective function should be constructed to reflect this requirement. Note that all numbers in the table are present values. There is no need to discount them any further. The net present value of each project should be the simple addition of each row over the years. That is,

Project	x_1	x_2	x_3	x_4	x_5
Net present value (NPV)	30	−5	38	49	26

If there is no capital rationing, and the selection is purely based on NPV ranking, only project x_2 would be rejected due to its negative NPV. If the initial budget is limited to \$175 million, the NPV ranking dictates the selection of only projects x_4 and x_3. By doing so, Computronics would generate a NPV of \$87 (i.e., $49+38$) million. These two projects require an initial investment of \$142 (i.e., $70+72$) million in year 0. Since the total initial budget cannot go beyond \$175 million, there is not enough money for acquiring any other project. However, if partial acquisition of a project is allowed, the company should invest the balance of the initial budget in x_1, because x_1 has the third highest NPV. The cost of x_1 at time 0 is \$90 million. With only \$33 million left after committing funds to x_4 and x_3, the investor can afford to acquire only 37% of project x_1. This combination of x_4, x_3, and $0.37x_1$ should bring in a total NPV of \$98 million.

Under capital rationing, the use of linear programming allows us to select a combination that is often different from selection based purely on the ranking of NPV. As illustrated below, this change in methodology ensures the maximization of NPV.

The objective function should include only the contributions by individual projects to the total net present value of the corporation. Therefore, we have

$$\text{Maximize NPV} = 30x_1 - 5x_2 + 38x_3 + 49x_4 + 26x_5.$$

The total investment at time 0 should not go beyond the budgeted limit of 175 million dollars. That is,

$$90x_1 + 85x_2 + 72x_3 + 70x_4 + 68x_5 \leq 175.$$

By the same token, budgetary constraints for year 1 and year 2 can be stated as

$$-30x_1 - 30x_2 + 15x_3 - 28x_4 + 20x_5 \leq 5,$$

and

$$-40x_1 + 20x_2 - 5x_3 + 10x_4 + 35x_5 \leq 5.$$

Since no budget has been assigned to these projects in years 3 and 4, the constrains for these years become

$$10x_1 - 30x_2 - 42x_3 - 46x_4 - 74x_5 \leq 0,$$

and

$$-60x_1 - 40x_2 - 38x_3 - 55x_4 - 75x_5 \leq 0.$$

The following constraints are introduced so that no project can be selected more than once.

$$x_1, x_2, x_3, x_4, x_5 \leq 1.$$

Finally, nonnegative constraints must be added so that the number of units of a selected project cannot be less than zero.

$$x_1, x_2, x_3, x_4, x_5 \geq 0.$$

Using the Solver, we obtain the following solutions.

Project	x_1	x_2	x_3	x_4	x_5
Number of units selected	0	0	1	1	0.49

It is worth emphasizing that only 0.49 units of x_5 have been included in the investment plan. This solution is admissible because the use of linear programming implies every project under evaluation is perfectly divisible. In this example, it means that the investor can choose to complete only 49%, not 100%, of project x_5. If a project or an asset is not completely divisible, the financial analyst should then switch to the technique of integer programming, which is a topic beyond the scope of this text. Given the above combination of projects, the corporation is expected to have its total net present value maximized at $99.62 million. This is higher than the total NPV of $98 million that resulted from the intuitive solution of x_4, x_3, and $0.37x_1$. ■

EXAMPLE 7.4.2 Speedy Pizza, a fast-food chain, is evaluating four home delivery systems. Currently, the company is expected to spend no more than $150,000 for improving home delivery. Cash flows generated by the new systems will be used to finance additional capital spending in the future. It is safe to assume that the interest rate will be stabilized at 10% over the whole planning horizon. The projected cash flows of individual delivery systems are as follows.

**Cash Flows of Home Delivery Systems
in Thousands of Dollars**

	Year			
Project	0	1	2	3
Delivery system 1 (x_1)	-75	40	40	40
Delivery system 2 (x_2)	-60	35	30	20
Delivery system 3 (x_3)	-58	-15	48	42
Delivery system 4 (x_4)	-43	-26	45	45

The objective of the company is to maximize the net present value of its investments.

Solution   Cash flows generated by individual systems must first be converted into present values. After discounting at the annual rate of 10%, the present values of these systems become

Present Values of Home Delivery Systems in Thousands of Dollars

Project	0	1	2	3
		Year		
Delivery system 1 (x_1)	−75	36.36	33.06	30.05
Delivery system 2 (x_2)	−60	31.82	24.79	15.03
Delivery system 3 (x_3)	−58	−13.64	39.67	31.55
Delivery system 4 (x_4)	−43	−23.64	37.19	33.81

Using the information in the above table, the net present value can be derived.

Project	x_1	x_2	x_3	x_4
Net present value (NPV)	24.47	11.64	−0.42	4.36

The capital rationing problem can be formulated as follows:

Maximize: $\text{NPV} = 24.47x_1 + 11.64x_2 - 0.42x_3 + 4.36x_4$,

Subject to: $75.00x_1 + 60.00x_2 + 58.00x_3 + 43.00x_4 \leq 150$,

$-36.36x_1 - 31.82x_2 + 13.64x_3 + 23.64x_4 \leq 0$,

$-33.06x_1 - 24.79x_2 - 39.67x_3 - 37.19x_4 \leq 0$,

$-30.05x_1 - 15.03x_2 - 31.55x_3 - 33.81x_4 \leq 0$,

$x_1 \qquad\qquad\qquad\qquad\quad \leq 1$,

$x_2 \qquad\qquad\qquad \leq 1$,

$x_3 \qquad\qquad \leq 1$,

$x_4 \leq 1$,

$x_1, x_2, x_3, x_4 \qquad\qquad \geq 0$.

To maximize the net present value, the company should acquire $1x_1$, $1x_2$, and $0.35x_4$. The resulting net present value from these investments is $37,630.93. ∎

PROBLEM SET 7.4

1. Given an initial budget of $90 million, without any additional future capital infusion and the following information, identify the combination of expansion projects so that the total net present value of XYZ Company is maximized.

XYZ Company
Present Values of Projected Annual Cash Flows
in Millions of Dollars

Project	0	Year 1	2	3
x_1	−40	15	20	10
x_2	−25	30	−20	40
x_3	−33	−15	25	42
x_4	−64	−28	80	46

2. In addition to the initial budget given in Problem 1, XYZ Company has now decided to allocate $5 million for Year 1, and $2.5 million for Year 2 to support the expansion. What is the new combination of projects for maximizing its total NPV? Comparing the solutions to Problems 1 and 2, are they different?

3. Java Island Café has experienced a growing demand for its gourmet coffee and pastry. A $100 million remodeling and expansion are planned. A financial consultant has been hired to review the following projects.

Present Values of Projected Annual Cash Flows
in Millions of Dollars

Project	0	Year 1	2	3
x_1	−32	25	20	10
x_2	−45	20	20	20
x_3	−28	−28	24	53
x_4	−53	−36	45	40
x_5	−50	−50	60	60

What should be the appropriate combination of projects for maximizing the net present value of Java Island Café?

4. If Java Island Café decides to infuse an additional $5 million for Year 1, and $2.5 million for Year 2 to support the remodeling and expansion projects, should the financial consultant change his or her original recommendation? Why or why not?

5. Well Wisher specializes in drilling deep wells in rural regions of New England. Because of the recent drought, its operations have extended to suburban areas. To capture the growing market, Well Wisher has budgeted $56 million for equipment, training, and labor in different locations. The forecasted annual cash flows in these locations are given below. The annual rate for discounting cash flows is 10%.

**Projected Annual Cash Flows
in Millions of Dollars
Year**

Location	0	1	2	3
x_1	−20	24	19	15
x_2	−27	19	20	21
x_3	−18	−17	35	35
x_4	−42	−25	41	46

Make sure the cash flows are converted into present values and that the net present value for each location is determined before the construction of the objective function.

6. Cloud Walkers, a chain which sells designer shoes in major metropolitan areas, plans to add new stores and/or to expand existing stores. Projected cash flows of the proposed stores are shown below. The initial budget of these stores is $120 million. The appropriate rate for discounting future cash flows is 8%.

**Projected Annual Cash Flows
in Millions of Dollars
Year**

Store	0	1	2	3	4
x_1	−79	25	34	−12	50
x_2	−74	20	−10	19	38
x_3	−61	−12	32	31	30
x_4	−59	23	−11	12	45
x_5	−46	−18	−24	63	64

Determine the best combination of stores for maximizing the net present value of Cloud Walkers.

7. As health consciousness sweeps across the country, the demand for wholesome bakery products grows by leaps and bounds. Facing this new trend, Hot Cross Buns is interested in broadening its customer base by increasing its stores in the metropolitan area of Boston. The following sites and their projected cash flows have been submitted for final consideration. The president of Hot Cross Buns has been advised by an investment banker that 12% is a reasonable annual rate for discounting cash flows. A line of credit of $100 million now, $60 million in Year 1, and $40 million in Year 2 has been arranged. As a financial consultant, you are asked to select the sites to optimize the net present value of Hot Cross Buns.

**Projected Annual Cash Flows
in Millions of Dollars
Year**

Site	0	1	2	3	4	5
x_1	−83	−22	35	35	35	35
x_2	−40	−20	−10	40	40	40
x_3	−60	−20	−10	30	50	70
x_4	−58	−24	−15	42	57	61
x_5	−20	−28	−18	33	45	27

8. Due to the recent currency deflation of many Asian currencies, Computronics has postponed its final decision on overseas expansion. New sites have now been added to the original proposal given in Example 7.4.1. The multiyear budget, however, stays the same—that is, an initial budget of $175 million, followed by an additional $5 million a year for the next two years. The up-to-date information is as follows:

**Present Values of Projected Annual Cash Flows
in Millions of Dollars
Year**

Project	0	1	2	3	4
x_1	−90	30	40	−10	60
x_2	−85	30	−20	30	40
x_3	−72	−15	45	42	38
x_4	−70	28	−10	46	55
x_5	−68	−20	−35	74	75
x_6	−63	30	−23	30	30
x_7	−56	22	22	22	22
x_8	−54	−18	−60	70	74

Select the appropriate combination of projects to maximize the net present value of Computronics.

9. If Computronics allocates an additional budget of $5 million a year to each of the last two years, would the extra funding change the outcome of Problem 8? Why or why not?

10. If the initial investment in the project increases from $85 million to $95 million, what would be the change in outcome? Why?

7.5 WORKING CAPITAL MANAGEMENT[2]

In finance, current assets refer to the sum of cash, marketable securities, accounts receivable, inventories, and other short-term assets. Working capital management is the management of current assets to cope with short-term uses of funds. Among current assets, it may take more time to convert inventories into cash. For this reason, a company often tries to maintain a quick ratio of higher than, say, 1. Here, quick ratio is defined as the ratio between current assets excluding inventories and current liabilities, or (current assets − inventories)/current liabilities. When this ratio is higher than one, the company is expected to have more than sufficient short-term funds to cover short-term debt. As a company adjusts its outputs, its current assets and liabilities change accordingly. For this reason, there is a need to adjust the output mix so that the company is able to meet the short-term obligations, and at the same time, maximize its profit.

EXAMPLE 7.5.1   Toy Wizard is a manufacturer of mechanical toys, unique toys tailor-made for high-income executives. Its three major products carry the codes of x_1, x_2, and x_3. For every unit of x_1 sold, the company gains an additional profit of $4,000. Similarly, the profit contribution of x_2 and x_3 is $2,000 and $3,000, respectively. The machine time needed for producing a unit of x_1 is one hour. The corresponding time requirements for x_2 and x_3 are two and three hours, respectively. The total amount of machine time for making all three toys is 14 hours per week. In addition, the amount of skilled labor is limited to 80 hours per week. The labor time for assembling x_1, x_2, and x_3 is 2.5, 3, and 1.2 hours, respectively. It is known that, with the exclusion of inventories, the three products build up current assets according to the formula $120 − 8x_1 − 3x_2 − 4x_3$. In the formula, the constant, 120, is the sum of cash, marketable securities, and accounts receivable, after covering interest payments, loan payments, and management salaries. $8x_1$ suggests that the variable cost for producing a unit of x_1 is $8. Similarly, the variable cost per unit for x_2 and x_3 is $3 and $4, respectively. Presently, Toy Wizard has current liabilities of $100. Determine the weekly output mix so that its weekly profit (P) can be maximized.

Solution   Based on the profit contribution, the company's objective function should be

$$\text{Maximize } P = 4{,}000x_1 + 2{,}000x_2 + 3{,}000x_3.$$

There are two types of production constraints: machine time constraint and labor time constraint. They can be stated as

$$1x_1 + 2x_2 + 3x_3 \leq 14,$$

and

$$2.5x_1 + 3x_2 + 1.2x_3 \leq 80.$$

The financial constraint in the form of quick ratio is

$$\frac{120 − 8x_1 − x_2 − 4x_3}{100} \geq 1,$$

---
[2]   I am also indebted to Dr. Hsi Li, Professor of Finance, Bryant College for contributing this section to the textbook.

or, after rearranging

$$8x_1 + 3x_2 + 4x_3 \geq 20.$$

Adding nonnegative constraints to the above, the full linear programming problem becomes

Maximize: $P = 4,000x_1 + 2,000x_2 + 3,000x_3,$

Subject to: $x_1 + 2x_2 + 3x_3 \leq 14,$

$2.5x_1 + 3x_2 + 1.2x_3 \leq 80,$

$8x_1 + 3x_2 + 4x_3 \leq 20,$

$x_1, x_2, x_3 \geq 0.$

Using the Solver, the weekly output-mix should be 0.2 units of x_1 and 4.6 units of x_3. For the sake of profit maximization, the production of x_2 should be discounted. By doing so, the weekly profit reaches $14,600. ∎

An extension of the above analysis is the use of current ratio as a financial constraint. This ratio is defined as current assets divided by current liabilities. Recall that in the process of deriving a quick ratio, inventory is removed from current assets before the difference is divided by current liabilities. When we calculate the current ratio, this step of removing inventory from current assets is no longer needed.

EXAMPLE 7.5.2   Assume that Toy Wizard has an inventory of $95, and the company requires its current ratio to be 2 or higher. Determine the best product mix for optimizing its profit. Based on the information given in Example 7.5.1, the following inequality has been used to calculate the quick ratio:

$$\frac{120 - 8x_1 - 3x_2 - 4x_3}{100} \geq 1,$$

where $(120 - 8x_1 - 3x_2 - 4x_3)$ represents the difference between current assets and inventory. To derive current assets, we simply add inventory back to this difference. That is, current assets of Toy Wizard reach the value of $(120 - 8x_1 - 3x_2 - 4x_3) + 95$. Dividing this by the current liabilities of $100, we obtain the current ratio. Since this ratio has to be two or higher, the financial constraint becomes

$$\frac{120 - 8x_1 - 3x_2 - 4x_3 + 95}{100} \geq 2.$$

Its simplification gives

$$8x_1 + 3x_2 + 4x_3 \leq 15.$$

Combining this financial constraint with the objective function and other constraints, we obtain the following linear programming problem:

Maximize: $P = 4,000x_1 + 2,000x_2 + 3,000x_3,$

Subject to: $x_1 + 2x_2 + 3x_3 \leq 14,$

$$2.5x_1 + 3x_2 + 1.2x_3 \le 80,$$
$$8x_1 + 3x_2 + 4x_3 \le 15,$$
$$x_1, x_2, x_3 \ge 0.$$

According to the results revealed by the Solver, Toy Wizard should produce only 3.75 units of x_3. The profit resulting from this output is $11,250.

PROBLEM SET 7.5

1. Precision Machinery, Inc. is a manufacturer of cutting-edge equipment for scientific research. Currently, it offers three models, x_1, x_2 and x_3. The weekly profit margins of x_1, x_2, and x_3 are $600, $300, and $500, respectively. The production manager reports that the making of these models is restricted by the availability of a special component of 15 units per week. It is also known that seven, one, and five units of this component must be installed in every unit of x_1, x_2, and x_3, respectively. According to the marketing analysis, it requires an average of four hours per week to sell a unit of x_1, while the amount of time for marketing of x_2 or x_3 takes only two hours. Restricted by the size of the sales force, a total of only 95 hours per week of sales personnel is available. After some mathematical manipulation, the finance department reports that if the condition of $7x_1 + 12x_2 + 6x_3 \le 60$ is met, then the company would have no problem to exceed the minimum quick ratio of one, as required by the board of directors. Given the above information, determine the combination of outputs so that the profit of Precision Machinery is maximized.

2. With the following profit functions and constraints given, determine the best output mix of x_1, x_2, x_3, and x_4.

Profit function:	$P = 2,000x_1 + 3,000x_2 + 1,500x_3 + 2,261x_4,$
Production constraint:	$65x_1 + 60x_2 + 70x_3 + 43x_4 \le 1,425,$
Marketing constraint:	$39x_1 + 28x_2 + 102x_3 + 36x_4 \le 730,$
Financial constraint:	$40x_1 + 72x_2 + 36x_3 + 55x_4 \le 1,670,$
Nonzero constraints:	$x_1, x_2, x_3, x_4 \ge 0.$

3. Vineyard Chia has just been licensed to bottle herbal tea drinks. Its products are marketed under three brand names, y_1, y_2, and y_3. The profit margin is $30 for every thousand cans of y_1; $40 for y_2; and $60 for y_3. The two basic ingredients are Chia syrup and herbal syrup. Each year, Vineyard has a quota to only import 800 gallons of Chia syrup and 700 gallons of herbal syrup. For every thousand cans of y_1, y_2, or y_3, two, three, or four gallons of Chia syrup must be used, respectively. The usage of herbal syrup for making a thousand cans of y_1, y_2, or y_3 is one, two, or four gallons, respectively. For the year, Vineyard has also bought 600 special filters for its bottling equipment. To ensure product quality, Vineyard Chia is expected to consume five, one, or two filters for bottling every thousand cans of y_1, y_2, or y_3, respectively. To prevent difficulty in cash flow, a quick ratio of one or higher is required. After covering all overhead cash expenses and debt payments, Vineyard Chia is estimated to have a cash flow of $120,000 for coping with current liabilities and variable cash expenses. The amount of current liabilities reaches $105,000, while variable cash expenses can be estimated by the formula $30y_1 + 35y_2 + 45y_3$. Find the best output mix for maximizing Vineyard Chia's profit.

4. Widget Warehouse stores and sells four types of widgets, x_1, x_2, x_3, and x_4. The annual contribution to corporate profit is $3,600 for 1,000 units of x_1. The contributions of 1,000 units of x_2, x_3, and x_4 are expected to be $4,200, $3,000, and $3,700. Widget Warehouse has a total of 940 square yards of display space. The areas required for displaying one thousand units of x_1, x_2, x_3, and x_4 are 43, 38, 62, and 51 square yards, respectively. The following table shows the annual labor requirements:

Labor Requirement Unit: in Thousands

| | | Minimum Hours Required for Each Widget | | | |
Type	Total hours available	x_1	x_2	x_3	x_4
Marketing and customer service	889	40	39	54	42
Shelving and maintenance	1,200	64	73	48	62

Finally, Widget Warehouse intends to maintain a quick ratio of one or higher. After covering all overhead cash expenses and debt payments, Widget Warehouse is estimated to have a cash balance of $286,000 for coping with current liabilities and variable cash expenses. The amount of current liabilities reaches $285,350, while variable cash expenses can be estimated by the formula $23x_1 + 31x_2 + 27x_3 + 36x_4$. Given the goal of profit maximization, find the best output mix for Widget Warehouse.

5. Gizmo Palace, a competitor of Widget Warehouse, has the following profit function and constraints:

Production function: $P = 245x_1 + 310x_2 + 290x_3 + 274x_4,$

Constraints: $55x_1 + 40x_2 + 74x_3 + 63x_4 \leq\ \ 780,$

$$60x_1 + 85x_2 + 74x_3 + 62x_4 \leq 1,100,$$

$$43x_1 + 52x_2 + 27x_3 + 40x_4 \leq\ \ 685,$$

$$28x_1 + 40x_2 + 32x_3 + 39x_4 \leq\ \ 590,$$

$$x_1, x_2, x_3, x_4 \geq\ \ \ 0.$$

Find the outputs of x_1, x_2, x_3, and x_4 to maximize the profit of Gizmo Palace.

6. As pointed out in Problem 4, Widget Warehouse is estimated to have a cash balance of $286,000, and the amount of current liabilities reaches $285,350. Note also that the variable cash expenses are estimated by the formula $23x_1 + 31x_2 +$

$27x_3 + 36x_4$. Assume the company's primary new financial concern is to reach a current ratio of two or above. It is known that its inventory has a value of $300,000. Determine the optimal outputs and the corresponding level of profit.

7. Given the information in Problems 4 and 6, if Widget Warehouse chooses to meet both a quick ratio of at least one and a current ratio of at least two, what should be the optimal outputs? At the optimal output mix, what is the profit? Show and explain the difference between the solutions of this problem and those of Problems 4 and 6.

8. Music Matters manufactures musical instruments. These instruments fall into four categories: x_1, x_2, x_3, and x_4. The profit margins per thousand units of x_1, x_2, x_3, and x_4 are $254,000, $256,000, $180,000, and $240,000, respectively. Music Matters faces the following constraints:

Type of constraint	Amount Required per Thousand Units of Output				
	x_1	x_2	x_3	x_4	Total Available
Machine hours	162,000	140,000	48,000	90,000	400,000
Shipping hours	60,000	72,000	42,000	69,000	300,000
Labor hours	80,000	60,000	45,000	56,000	600,000

In addition, Music Matters is required by its bank to maintain a current ratio of at least two. According to its balance sheet, the sum of cash, marketable securities, and accounts receivable is $8,500,000, and current liabilities are $3,000,000. The unit value of x_1, x_2, x_3, and x_4 is $560, $670, $470, and $540, respectively. Identify the optimal outputs for maximizing its profit.

Note that inventory value equals to $560x_1 + 670x_2 + 470x_3 + 540x_4$, and current assets is the sum of cash, marketable securities, and accounts receivable. As mentioned before, current ratio = (current assets)/(current liabilities).

9. Ye Olde Cheese Shoppe collects milk from local dairy farms to make gourmet cheese. The average daily production and financial constraints are as follows:

Type of constraint	Amount Required per Unit of Output				
	x_1	x_2	x_3	x_4	Total Available
Milk supply	38	20	41	34	175
Labor supply	32	36	42	32	150
Equipment	40	30	32	28	160
Quick ratio	30	34	24	27	125

Find the best combination of cheese products so that the profit function, $P = 52x_1 + 53x_2 + 65x_3 + 42x_4$, can be maximized.

The linear program for Ye Olde Cheese Shoppe can be stated as

$$\text{Maximize:} \quad P = 52x_1 + 53x_2 + 65x_3 + 42x_4,$$

$$\text{Subject to:} \quad 38x_1 + 20x_2 + 41x_3 + 34x_4 \leq 175,$$
$$32x_1 + 36x_2 + 42x_3 + 32x_4 \leq 150,$$
$$40x_1 + 30x_2 + 32x_3 + 28x_4 \leq 160,$$
$$30x_1 + 34x_2 + 24x_3 + 27x_4 \leq 125,$$
$$x_1, x_2, x_3, x_4 \qquad \geq \quad 0.$$

10. Assume that Ye Olde Cheese Shoppe has its quick ratio constraint derived from the following inequality:

$$\frac{2,250 - (30x_1 + 34x_2 + 24x_3 + 27x_4)}{2,125} \geq 1.$$

In addition, the company requires its current ratio to be 2 or higher. To determine the inventory value, Ye Olde Cheese Shoppe has assigned a unit value of $82 to x_1, $85 to x_2, $98 to x_3, and $72 to x_4. With the imposition of this additional financial constraint, what should be the new product mix and profit? (Hint: To derive current assets, add the inventory (i.e., $82x_1 + 85x_2 + 98x_3 + 72x_4$) to the sum of cash, marketable securities, and accounts receivable (i.e., $2,250 - (30x_1 + 34x_2 + 24x_3 + 27x_4)$)).

CHAPTER

8

Breakeven Models

A breakeven model gives a business an approximate point when a profit will begin to be realized on an investment. It may be the point when total cost will have been recovered by revenue from sales in a manufacturing process, or it may be the point in the future when the choice between two investment choices currently being analyzed would be influenced.

8.1 FUNDAMENTAL BREAKEVEN ANALYSIS

There are two components of cost in manufacturing a particular product: fixed costs and variable cost. *Fixed costs* are those costs that are not associated with the number of units of production. They include such items as rent, loan payments, insurance, depreciation cost of machinery, and utilities. *Variable cost* is the cost that is associated with production quantities and includes such items as material, labor, warehouse facilities, and shipping costs. Letting:

C be the total cost,

c be the variable cost per unit,

C_0 be the fixed cost, and

q be the quantity produced,

the total cost equation, as a linear function of the quantity produced, is given by

$$C(q) = cq + C_0. \tag{8.1-1}$$

Revenue is typically the product of the selling price per unit and the number of units that have been sold. Letting R be the revenue and s be the selling price per unit, the total revenue is given by

$$R(q) = sq. \tag{8.1-2}$$

It is axiomatic that the selling price per unit must be greater than the variable cost per unit. Then, the slope of the revenue function must be greater than the slope of the cost function. These functions are sketched in Figure 8-1.

Profit (P) is defined as the difference between revenue and cost, or

$$P(q) = R(q) - C(q), \tag{8.1-3}$$

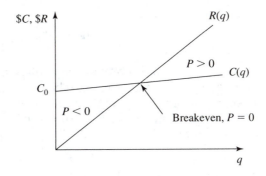

FIGURE 8-1 Fundamental Breakeven

and Figure 8-1 shows that profit equals zero at the breakeven point. Equation (8.1-3) reduces to

$$R(q) = C(q),$$
$$(8.1\text{-}4)$$

for $P(q) = 0$.

EXAMPLE 8.1.1 In order to manufacture an electronic device, a company has incurred $100,000 of fixed costs. Cost accountants have established a variable cost per unit of $50. The device is priced to sell for $100 each. Determine the breakeven point.

Solution   For Equation (8.1-1) and (8.1-2), the cost and revenue functions are $C(q) = 50q + 100,000$ and $R(q) = 100q$. For breakeven, Equation (8.1-4) gives

$$100q = 50q + 100,000,$$

$$50q = 100,000,$$

and

$$q = 2,000.$$

Then, the number of units to breakeven is 2,000. However, in mathematics, the identification of a point requires an ordered pair, in this case (q, R) or (q, C). The cost or revenue to realize the breakeven number of units is

$$C(2,000) = R(2,000) = 100(2,000) = \$200,000,$$

and the breakeven point is given by (2,000, 200,000). ∎

Since money has a time value, it may be important to know the point in time when breakeven will occur. The number of units that can be produced in a given amount of time is the product of a production rate and time. Letting:

q' be the production rate, and

t be the time,

the time required to produce a given number of units is given by

$$q = q't,$$
$$(8.1\text{-}5)$$

from which the time t can be determined. Then, in Example 8.1.1, if the production rate is 100 units per day, the time to breakeven would be determined by Equation (8.1-5) where,

$$2,000 = 100t,$$

from which $t = 20$ days.

The foregoing analysis for breakeven implies the assumption that the revenue cash flow is coincident with the variable cost cash flow. Typically, there is a lag between the production of an item and the sale of that item. This time lag is labeled t_R in Figure 8-2. Substituting Equation (8.1-5), transforms Equation (8.1-1) into

$$C(t) = cq't + C_0. \tag{8.1-6}$$

Letting t_R be the time to the beginning of the revenue, substituting Equation (8.1-5) transforms Equation (8.1-2) into

$$R(t) = sq'(t - t_R). \tag{8.1-7}$$

Equating Equations (8.1-6) and (8.1-7) and solving for the time to breakeven gives

$$t_B = \frac{C_0 + sq't_R}{(s - c)q'}. \tag{8.1-8}$$

This breakeven analysis with a revenue time lag is shown in Figure 8-2.

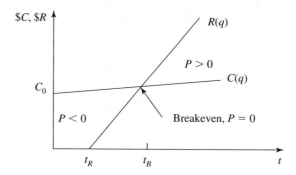

FIGURE 8-2 Breakeven with a Revenue Time Lag

EXAMPLE 8.1.2 In order to manufacture an electronic device, a company has incurred $100,000 of fixed costs. Cost accountants have established a variable cost per unit of $50, and the manufacturing engineers have determined that the production rate will be 100 units per day. The device is priced to sell for $100 each, and the sales manager estimates revenues to begin 30 days from the start of production. Determine the breakeven point.

Solution From the given information, Equation (8.1-6) gives the cost function as

$$C(t) = 50(100)t + 100,000,$$

and Equation (8.1-7) gives the revenue function as

$$R(t) = 100(100)(t - 30).$$

Setting $R(t) = C(t)$ for breakeven gives

$$10,000(t - 30) = 5,000t + 100,000.$$

This reduces to

$$5,000t = 400,000,$$

from which the time to breakeven is determined to be 80 days from the beginning of production. The associated cost or revenue can be determined from either the cost or revenue functions as

$$C(80) = R(80) = 100(100)(80 - 30) = \$500,000.$$

Then, the ordered pair (t, C) or (t, R) is (80, 500,000). Equations (8.1-7) and (8.1-8) can be programmed into a spreadsheet in order to solve for the breakeven point, thereby foregoing the repeated algebraic solution. ■

If we assume 250 production days in a year, 100 units per day will generate 25,000 units per year. The profit that would have been projected at the end of one year in Example 8.1.1 would be determined by Equation (8.1-3) as

$$P(25,000) = 100(25,000) - 50(25,000) - 1,000,000 = \$1,150,000.$$

The profit that would have been projected at the end of one year in Example 8.1.2 would be determined by Equation (8.1-3) as

$$P(250) = 100(100)(250 - 30) - 50(100)(250) - 100,000 = \$850,000.$$

The \$300,000 difference is due to the 30-day time lag and would be a receivable. However, the company would have to plan for this shortfall in the current year.

A further consideration in the determination of breakeven is the fact that in many, if not most, cases, the company will borrow, until the start of revenues, the amount of the fixed cost and the amount of the variable costs that are necessary. The loan may be either a short-term loan or a longer-term debt depending upon the amount. There is also a time lag from the date of the loan until production can begin. This situation is shown in Figure 8-3.

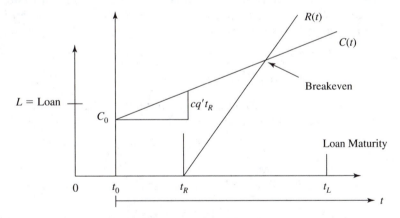

FIGURE 8-3 Breakeven with a Loan

The amount of the loan would be $C_0 + cq't_R$. Letting r be the simple interest rate, the interest on the loan would be determined by

$$I = (C_0 + cq't_R)rt_L. \tag{8.1-9}$$

EXAMPLE 8.1.3   In order to manufacture an electronic device, $100,000 of fixed costs has been incurred. Cost accountants have established a variable cost per unit of $50, and the manufacturing engineers have determined that the production rate will be 100 units per day. The device is priced to sell for $100 each, and the sales manager estimates revenues to begin 30 days from the start of production. A simple interest loan was negotiated 45 days prior to production. The interest rate on the loan is 8%, and the maturity date is one year from the date of the loan. Determine the interest cost and the effect it will have on profit one year from the starting date of production.

Solution   The amount of the loan is fixed cost plus the variable cost to the beginning of revenue income. That is,

$$L = C_0 + cq't_R$$

$$= 100,000 + 50(100)(30)$$

$$= 250,000.$$

The interest cost one year from the date of the loan would be

$$I = 250,000(0.08)(1) = \$20,000.$$

Example 8.1.2 shows that the profit, one year from the start of production, to be $850,000. Then, the interest cost would be subtracted from this amount for a net profit of $830,000.   ■

EXAMPLE 8.1.4   The company of Example 8.1.3 wishes to liquidate the loan at the earliest possible date in order to maximize the end-of-year profit. Determine that date.

Solution   In order to liquidate the loan at the earliest possible date, the profit must equal the interest cost or $P = I$. From the definition of profit using Equations (8.1-3) and (8.1-9),

$$I = R(t) - C(t).$$

$$(C_0 + cq'tR)r(t + t_0) = sq'(t - t_R) - (cq't + C_0). \qquad (8.1\text{-}10)$$

The value of t would be the date on which the profit equals the interest cost and can be determined by Equation (8.1-10) or by Equation (8.1-11).

$$t = \frac{(C_0 + cq't_R)rt_0 + sq't_R + C_0}{(s - c)q' - (C_0 + cq't_R)r}. \qquad (8.1\text{-}11)$$

Then using a daily simple interest rate and referring to Figure 8-3,

$$[100,000 + 50(100)(30)]\left(\frac{0.08}{365}\right)(t + 45) = 100(100)(t - 30) - [50(100)t + 100,000]$$

$$54.79t + 2,465.55 = 10,000t - 300,000 - 5,000t - 100,000$$

$$4,945.21t = 400,000 - 2,465.55$$

$$t = 80.38 \text{ days.}$$

Then, on the 81st day from the start of production, the interest cost may be liquidated with a small amount remaining. A check on this will show that the interest would be \$4,428.36, and the profit would be \$5,000 on that 81st day. ∎

PROBLEM SET 8.1

Problems 1–4 represent first year cost and revenue functions. Determine the fundamental breakeven points and the time to breakeven if production is 100 units per day for the following revenue and cost functions.

1. $R(q) = 15q,$ $C(q) = 5q + 100,000$

2. $R(q) = 21q,$ $C(q) = 15q + 120,000$

3. $R(q) = 0.40q,$ $C(q) = 0.20q + 12,000$

4. $R(q) = 300q,$ $C(q) = 150q + 200,000$

Problems 5–8 represent first year cost and revenue functions where the revenue begins at the indicated number of days after production begins. Determine the time to breakeven and the breakeven cost if production is 100 units per day for the following revenue and cost functions.

5. $R(q) = 15q,$ $C(q) = 5q + 100,000,$ $t_R = 30$ days

6. $R(q) = 21q,$ $C(q) = 15q + 120,000,$ $t_R = 28$ days

7. $R(q) = 0.40q,$ $C(q) = 0.20q + 12,000,$ $t_R = 32$ days

8. $R(q) = 300q,$ $C(q) = 150q + 200,000,$ $t_R = 40$ days

Problems 9–12 represent first year cost and revenue functions where the revenue begins at the indicated number of days, at 100 units per day, after production begins. The initial fixed costs plus the variable cost to the first day of revenue income must be borrowed using a short-term note to be paid one year after the date of the loan at the indicated interest rates. Determine the interest costs.

9. $R(q) = 15q$ $C(q) = 5q + 100,000,$ $t_R = 30$ days, $r = 8\%$

10. $R(q) = 21q,$ $C(q) = 15q + 120,000,$ $t_R = 28$ days, $r = 9\%$

11. $R(q) = 0.40q,$ $C(q) = 0.20q + 12,000,$ $t_R = 32$ days, $r = 10\%$

12. $R(q) = 300q,$ $C(q) = 150q + 200,000,$ $t_R = 40$ days, $r = 12\%$

Problems 13–16 represent first year cost and revenue functions where the revenue begins at the indicated number of days, at 100 units per day, after production begins. The initial fixed costs plus the variable cost to the first day of revenue income must be borrowed using a short-term note. Determine the number of days to the time when the profit will just equal the interest cost if the loans were negotiated on the indicated number of days prior to the beginning of production.

13. $R(q) = 15q,$ $C(q) = 5q + 100,000,$ $t_R = 30$ days, $r = 8\%,$ $t_0 = 20$ days

14. $R(q) = 21q,$ $C(q) = 15q + 120,000,$ $t_R = 28$ days, $r = 9\%,$ $t_0 = 30$ days

15. $R(q) = 0.40q,$ $C(q) = 0.20q + 12,000,$ $t_R = 32$ days, $r = 10\%, t_0 = 40$ days

16. $R(q) = 300q,$ $C(q) = 150q + 200,000,$ $t_R = 40$ days, $r = 12\%, t_0 = 60$ days

17. In order to tool up for and begin production of a new model sports car, an automobile company issues \$100,000,000 worth of 10-year, 6%, semiannual coupon bonds to yield the investor 6%. The company establishes a sinking fund that earns 10% annually in order to accumulate the face value of the bonds at the end of the 10 years. Thus, the annual fixed cost will be the coupon interest payment plus the semiannual sinking fund contribution amount. The variable cost per car is estimated to be \$18,000, and the car manufacturer will receive \$24,000 per car. Ignoring any revenue lag, how many units must be produced and sold during each of the 10 years to breakeven?

8.2 BREAKEVEN BASED ON SUPPLY AND DEMAND

In a "perfect" economy, the breakeven point for a product is that point when the supply of the product will equal the demand for the product. At that point, the market economists consider the market to be in equilibrium, and the breakeven point is called the "equilibrium" point. Letting:

P be the price on a demand or supply equation as appropriate,

P_0 be the intercept on the demand or supply axis as appropriate,

p' be the slope of either a demand or supply equation as appropriate,

q be the quantity in demand or supply,

S represent supply, and

D represent demand,

a linear demand equation would be of the form

$$P_D = p'q + P_{0D}. \qquad (8.2\text{-}1)$$

The value of P_{0D} is a theoretical value that specifies the price if the demand is zero. In reality, if the price of a product is relatively high, the demand for the product will be small. Similarly, if the price of a product is too low, the demand could be beyond the level of the suppliers to produce. For example, the demand for the Porsche automobile is much smaller than the demand for Ford automobiles because the price of the Porsche is much greater than the price of a Ford. Since there is a decreasing price for increasing demand, the slope of a demand function will be negative. Similarly, a linear supply would be of the form

$$P_S = p'q + P_{0S}. \qquad (8.2\text{-}2)$$

The value of P_{0S} is the theoretical threshold price in that no matter how small a supply is needed, the price will never be less than the threshold price. Suppliers will offer quantities that are sufficient to meet the price that people are willing to pay. For example, manufacturers will produce as many high-priced automobiles as there are people willing to pay the high prices. Consequently, manufacturers will produce greater and greater quantities at higher and higher prices. The slope of the supply function will be positive. Figure 8-4 shows the equilibrium point for linear supply and demand functions.

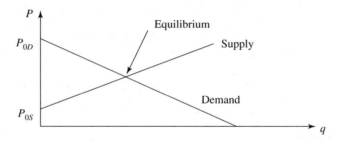

FIGURE 8-4 Supply and Demand Breakeven — Equilibrium

EXAMPLE 8.2.1   The purchase and sale of stocks in the stock market relies on the connection of willing buyers and sellers. Suppose that 1,000 people bid on a stock when its price was $49 per share, and 700 people bid when the price was $55 per share. At the same time, suppose 500 people offered the stock when the price was $49 per share, and 2,000 people offered the stock when the price was $55 per share. If both the demand (bid) and the supply (offer) are linear functions, determine the equilibrium price of the stock.

Solution   The ordered pairs for the development of the linear demand function are (q, P) in general and specifically for this example, $(1,000, 49)$ and $(700, 55)$. Then the slope of the demand equation would be determined by

$$p' = \frac{P_2 - P_1}{q_2 - q_1}$$

$$= \frac{55 - 49}{700 - 1,000} = -0.02.$$

From Equation (8.2-1),

$$P_D = -0.02q + P_{0D},$$

or

$$49 = -0.02(1,000) + P_0.$$

Then, $P_{0D} = 69$, and the demand equation becomes

$$P_D = -0.02q + 69.$$

The slope of the supply equation would be determined by

$$p' = \frac{P_2 - P_1}{q_2 - q_1}$$

$$= \frac{55 - 49}{2,000 - 500} = 0.004.$$

From Equation (8.2-2),

$$P_S = 0.004q + P_{0S},$$

or

$$49 = 0.004(500) + P_{0S}.$$

Then, $P_{0S} = 47$ and the supply equation becomes
$$P_S = 0.004q + 47.$$

The equilibrium point occurs when the supply equation equals the demand equation, or
$$0.004q + 47 = -0.02q + 69,$$
$$0.024q = 22.$$

Solving gives
$$q = 916.67.$$

Thus, the equilibrium number of shares would be 917 if fractional shares cannot be purchased. At the bid and offer number of 917, the per share price would be
$$P = 0.004(917) + 47 = \$50.67. \quad \blacksquare$$

If a demand function is truly linear, that linear demand can generate diminishing returns. Revenue from a sale is the product of the price times the quantity sold or $R = pq$. Then if a demand function is linear, revenue would be given by
$$R(q) = (p'q + P_{0D})q,$$

which multiplies to
$$R(q) = p'q^2 + (P_{0D})q. \tag{8.2-3}$$

In a general quadratic of the form, $y = ax^2 + bx + c$, the graph is a parabola. The vertex is at the value, x_v, given by
$$x_V = -\frac{b}{2a}.$$

If the coefficient of the quadratic term, a, is negative the parabola opens downward, and the vertex is the point of maximum y. Also, if the coefficient of the linear term, b, is positive, the x-coordinate of the vertex is positive. Both conditions hold for revenue based on linear demand; p' is negative, and P_0 is positive, for the revenue function of Equation (8.2-3). Therefore, the quantity for maximum revenue generated by a linear demand function is given by
$$q_{V(\text{Max } R)} = -\frac{P_{0D}}{2p'}. \tag{8.2-4}$$

The maximum revenue is determined by substituting q_v into $R(q)$. If a linear cost function is superimposed on the quadratic revenue function, two breakeven points are possible as shown in Figure 8-5.

EXAMPLE 8.2.2   Suppose that at the selling price of \$100 per unit in Example 8.1.1, the demand for the product is 100 units per day. Also suppose that if the selling

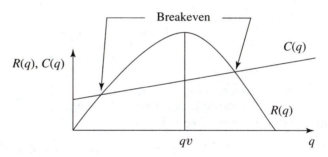

FIGURE 8-5 Revenue Based on Linear Demand

price had been set at $110, the demand would have been 95 units per day. Based on the assumption that the price for this product is a linear function of demand, determine the point of maximum revenue.

Solution   The ordered pairs, (q, p), for this function are (100, 100) and (95, 110). The demand function would be given by Equation (8.2-1) as

$$P_D = p'q + P_{0D}.$$

The slope of the equation is determined by

$$p' = \frac{P_2 - P_1}{q_2 - q_1}$$

$$= \frac{110 - 100}{95 - 100} = -2.$$

To find P_{0D},

$$110 = -2(95) + P_{0D},$$

$$P_{0D} = 300.$$

Then, the demand function is determined to be

$$P_D = -2q + 300.$$

Since revenue is the product of price and quantity,

$$R(q) = (-2q + 300)q$$

$$= -2q^2 + 300q.$$

This is a quadratic of the form $y = ax^2 + bx$, and since a is negative, the parabola opens downward giving the point of maximum revenue at the vertex. Then, from Equation (8.2-4) the number of units to be sold for maximum revenue is determined to be

$$q_V = -\frac{300}{2(-2)} = 75.$$

If the demand function is based on a daily number of units produced and sold, substituting 75 into the revenue equation would determine the maximum daily revenue

$$R(75) = -2(75)^2 + 300(75)$$
$$= \$11{,}250. \quad \blacksquare$$

For the breakeven points when cost is a linear function of the number of units produced, profit is determined by subtracting the linear cost function from the quadratic revenue function, or

$$P(q) = p'q^2 + P_{0D}q - (cq + C_0).$$

This reduces to

$$P(q) = p'q^2 + (P_0 - c)q - C_0. \tag{8.2-5}$$

The point of maximum profit occurs at the vertex of this quadratic profit equation or

$$q_{v(\text{Max } P)} = -\frac{P_{0D} - c}{2p'}. \tag{8.2-6}$$

Note that the numerator in Equation (8.2-4) will be less than the numerator of Equation (8.2-6). Since the denominators are the same, the difference indicates that when revenue is based on a linear demand function and cost is a linear function of quantity, the quantity for maximum profit is less than the quantity for maximum revenue.

EXAMPLE 8.2.3   Suppose that the product of Example 8.2.2 has been successfully selling at $100 per unit. The initial costs were recovered during the first year of production and now annual fixed costs are $2,800. The variable cost is $50 per unit. Determine the daily demand and price that will be necessary to generate maximum profit and determine the maximum daily profit.

Solution   The revenue function was determined in Example 8.2.2 as
$$R(q) = -2q^2 + 300q.$$

From the information in this problem, the cost function is
$$C(q) = 50q + 2{,}800.$$

Subtracting the cost equation from the revenue equation gives the profit function as
$$P(q) = -2q^2 + 250q - 2{,}800.$$

The production quantity that is needed for maximum daily profit is the vertex of the profit equation and is determined as
$$q_{V(\text{Max } P)} = -\frac{250}{-2(2)} = 62.5.$$

Then, the production of 62 complete units and one-half of the 63rd unit each day will generate maximum profit. But since only a whole number of units can be shipped daily, the maximum daily profit would be the profit resulting from producing 62 units each day and selling them at the price that is indicated by the demand equation. The selling price per unit can be determined by the demand equation, Equation (8.2-1), as

$$P = -2(62) + 300$$

$$= \$176.$$

Whether or not the popularity of this product could sustain a 76% increase is problematic. That is, would there be a demand for 62 units daily at a price of $176? However, it is the condition for the maximum daily profit that would be determined by substituting 62 units into the profit equation. Then,

$$P(62) = -2(62)^2 + 250(62) - 2,800 = \$5,012. \quad \blacksquare$$

In order to determine the breakeven quantities, the quadratic formula may be applied to the profit function after setting it equal to zero. Since it is quadratic in the form,

$$y = ax^2 + bx + c,$$

and recalling the quadratic formula

$$x = \frac{-b \pm \sqrt{b^2 - 4ac}}{2a},$$

substituting the appropriate quantities from the profit equation gives

$$q = \frac{-250 \pm \sqrt{(250)^2 - 4(-2)(-2,800)}}{2(-2)},$$

the solution of which gives the breakeven demand quantities as 12.44 and 112.56 units. The appropriate values are shown in Figure 8-6.

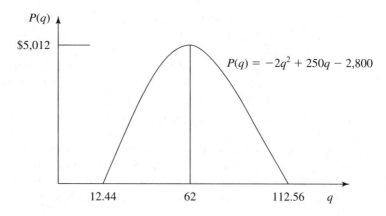

FIGURE 8-6 Graph for Example 8.2.3

PROBLEM SET 8.2

Problems 1–4 represent supply and demand functions. Determine the equilibrium points.

Demand Functions	Supply Functions
1. $p = -\left(\frac{4}{3}\right)q + 20$	$p = \left(\frac{5}{6}\right)q + 3$
2. $p = -\left(\frac{2}{7}\right)q + 8$	$p = -\left(\frac{3}{11}\right)q + 4$
3. $p = -2q + 22$	$p = 3q + 12$
4. $P = -0.3q + 6$	$p = 0.15q + 1.5$

In Problems 5–8, determine the revenue functions and points of maximum revenue.

Demand Function

5. $p = -\left(\frac{4}{3}\right)q + 20$

6. $p = -\left(\frac{2}{7}\right)q + 8$

7. $p = -2q + 22$

8. $p = -0.3q + 6$

In Problems 9–12, develop the profit equations and determine (a) the breakeven quantities and (b) the points of maximum profit.

Demand Functions	Cost Functions
9. $p = -1.33q + 20$	$C(q) = 1.67q + 50$
10. $p = -0.22q + 8$	$C(q) = 0.67q + 47$
11. $p = -1.58q + 22$	$C(q) = 3q + 45$
12. $p = -0.3q + 6$	$C(q) = 1.5q + 10$

13. The monthly demand for a particular model of television set is 250 when the price is $140; however, when the set is advertised at $110, the monthly demand increases to 1,000. If the unit price to the suppliers is $60, they are willing to supply 750 sets per month, but if the unit price to the suppliers increases to $80, they are willing to supply 2,240 sets. Determine the equilibrium quantity to the nearest whole number and the associated price.

14. At one point during a trading day, 1,000 investors bid $120 to buy each share of a particular stock, but 2,000 people bid to buy each share at $115. When the bid was $120, there were 3,000 offers to sell the stock, and when the bid was $115, there were only 1,200 offers to sell. Assuming the market supply and demand are linear functions of quantity, determine the equilibrium quantity to the nearest whole number and the associated price.

8.3 BREAKEVEN BASED ON FINANCE

Breakeven points in investments can be useful in decision making. For example the purchase of a stock in a company requires the payment of a commission unless the stock is purchased directly from the company. When the stock is sold using a brokerage house, a commission is charged on the sale. It may be useful to know the breakeven price, after which the sale of the stock will yield a profit on the investment. Letting:

N be the number shares,

c be the purchase cost per share,

s be the selling price per share,

r_c be the commission rate on purchases,

r_s be the commission rate on sales,

C be the total cost, and

R be the net revenue,

the total cost to purchase N shares of stock in a company such as IBM would be determined as

$$C = cN + r_c cN,$$ (8.3-1)

or

$$C = (1 + r_c)cN.$$ (8.3-2)

The net revenue from the sale of the stock would be determined as

$$R = sN - sr_s N$$ (8.3-3)

or

$$R = (1 - r_s)sN.$$ (8.3-4)

The breakeven selling price would be determined when the net revenue equals the total cost. Equating Equations (8.3-2) and (8.3-4) and solving for the breakeven selling price gives

$$s = \left(\frac{1 + r_c}{1 - r_s}\right)c.$$ (8.3-5)

Commission rates may be graduated based on the dollar amount of the sale. If the dollar amount places the sale in a different commission bracket, the rate in the denominator will be different than the rate in the numerator.

EXAMPLE 8.3.1   A person buys 100 shares of IBM at a price of $105 per share. Determine the breakeven selling price if the commission scale below is in effect.

Sale Amount	Commission Rate
Less than $5,000	3.0%
At least $5,000 but less than $50,000	2.5%
At least $50,000 but less than $500,000	2.0%
At least $500,000	1.5%

Solution   The cost of 100 shares of IBM at $105 per share would be $10,500. From the rate scale above, the commission rate would be 2.5%. Then, from

Equation (8.3-5), the breakeven selling price would be

$$s = \left(\frac{1+r}{1-r}\right)c$$

$$= \left(\frac{1+0.025}{1-0.025}\right)(105) = 110.3846.$$

At a selling the price of $110.3846 per share, the total revenue for the IBM stock would be $11,038.46. At this sale amount, the commission rate is 2.5% and use of Equation (8.3-5) is valid. The commission would be $275.96 leaving a net revenue of $10,762.50. The cost to purchase the 100 shares would have been $10,500, and the commission on the purchase would have been $262.50 for a total cost $10,762.50. Hence, at a selling price of $110.3846 breakeven will have occurred. ■

Another possible necessity for a breakeven analysis could be the determination of the point in time when total revenues or total costs are equal for different alternatives currently being considered. An example of this type of analysis is the increasing number of people choosing to retire at a relatively early age. Normally, when a person retires at an age that is earlier than a "normal" retirement age, the annual pension amount will be reduced from the amount that would have been paid at the "normal" age. Before people elect a retirement date strategy, it might be a good idea for them to determine the breakeven point where the total revenues would be equal. Figure 8-7 is a diagram that represents the situation.

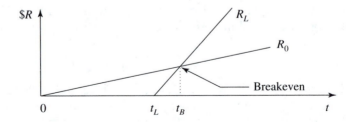

FIGURE 8-7 Retirement Age Breakeven

Letting:

P_0 be the periodic amount at an early retirement age,

P_L be the periodic amount at a later retirement age,

t be the number of periods from the early retirement date,

t_L be the number of periods to the later start date,

R_0 be the total revenue from the start of early retirement, and

R_L be the total revenue from the start of later retirement,

the total revenue from early retirement would be

$$R_0 = P_0 t. \tag{8.3-6}$$

The total revenue from the normal retirement date would be

$$R_L = P_L(t - t_L). \tag{8.3-7}$$

Equating Equations (8.3-6) and (8.3-7) and solving for t, give the time to breakeven as

$$t_B = \frac{P_L t_L}{P_L - P_0}. \tag{8.3-8}$$

EXAMPLE 8.3.2   The pension plan of a company will provide a particular employee an annual amount of $50,000 for retirement at age 65. It allows early retirement at age 62 at a reduced amount of $40,000 per year. If the employee chooses to retire at age 62, at what age will the total revenue from early retirement equal the total revenue had the employee waited to age 65 to retire?

Solution   The number of years of early retirement is three—(ages 65 − 62). From Equation (8.3-8), the number of years to breakeven is determined to be

$$t_B = \frac{P_L t_L}{P_L - P_0}$$

$$= \frac{(50,000)(3)}{50,000 - 40,000} = 15.$$

Then at age 77—(age 62 + 15) the total revenues from either retirement age will be the same. If the employee is in good health and lives beyond age 77, he or she would have made a mistake, relative to total revenues, to retire early. If the employee is in poor health, it might have been to his or her advantage to retire early. ∎

Some pension plans and Social Security have provisions for annual increases in the pension due to increases in cost of living. Equation (8.3-8) is not valid for those types of plans; however if the annual benefit is increased by a constant annual rate, r, the number of years to breakeven can be determined by Equation (8.3-9) as

$$t_B = \frac{\mathrm{Ln}\left[\dfrac{P_L - P_0}{P_L(1+r)^{-t_L} - P_0}\right]}{\mathrm{Ln}(1+r)}. \tag{8.3-9}$$

The derivation of the formula to accommodate annual increases in pension amounts is left as an exercise for the student as Problem 15b below.

EXAMPLE 8.3.3   The pension plan of a company will provide a particular employee an annual amount of $50,000 for retirement at age 65. It allows early retirement at age 62 at a reduced amount of $40,000 per year. For either choice, the company pension plan allows for a 2% annual increase in the pension amount. If the employee chooses to retire at age 62, at what age will the total revenue from early retirement equal the total revenue had the employee waited to age 65 to retire?

Solution The number of years of early retirement is three—(ages 65 − 62). From Equation (8.3-9), the number of years to breakeven is determined by

$$t_B = \frac{Ln\left[\dfrac{50,000 - 40,000}{50,000(1.02)^{-3} - 40,000)}\right]}{Ln(1.02)} = 17.2$$

years or approximately age 79 years and two months. ■

PROBLEM SET 8.3

Use the following commission rate table for problems 1–10 to determine the breakeven prices.

Sale Amount	Commission Rate
Less than $5,000	3.0%
At least $5,000 but less than $50,000	2.5%
At least $50,000 but less than $500,000	2.0%
At least $500,000	1.5%

	Company	Number of Shares	Price per Share
1.	Abbot Laboratories	1,000	$31.1875
2.	Bell Atlantic Corporation	2,500	$60.0625
3.	Ford Motor Corporation	5,000	$50.4375
4.	IBM	200	$110.1250
5.	Intel Corporation	500	$97.9375
6.	J. P. Morgan & Company	1,500	$117.6250
7.	Lucent Technologies	750	$52.7500
8.	Merck & Company	3,000	$72.4375
9.	Microsoft Corporation	250	$103.7500
10.	Morgan Stanley Dean Witter	3,000	$66.2500

In problems 11–14, determine the age at which total revenues will become equal for the indicated retirement ages and annual incomes.

	Retirement Age	Annual Income	Retirement Age	Annual Income
11.	55	$25,000	65	$40,000
12.	60	$27,500	65	$50,000
13.	62	$30,000	65	$55,000
14.	65	$50,000	70	$70,000

15. a) If a person had worked under Social Security in the United States and had earned the maximum Social Security taxable amount from age 22

through age 62 in the year 2001, that person would receive a monthly Social Security Benefit of $1,335 at age 62. If that person elected to wait to age 65 to retire, the monthly benefit would have been approximately $1,904. Ignoring the fact that there are annual cost of living increases in the monthly benefit, determine the age at which the total revenues from either retirement age would have been the same. That is, determine the breakeven point.

b) As indicated in Problem 15a, the monthly benefit is increased annually. If it is assumed that the future annual increases will be 3.5% per year, derive Equation (8.3-9) and determine the breakeven point for Problem 15a. Use annual amounts.

CHAPTER

9

Life Insurance

Life insurance combines the time-value of money and mortality. That is, a life insurance company uses the probability that a policyholder will die within a certain number of years. This probability helps to establish a price that the individual will have to pay in order to buy a life insurance policy. That price will depend upon an assumption of future interest rates. For both mortality and future interest rate assumptions, the insurance company will use a conservative approach by assuming higher probabilities of death and lower interest rates than indicated by the company's experience. For life insurance, there are two fundamental concepts from the company's perspective.

1. The contract will be honored at the death of the insured. This is the conventional life insurance concept. Upon the death of an individual, the face amount of the insurance policy will be paid to a designated beneficiary. This type of insurance is favored by people who think that they will not survive beyond the life expectation used by the insurance company or by people who wish to use life insurance in estate planning.

2. The contract will be honored if the insured is alive at a specified future date. Basically, this type of insurance contract will pay the insured a lump sum or a periodic amount beginning at a specified future date and will terminate at the death of the individual or after a specified number of payments. Some modifying options can be included in the contract such as an annuity, payable as long as one of the spouses of a married couple is alive. This is, obviously, favored by people who think that one of the spouses will survive longer than the life expectation used by the insurance company.

It would be a reasonable assumption that the relative cost of 1) above would be greater than the relative cost of 2) above. There is also the possibility of combining a pure life insurance contract with a life annuity contract at a commensurate cost. This chapter presents some of the elementary aspects of life insurance, or more generally, life contingencies. It is intended to introduce you to the topic and to give you a "vocabulary" of the subject.

9.1 ELEMENTS OF PROBABILITY

Probability is defined by a fraction in which the numerator is the number of times an event occurs and the denominator is the total number of ways that the event can occur. An event is simply a specified outcome such as the result of flipping a coin being a head (or tail). Also, an event is the selection of a particular card out of a deck of cards. It can also be the death of a person in a given year.

Using the notation, $P(E)$, to represent the probability of an event, the mathematical definition of probability would be given by

$$P(E) = \frac{n}{N}, \qquad (9.1\text{-}1)$$

where n is the number of times an event occurs, and N is the total number of ways that the event can occur. For example, there are only two possibilities in the toss of an honest coin (head or tail).

We discount the extremely rare chance that the coin could land on edge. Therefore, the probability of tossing either a head or a tail is $\frac{1}{2}$. Now, if you toss a coin 10 times, you might get any combination of heads and tails. However, if you toss the coin 100,000 times, the number of heads and tails would be close to 50,000 each. That is, as the number of tosses of a coin increases, the actual probability of the result will approach the theoretical probability of $\frac{1}{2}$. This is known as the **law of large numbers**. As another example; there are four aces in a conventional deck of 52 playing cards. If a person selects one card from such a deck, the probability that it will be an ace is given by Equation (9.1-1) as $\frac{4}{52}$ or $\frac{1}{13}$.

MULTIPLE EVENTS—ADDITION RULE

If there is a possibility of several events but only one can occur, the probability that one of the events will occur is the sum of the probabilities of the occurrence of each of the possible events. If the occurrence of one event does not have an effect on the occurrence of another event, the events are **independent**. Given the possibility of independent events A and B, the probability of either A happening or B happening, but not both, in one effort is the sum of the two probabilities.

$$P(A \cup B) = P(A) + P(B), \qquad (9.1\text{-}2)$$

where,

$P(A \cup B)$ is the probability of A or B,

$P(A)$ is the probability of A, and

$P(B)$ is the probability of B.

EXAMPLE 9.1.1 In a conventional deck of 52 playing cards, there are four aces and four kings. In one selection, what is the probability of selecting an ace or a king?

Solution If we chose to use Equation (9.1-1), there are eight possible card choices from the 52. Then the P(an ace or a king) $= \frac{8}{52} = \frac{2}{13}$. If we chose to use

Equation (9.1-2), then

$$P(\text{an ace or a king}) = P(\text{an ace}) + P(\text{a king})$$

$$= \frac{4}{52} + \frac{4}{52}$$

$$= \frac{1}{13} + \frac{1}{13}$$

$$= \frac{2}{13}. \quad \blacksquare$$

Life contingencies are similar to the toss of a coin in that only two possibilities exist each year and only one of them will occur. A person will either be alive at the end of a year or the person will die during the year. If experience has shown that out of every 35,046 people who reach age 84, 3,154 of them will die within the year, the probability of a person at age 84 dying before age 85 is given by Equation (9.1-1) as

$$P(\text{Dying}) = \frac{3,154}{35,046}$$

$$= 0.09,$$

after rounding to two decimal places.

RATE OF MORTALITY

The foregoing has been introduced in order to define, formally, a rate of mortality. Let:

q_x = the rate of mortality at age x,
d_x = the number dying before reaching age $x + 1$, and
l_x = the number living at the beginning of age x.

Then the **rate of mortality** is given by Equation (9.1-1) as

$$q_x = \frac{d_x}{l_x}. \tag{9.1-3}$$

Similarly, if 3,154 of the people of the group die during the year, 31,892 will be alive at the end of the year. Then, the probability that a person age 84 will survive to age 85 is given by Equation (9.1-1) as

$$P(\text{Surviving}) = \frac{31,892}{35,046}$$

$$= 0.91,$$

after rounding to two decimal places. Since a person will either die during a year or will survive the year, the probability that either will happen is one. Thus, the probability that a person age 84 will either die before reaching age 85 or will survive to age 85 is $0.09 + 0.91 = 1.00$. If $P(S)$ designates the probability of surviving, $P(D)$ designates the probability of dying, and the notation $P(D \cup S)$ designates the probability of dying or surviving, the addition rule gives

$$P(D \cup S) = P(D) + P(S). \qquad (9.1\text{-}4)$$

If a probability of dying is given as 0.06225, the probability of surviving could be determined from Equation (9.1-4) as

$$P(S) = 1 - P(D) = 1 - 0.06225$$
$$= 0.93775.$$

When the sum of two probabilities equals one, the probabilities are called complementary.

MULTIPLE EVENTS—MULTIPLICATION RULE

If several events can occur independently of each other, the probability that all the events will occur is the product of the probability of the occurrence of each of the events. Again, events are independent of each other if the occurrence of one has no effect on the occurrence of the other. Mathematically, if the probability of an independent event A is $P(A)$ and the probability of an independent event B is $P(B)$, the probability of both events occurring $P(A$ and $B)$ is given by

$$P(A \cap B) = P(A) \times P(B). \qquad (9.1\text{-}5)$$

EXAMPLE 9.1.2 In two tosses of a coin, what is the probability of two heads, three heads?

Solution Since the probability of a head is $\frac{1}{2}$, the probability of two heads is $\left(\frac{1}{2}\right)\left(\frac{1}{2}\right) = \frac{1}{4}$.

The probability of tossing three heads is $\left(\frac{1}{2}\right)\left(\frac{1}{2}\right)\left(\frac{1}{2}\right) = \frac{1}{8}$. ■

In life contingencies, it is often necessary to know the probability that both husband and wife will survive to a particular age. Since the survival of each is independent of the survival of the other, the multiplication rule would apply.

EXAMPLE 9.1.3 If the probability that a male will live to age 81 is 0.44232 and if the probability a female will live to age 81 is 0.44232, what is the probability that both a husband and wife, both age 81, will live to their respective life expectancies?

Solution The survival of each is independent of the survival of the other. The probability that a husband and wife will survive to their respective life expectancies is the product of the probability that he will live to age 81 and that she will live to age 81. The multiplication rule gives the probability as (0.44232)(0.44232) and is equal to 0.19565. ■

Experience has shown that the life expectancy of females is greater than that of males, but for this elementary presentation of life contingencies, it will be assumed that the life expectancies of males and females are the same.

PROBLEM SET 9.1

1. What is the probability of getting a four when one ordinary 6-faced die is thrown one time?

2. What is the probability of getting a six when one ordinary 6-faced die is thrown one time?

3. If a red die and a green die are each thrown once, what will be probability that the red die will show a three and the green die will show a four?

4. If a red die and a green die are thrown one time each, what is the probability that the red die will show a two and the green die will show a five?

5. A pair of ordinary 6-faced dice is rolled once. What is the probability that the sum of the faces will be seven?

6. Given the following probabilities: $P(\text{Twins}) = 0.0114$, $P(\text{Triplets}) = 0.0017$, $P(\text{Quadruplets}) = 0.0003$, and $P(\text{Births} > 4) = 0$. What is the probability of a multiple birth by a pregnant woman selected at random?

7. If the probability that a person age 35 will survive to age 36 is 0.999 and the probability that a person age 36 will die before reaching age 37 is 0.0008, what is the probability that a person who is 35 years of age will live to age 36 and then die before reaching age 37?

8. A standard deck of 52 playing cards is comprised of four suits (clubs, diamonds, hearts, and spades) of 13 cards each. In a single selection, what is the probability of selecting an ace or a club?

9. If a group of 1,000 people 25 years of age is observed for one year and two of them die during the year, what is the probability that a person age 25 will survive to age 26?

10. If the probability that a newborn child will be a girl is 0.4961, what is the probability that a newborn child will be a boy?

11. If the probability of dying at a particular age is 0.952, what is the probability of surviving to the next birthday?

9.2 MORTALITY

A **mortality table** is simply a tabulation of the number of people living and dying at each age from 0 to an arbitrary upper limit such as 110 or 115 years. The population at age 0, which serves as a base population for computing the table is called a **cohort**. In addition to the actual numbers of the cohort who die at each age, the table may also include the rates of mortality and other information.

Depending upon the data that are used, there are two principal types of mortality tables.

1. Tables that are derived from population statistics and that are generally prepared by the National Center for Health Statistics. The data are

based on information gathered during a regular census. Such a table is Table 80CNSMT, *1980 Commissioners Standard Mortality Table*. It is included as Table A-6 in the Appendix.

2. Tables that are derived from data on the lives of the insured and are classified into two categories.

   (a) Annuity mortality tables that specify the present values of contingent life annuities. Examples of these are the 1983 Individual Annuity Mortality Table and UP-84 Unisex Annuity Mortality Table. The UP represents "Uninsured Pensions." However, for the purposes of this text, a table that is derived from Table 80CNSMT will be used and is included as Table A-7 in the Appendix.

   (b) Insurance mortality tables that specify the life expectancy of an individual and/or the joint life expectancies of a couple. The latter is used for life insurance contracts, which specify the "last survivor." Examples of these types of tables have been extracted from Publication 939 of the U.S. Treasury Department and are included in the Appendix as Table A-8 for single life expectancies, Table A-9 for joint life expectancies with a 10-year age difference, and Table A-10 for joint life expectancies as published in Table VI of Publication 939.

Mortality tables such as life tables include columns for the number of people living and dying at each age from a cohort at age zero. From the definition of the rate of mortality, a column for q_x is also included. Typically, there will be columns headed D_x, N_x, C_x, M_x, and life expectancy, all of which will be explained later. Table A-6 uses a cohort of 100,000 people who were born in 1990. Of that number 1,260 died before reaching age one. The rate of mortality would be determined by

$$q_0 = \frac{1,260}{100,000}$$
$$= 0.0126.$$

Consequently, the number of newborns who survived to age one is given by $100,000 - 1,260 = 98,740$, and in general

$$l_{x+1} = l_x - d_x, \tag{9.2-1}$$

where,

l_x is the number alive at age x,

d_x is the number who die at age x, and

l_{x+1} is the number alive at age $x + 1$.

Similarly, given the rate of mortality, the number of people who will die in the year they become age x may be determined by

$$d_x = l_x q_x, \tag{9.2-2}$$

and consequently, the rate of mortality can be determined, as defined by Equation (9.1-3), by

$$q_x = \frac{d_x}{l_x}.$$ (9.2-3)

This is referred to as the **rate of mortality**.

EXAMPLE 9.2.1   If the rate of mortality at age 30 is 0.0014 and if, from a cohort of 100,000 people, 97,077 were alive at the beginning of year, how many persons can be expected to die during the year?

Solution   From Equation (9.2-2),

$$d_{30} = (0.0014)(97,007)$$

$$= 136. \quad \blacksquare$$

EXAMPLE 9.2.2   If 214 people who are age 40 die, and, from a cohort of 100,000 people, 95,382 persons have attained age 40, what is the rate of mortality at age 40?

Solution   From Equation (9.2-3),

$$q_{40} = \frac{214}{95,382} = 0.00224. \quad \blacksquare$$

A life mortality table can be constructed in two ways. If the number alive at the beginning of each year is used, then the number expected to die can be determined by the rearrangement of Equation (9.2-1) to

$$d_x = l_x - l_{x+1}.$$ (9.2-4)

Given the numbers of living and dying, the rate of mortality at each age can be determined by using Equation (9.2-3).

PROBABILITIES OF LIVING AND DYING

The probability that a person age x will survive to age $x + 1$ is defined as the ratio of the number of people alive at each age, or

$$p_x = \frac{l_{x+1}}{l_x}.$$ (9.2-5)

One of the absolute certainties in insurance is that a person will either live through one year or will die during the year. Thus, the probabilities of living and dying are complementary, and

$$p_x + q_x = 1.$$ (9.2-6)

EXAMPLE 9.2.3   Given the following from a mortality table. What is the probability that a person age 82 will survive to age 83? What is the probability that the person will die before reaching age 83?

Age	80	81	82	83	84
l_x	2,410	2,180	1,955	1,737	1,528

Solution   From Equation (9.2-5),

$$p_{82} = \frac{l_{83}}{l_{82}}$$

$$= \frac{1,737}{1,955}$$

$$= 0.8885. \quad \blacksquare$$

From the complementary nature of the probabilities, Equation (9.2-6) may be solved for q_{82} as 0.1115.

PROBABILITY OF LIVING n YEARS

If a person survives to age x, the probability that he or she will survive to age $x + n$ is denoted by $_np_x$ and is read as "*npx*." Thus, a notation $_{15}p_{65}$, which is read "15*p*65," is asking for the probability that a person age 65 will live 15 years to age 80 $(= 65 + 15)$. If it is desired to determine the probability of surviving to age 80 if a person is 65 now, the value of $n = (x + n) - x = 80 - 65 = 15$. The probability that a person age x will survive to age $(x + n)$ is the ratio of the number of people expected to be alive at age $(x + n)$ to the number of people alive at age x. Then,

$$_np_x = \frac{l_{x+n}}{l_x}. \tag{9.2-7}$$

EXAMPLE 9.2.4   From the table in Example 9.2.3, what is the probability that a person age 80 will live four years?

Solution   From Equation (9.2-7),

$$_4p_{80} = \frac{l_{80+4}}{l_{80}}$$

$$= \frac{l_{84}}{l_{80}}$$

$$= \frac{1,528}{2,410}$$

$$= 0.6340. \quad \blacksquare$$

From the complementary nature of the probabilities of living and dying,

$$_np_x + {_nq_x} = 1. \tag{9.2-8}$$

EXAMPLE 9.2.5   For the 80-year-old person in Example 9.2.4, what is the probability that he or she will not survive to age 84?

Solution   Obviously, not surviving implies dying between ages 80 and 84. From Equation (9.2-8),

$$_4q_{80} = 1.0000 - 0.6340$$

$$= 0.3660. \quad \blacksquare$$

PROBLEM SET 9.2

1. Using Table A-6 of the Appendix, what is the probability that a person age 60 will die before reaching age 61?

2. According to Table A-6, what is the rate of mortality at age 82?

3. If the rate of mortality at a certain age is 0.00742, and the number of persons living at that age is 107,412, how many of them may be expected to die within the year?

4. If a mortality table shows $l_{18} = 994,831$ and $d_{18} = 1,094$, calculate l_{19}.

5. If a mortality table shows $l_{36} = 951,003$ and $q_{36} = 0.0022$, calculate the value of d_{36}.

6. If a mortality table shows $l_{75} = 4,940,810$ and $d_{75} = 361,490$, calculate the value of q_{75}.

7. What is the meaning of the notation $_n p_x$?

8. What is the probability that a person age 20 will live to reach age 21?

9. What is the probability that a person age 20 will live 25 years?

10. What is the probability that a person age 30 will be alive at age 50?

11. Calculate the probability that a person age 70 will live to age 80.

12. Of 10,000 persons age 35, how many can be expected to live to age 65?

13. Fill in the missing items in the following mortality table.

Age	l_x	d_x	q_x
50	92,637	_____	_____
51	91,941	_____	_____
52	91,195	_____	_____
53	90,481	_____	_____

14. The life expectancy of females is approximately six years longer than the life expectancy of males. If Table A-6 is assumed to be a male mortality table, then female mortality can be determined from the same table by using ages six years less than indicated in the table. That is, for a female age 30, use the male age of 24. This process is called a six-year setback. Using a six-year setback for females, calculate how many of 100,000 females age 27 can be expected to die within 10 years.

15. What is the probability that a person age 30 will live for 25 additional years?

16. What is the probability that a person age 30 will die before reaching age 65?

17. What is the probability that two persons, ages 30 and 40, will die within the next 10 years?

9.3 COMMUTATION FUNCTIONS

Commutations symbols are given by the columns headed by D_x, N_x, C_x, and M_x in Table A-6 and are used to simplify calculations in life insurance. We will define the terms in this section but leave their use until later. All references to the commutation functions are to be understood to be per 100,000 people, the size of the cohort used in this book.

The **commutation symbol** D_x represents the present value, at age 0, of $1.00 for each person of the cohort who is age x and alive. It is, therefore, defined by multiplying the number of people who are age x by the present value of $1.00 due in x years at a specified interest rate. Mathematically,

$$D_x = l_x v^x, \qquad (9.3\text{-}1)$$

where the notation v^x represents the present value of $1.00, $(1 + i)^{-x}$. By defining v in this manner, only positive exponents appear in equations involving life insurance. This is the symbol that is standard notation of the Society of Actuaries.

EXAMPLE 9.3.1 Find D_x in Table A-6 at a 5% interest rate for a person age 25, and confirm its value using Equation (9.3-1).

Solution Table A-6 shows that $l_{25} = 97{,}110$ and $D_{25} = 28{,}676.85$. From Equation (9.3-1),

$$D_{25} = 97{,}110(1.05)^{-25}$$
$$= 97{,}110(0.295303)$$
$$= 28{,}676.87. \quad \blacksquare$$

If D_x represents the present value of $1.00 for each person of the cohort age x and alive, then D_{x+n} represents the present value, at age zero, of $1.00 for each person of the cohort age $x + n$ and alive. The amounts represented by D_x and D_{x+n} are referred to as the present value with **benefit of survivorship**.

Instead of a single payment with benefit of survivorship at the end of n years, the obligation of the insurance company could be in the form of an annuity of payments as long as the person is alive to receive the payments. N_x denotes the present value of such an annuity. It represents the amount of money needed at age x for all persons who are alive at each age from x to infinity (or the end of the mortality table). Thus,

$$N_x = \sum_{t=x}^{\infty} D_t. \qquad (9.3\text{-}2)$$

EXAMPLE 9.3.2 What is the present value with a 5% interest rate, at age zero, of an annuity of $1,000 per year for each person of the cohort at age 65 and living?

Solution   From Table A-6, $N_{65} = 35,523.27$. Then, the present value, at age zero, for $1,000 per year is the same as $(1,000)N_{65}$, or $(1,000)(35,523.27) = \$35,523,270$, per 100,000 people. ■

When there is no benefit of survivorship, the obligation of the insurance company may be to pay a contracted amount of money at the death of an insured person. Then, the present value at age zero for the payment of $1.00 to the beneficiaries of those in the cohort who will die at age x is denoted by C_x and is defined by

$$C_x = d_x v^{x+1}. \tag{9.3-3}$$

In Equation (9.3-3), it is assumed that the payment will be made at the end of the year of death. In practice, the payment is usually made shortly after death and is referred to as the **moment of death**. The calculations for the present value at the "moment of death" are beyond the level of this text. The number of people at age 65 is found in Table A-6 as 1,587. The present value at age 0 for the payment of $1.00 to the beneficiaries of all who will die at age 65 is,

$$(1,587)(1.05)^{-66} = (1,587)(0.039949)$$

$$= 63.40,$$

as seen in Table A-6 opposite age 65. The present value at age 0 for the payment of $1.00 at the death of all persons, of the cohort, who are alive at age 65 is determined from Table A-6 under column headed by M_x as $1,542.78. It is analogous to N_x and is defined by

$$M_x = \sum_{t=0}^{\infty} C_t. \tag{9.3-4}$$

Note that D_x and C_x are used when persons at age x are alive and dead, respectively, whereas N_x and M_x are accumulations, which begin at age x, of C_x and D_x, respectively.

Lastly, Table A-6 includes a column headed by "Life Exp," which is the life expectancy at age x. Thus, the life expectancy of a person who is alive at age 65 is 17.0 years.

METHODS USED IN PRACTICE

It is highly unlikely that the same number of people reach age x each year. Therefore, the rate of mortality for age x would be different each year. In order to compensate for the differences, actuaries observe the mortality rate of each cohort over a long period of time to arrive at the annual changes in the mortality rates. Adjustments in the table are also needed due to the facts that not everyone reaches age x on the same day of the year and everyone who dies in a given year is not of the same age. Further, the actuaries will make conservative mortality assumptions to allow for unpredictable events, such as hurricanes, earthquakes, and floods that can be the cause of some deaths.

The assumptions and methods used in the construction of mortality tables are beyond the level of this text, but the student should be aware that the numbers are not absolute. An insurance company may use tables based on its own experience, publications of the U.S. Government, or on estimates of the U.S. Internal Revenue Service. Table A-6 is based on Table 80CNSMT as published in the Regulations of the Internal Revenue Service Regulations Section 20.2031-7A.

PROBLEM SET 9.3

Use Table A-6 of the Appendix for the following problems.

1. Find the number of people who will be alive at age 30.

2. Find the number of people who will be alive at age 65.

3. Find the number of people who will die at age 30.

4. Find the number of people who will die at age 65.

5. Find the number of people who will die between ages 30 and 31.

6. Find the number of people who will die between the ages of 30 and 65.

7. How much money will be necessary, at age zero, to pay $1,000 to everyone alive at age 22?

8. How much money will be necessary, at age zero, to pay $1,000 to everyone alive at age 65?

9. How much money will be necessary, at age zero, to pay $1,000 per year to everyone alive at age 65?

10. Verify that $N_{100} = 179.95$ at 3%.

11. Verify that $C_{100} = 16.92$ at 3%.

12. Verify that $M_{100} = 54.60$ at 3%.

9.4 LIFE ANNUITIES

A **life annuity** is a periodic amount paid to a living individual. One example of a life annuity is the Social Security benefit whereby the U.S. Government pays a retiree a monthly stipend for as long as the individual is alive. In the insurance industry, this type of annuity is referred to as **payments *with the benefit of survivorship*** or **contingent annuities** (as opposed to annuities certain, where the number of payments is fixed).

For a future payment of $\$P$, the present value at age x for a payment at age $x + n$ with the benefit of survivorship is determined by combining the present value of $1.00 due in n periods with the probability of a person age x surviving to age $x + n$ as given by Equation (9.2-7). Then,

$$A_x = P \left[\frac{l_{x+n}}{l_x} \right] v^n. \tag{9.4-1}$$

Using the definition of D_x as given by Equation (9.3-1), Equation (9.4-1) may be redefined as

$$A_x = P \left[\frac{D_{x+n}}{D_x} \right]. \tag{9.4-2}$$

EXAMPLE 9.4.1   Find the present value with the benefit of survivorship for a payment of $50,000 at age 65 to a person at age 40 if the interest rate is 5%.

Solution   From Table A-6, we find $l_{40} = 94,926$ and $l_{65} = 77,107$. Then, from Equation (9.4-1)

$$A_{40} = 50,000 \left[\frac{77,107}{94,926} \right] (1.05)^{-(65-40)}$$

$$= \$11,993.51,$$

or, from Equation (9.4-2),

$$A_{40} = 50,000 \left[\frac{3,234.37}{13,483.83} \right]$$

$$= \$11,993.50.$$

The 1-cent difference is due to rounding.   ■

EXAMPLE 9.4.2   What would be the present value of $50,000, with the benefit of survivorship, at age 65 for a person age 40 if the interest rate is 3%?

Solution   Using the definition, given by Equation (9.4-1), the present value at age 40 is determined as

$$A_{40} = 50,000 \left[\frac{77,107}{94,926} \right] (1.03)^{-(65-40)}$$

$$= \$19,397.60.$$

This may be confirmed using Equation (9.4-2) and the 3% rate of Table A-6. From Equation (9.4-2) and the table

$$A_{40} = P \left[\frac{D_{65}}{D_{40}} \right]$$

$$= 50,000 \left[\frac{11,289.49}{29,100.21} \right]$$

$$= \$19,397.60.   ■$$

The present values in Examples 9.4.1 and 9.4.2 indicate that the present values with the benefit of survivorship vary inversely as the interest rate. Thus, insurance companies tend to use low, and thus conservative, interest rates in the determination of the present values.

PRESENT VALUE OF A TEMPORARY LIFE ANNUITY

The concept of a life annuity is that the payment will be made as long as the individual is alive to receive it. However, under some conditions, a fixed number of payments will be guaranteed. When this is the situation, the annuity is referred to as a **temporary life annuity**. Recall that in Chapters 3–5, when payments occurred at the ends of payment periods the annuity was referred to as an ordinary annuity. However, the actuarial and insurance standard is to refer to the ordinary annuity as an **Annuity immediate** or an **Immediate annuity**. When the payment is at the beginning of a payment period, the annuity is referred to as an **annuity due** as previously explained in Chapter 4. The notation we shall use for the present value of an n-payment temporary life annuity immediate is $A_{x:\overline{n}|}$, for an n-payment temporary life annuity due is $\ddot{A}_{x:\overline{n}|}$.

Consider a 3-payment (for brevity) life annuity beginning in one year as shown in Figure 9-1.

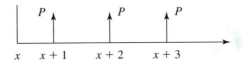

FIGURE 9-1 A 3-Payment Temporary Life Annuity

Using Equation (9.4-1) the present value at age x would be the sum of the contingent present values of each of the future payments, or

$$A_{x:\overline{3}|} = P\left(\frac{l_{x+1}}{l_x}\right)v^{(x+1)-x} + P\left(\frac{l_{x+2}}{l_x}\right)v^{(x+2)-x} + P\left(\frac{l_{x+3}}{l_x}\right)v^{(x+3)-x}$$

or in general

$$A_{x:\overline{n}|} = P\sum_{t=1}^{x+n}\left(\frac{l_{x+t}}{l_x}\right)v^t. \tag{9.4-3}$$

In actual calculations, l_x may be factored from the denominator of Equation (9.4-3).

EXAMPLE 9.4.3   What is the present value of a 3-payment temporary life annuity of $10,000 per year beginning at age 65 for a person age 64 if the interest rate is 5%?

Solution   Using Table A-6 for the number living at each age and after factoring l_x, Equation (9.4-3) gives present value as

$$A_{64:\overline{3}|} = \left[\frac{10,000}{78,609}\right]\left(77,107(1.05)^{-(65-64)} + 75,520(1.05)^{-(66-64)}\right.$$

$$\left. +73,846(1.05)^{-(67-64)}\right)$$

$$= \$26,170.68.$$

For a payment of $26,170.68 at age 64, this individual will receive $10,000 at ages 65, 66, and 67, if alive. ■

The use of the commutation function N_x facilitates these calculations. From its definition as given by Equation (9.3-2)

$$N_x = \sum_{t=x}^{\infty} D_t.$$

Now,

$$A_{64:\overline{3}|} = P\left[\frac{D_{65} + D_{66} + D_{67}}{D_{64}}\right].$$

Since N_x is the sum of the D_x's to the end of a mortality table, the present value is determined as follows:

$$N_{65} = D_{65} + D_{66} + D_{67} + D_{68} + D_{69} + \cdots \text{ to the end of the table.}$$

$$N_{68} = \qquad\qquad\qquad\quad D_{68} + D_{69} + \cdots \text{ to the end of the table.}$$

Subtracting N_{68} from N_{65} gives

$$N_{65} - N_{68} = D_{65} + D_{66} + D_{67}.$$

Using this concept for Example 9.4.3,

$$A_{64:\overline{3}|} = P\left[\frac{N_{65} - N_{68}}{D_{64}}\right],$$

and from Table A-6,

$$A_{64:\overline{3}|} = 10,000\left[\frac{35,523.27 - 26,462.35}{3,462.24}\right]$$

$$= \$26,170.68,$$

as previously determined. The use of the commutation function N may be generalized as follows. Let:

x be the age at which the present value is to be determined,
t be the number of periods (e.g., years) to the first payment, and
n be the number of payments.

Then,

$$A_{x:\overline{n}|} = P\left[\frac{N_{x+t} - N_{x+t+n}}{D_x}\right]. \qquad (9.4\text{-}4)$$

EXAMPLE 9.4.4   What is the present value of a 3-payment temporary life annuity of $10,000 per year to a person now age 65, with the first payment due at age 65, if the interest rate is 5%?

Solution   The present value is desired for age 65. Therefore, $x = 65$, and since the first payment is due at this age, $t = 0$, and the annuity is an annuity due. For three payments, $n = 3$ and $x + t + n = 68$. Then, Equation (9.4-4) gives

$$\ddot{A}_{65:\overline{3}|} = 10,000\left[\frac{N_{65} - N_{68}}{D_{65}}\right].$$

Substituting from the 5% rate of Table A-6,

$$\ddot{A}_{64:\overline{3}|} = 10,000 \left[\frac{35,523.27 - 26,462.35}{3,234.37} \right]$$

$$= \$28,014.48. \quad \blacksquare$$

Note that in both Examples 9.4.3 and 9.4.4, the first payment was to be at age 65, but in Example 9.4.3, the present value was requested at age 64, one year prior to the first payment. In Example 9.4.4, the first payment was on the present value date. Thus Equation (9.4-4) represents an annuity immediate if $t = 1$ and an annuity due if $t = 0$.

PRESENT VALUE OF A WHOLE LIFE ANNUITY

It is often necessary to plan an annuity for the remaining life of an individual. In this capacity, the number of payments is not known for certain. Therefore, the benefit of survivorship must be accounted for until everyone from the original cohort is dead—that is, from a current age to the end of a mortality table. For a whole life annuity, n becomes "infinite," and the notation we will use is A_x for a whole life annuity immediate and $\ddot{A}_x$ for a whole life annuity due.

As a trivial example to demonstrate the concept, suppose we desire the present value of a whole life annuity immediate of $10,000 per year for a person age 106. Since Table A-6 indicates that the number living at age 110 is 0, the present values with the benefit of survivorship are needed for ages 107, 108, and 109. From Equation (9.4-3) and an interest rate of 5%,

$$A_{106} = 10,000 \left[\left(\frac{l_{107}}{l_{106}} \right) v + \left(\frac{l_{108}}{l_{106}} \right) v^2 + \left(\frac{l_{109}}{l_{106}} \right) v^3 \right]$$

$$= 10,000 \left[\left(\frac{78}{119} \right) (1.05)^{-1} + \left(\frac{51}{119} \right) (1.05)^{-2} + \left(\frac{33}{119} \right) (1.05)^{-3} \right]$$

$$= 10,000[0.62 + 0.39 + 0.24]$$

$$= \$12,500.$$

To use commutation functions, Equation (9.4-4) gives

$$A_{106} = 10,000 \left[\frac{N_{107} - N_{110}}{D_{106}} \right],$$

and after substitution from Table A-6

$$A_{106} = 10,000 \left[\frac{0.85 - 0}{0.68} \right]$$

$$= 10,000[1.25]$$

$$= \$12,500,$$

as above. Then, for a whole life annuity,

$$A_x = P \left[\frac{N_{x+t}}{D_x} \right], \tag{9.4-5}$$

where $t = 0$ for a whole life annuity due, where A_x is replaced by $\ddot{A}_x$, and $t = 1$ for a whole life annuity immediate.

EXAMPLE 9.4.5   Find the present value of a whole life annuity immediate that pays $10,000 per year for a person at age 64 and the present value of a whole life annuity due that pays $10,000 per year for a person at age 65. The interest rate is 5%.

Solution   For a whole life annuity immediate $t = 1$ and $x + t = 65$. From Equation (9.4-5) and Table A-6,

$$A_{64} = 10{,}000 \left[\frac{N_{65}}{D_{64}} \right] = 10{,}000 \left[\frac{35{,}523.27}{3{,}462.24} \right]$$

$$= \$102{,}601.98.$$

For a whole life annuity due, $t = 0$ and $x + t = 65$.

$$\ddot{A}_{65} = 10{,}000 \left[\frac{N_{65}}{D_{65}} \right]$$

$$= 10{,}000 \left[\frac{35{,}523.27}{3{,}234.37} \right]$$

$$= \$109{,}830.57. \quad \blacksquare$$

The foregoing analyses for the present values of temporary and whole life annuities were for annuities where the first payment is now or at the end of one year. It is often necessary to determine the present values of temporary and whole life annuities long before the first payment. Those types of annuities are referred to as deferred life annuities.

PRESENT VALUE OF A DEFERRED LIFE ANNUITY

In annuities certain (without the benefit of survivorship), the period of deferment is defined as the number of interest conversion periods to the beginning of an ordinary annuity. The beginning of an ordinary annuity is one interest conversion period before the first payment. For deferred life annuities, it is more convenient to define the period of deferment as the number of interest conversion periods to the first payment. The resulting annuity becomes an annuity due as indicated in Figure 9-2.

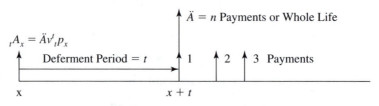

FIGURE 9-2 Time Diagram for Deferred Life Annuities

Again, to illustrate by example, consider the present value, at age 20, of a 3-payment temporary life annuity with the first payment of P deferred 22 years. The interest rate is 5%. The time diagram is shown in Figure 9-3.

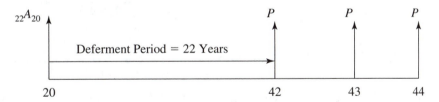

FIGURE 9-3 Time Diagram for a 3-Payment Deferred Life Annuity

From Equation (9.4-3),

$$_{22}A_{20:\overline{3}|} = P\left[\left(\frac{l_{42}}{l_{20}}\right)v^{22} + \left(\frac{l_{43}}{l_{20}}\right)v^{23} + \left(\frac{l_{44}}{l_{20}}\right)v^{24}\right].$$

From Equation (9.4-2),

$$_{22}A_{20:\overline{3}|} = P\left[\frac{D_{42} + D_{43} + D_{44}}{D_{20}}\right],$$

and in terms of the commutation function, N,

$$_{22}A_{20:\overline{3}|} = P\left[\frac{N_{42} - N_{45}}{D_{20}}\right].$$

Then, in general terms, the present value of an n-payment deferred life annuity is

$$_tA_{x:\overline{n}|} = P\left[\frac{N_{x+t} - N_{x+t+n}}{D_x}\right], \tag{9.4-6}$$

which is the same as Equation (9.4-4). Equation (9.4-6) is used directly for an n-payment life annuity that is deferred t years. For a deferred whole life annuity, the number of payments, n, is theoretically infinite, and therefore $N_{x+t+n} = 0$ and Equation (9.4-5) becomes

$$_tA_x = P\left[\frac{N_{x+t}}{D_x}\right]. \tag{9.4-7}$$

which is the same as Equation (9.4-5).

EXAMPLE 9.4.6   What is the present value at age 35 of a 3-year temporary life annuity of $10,000 per year deferred 30 years if the interest rate is 5%?

Solution   Since the current age, x, is 35, and the number of years of deferment is 30, the starting age of the annuity is $x+t = 35+30$, or $x = 65$. Since $n = 3$, $x+t+n = 68$. From Equation (9.4-6),

$$_{30}A_{35:\overline{3}|} = 10,000 \left[\frac{N_{65} - N_{68}}{D_{35}} \right]$$

$$= 10,000 \left[\frac{35,523.37 - 26,462.35}{17,369.06} \right]$$

$$= \$5,216.76. \quad \blacksquare$$

EXAMPLE 9.4.7   What is the present value at age 35 of a deferred whole life annuity of $10,000 per year beginning at age 65 if the interest rate is 5%?

Solution   Since this is a whole life annuity, $N_{x+t+n} = 0$. Then, from Equation (9.4-7),

$$_{30}A_{35} = 10,000 \left[\frac{N_{65}}{D_{35}} \right]$$

$$= 10,000 \left[\frac{35,523.37}{17,369.06} \right]$$

$$= \$20,452.10.$$

This result may be verified as the present value, at age 35, of a whole life annuity at age 65. The life annuity at age 65 is given by Equation (9.4-7) as

$$A_{65} = P \left[\frac{N_{65}}{D_{65}} \right]$$

$$= 10,000 \left[\frac{35,523.27}{3,234.27} \right]$$

$$= \$109,833.57,$$

as shown in Example 9.4.5. The present value of this annuity due at age 35 is determined by Equation (9.4-1) as

$$A_{35} = A_{65} v^{30} \frac{l_{65}}{l_{35}}$$

$$= 109,830.57(1.05)^{-30} \left(\frac{77,107}{95,808} \right)$$

$$= \$20,452.58. \quad \blacksquare$$

TWO INTEREST RATES

Because long-term interest rates are not known with certainty, a relatively high interest rate may be given for a limited number of years, and a lower interest rate is used for the remaining term. The time diagram is shown in Figure 9-4 for

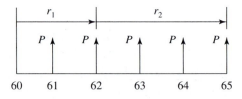

FIGURE 9-4 Contingent Present Value at Two Interest Rates

the present value at age 60 for a 2-payment temporary life annuity immediate at one interest rate followed by a 3-payment temporary life annuity immediate at a second interest rate.

The present value at age 60 for the 2-payment temporary life annuity immediate is given by Equation (9.4-6) as

$$A_{60:\overline{2}|} = P \left[\frac{N_{61} - N_{63}}{D_{60}} \right]_{r_1}.$$

The present value at age 60 of a deferred 3-payment temporary life annuity immediate at age 62 is given by Equation (9.4-6) as

$$A_{60:\overline{3}|} = P \left[\frac{N_{63} - N_{66}}{D_{60}} \right]_{r_2}.$$

Then, the present value of the two-interest life annuity described in Figure 9-4 is the sum of the two present values at age 60. The concept can be generalized to

$$A_x = P_1 \left[\frac{N_{x+t} - N_{x+t+n_1}}{D_x} \right]_{r_1} + P_2 \left[\frac{N_{x+t+n_1} - N_{x+t+n_1+n_2}}{D_x} \right]_{r_2}. \qquad (9.4\text{-}8)$$

If the second interest rate is to be applied to a deferred whole life annuity due, Equation (9.4-8) reduces to

$$A_x = P_1 \left[\frac{N_{x+t} - N_{x+t+n_1}}{D_x} \right]_{r_1} + P_2 \left[\frac{N_{x+t+n_1}}{D_x} \right]_{r_2}. \qquad (9.4\text{-}9)$$

Normally, the annual payment remains the same, but Equations (9.4-8) and (9.4-9) allow for a different annual payment when the second interest rate is imposed.

EXAMPLE 9.4.8 Find the present value at age 60 of a 2–payment temporary life annuity immediate of $1,000 per year at an interest rate of 5% followed by a 3–payment temporary life annuity immediate of $1,000 per year at an interest rate of 3%.

Solution For the present value at age 60 of the 2-payment temporary life annuity of $1,000 per year $x + 1 = 60 + 1$, and $x + 1 + n_1 = 60 + 1 + 2 = 63$. From Equation (9.4-8),

$$A_{60:\overline{2}|} = 1{,}000 \left[\frac{N_{61} - N_{63}}{D_{60}} \right]_{5\%}$$

$$= 1{,}000 \left[\frac{50{,}846.91 - 42{,}686.30}{4{,}482.32} \right]$$

$$= 1{,}820.62.$$

For the present value at age 60 of a deferred 3-payment temporary life annuity due of $1,000 per year, $x + 1 + n_1 = 63$ and $x + 1 + n_1 + n_2 = 60 + 1 + 2 + 3 = 66$. From Equation (9.4-8)

$$A_{60:\overline{3}|} = 1,000 \left[\frac{N_{63} - N_{66}}{D_{60}} \right]_{3\%}$$

$$= 1,000 \left[\frac{169,503.00 - 133,928.72}{14,211.07} \right]$$

$$= 2,503.28.$$

The present value at age 60 is the sum of the present values at the two interest rates, $4,323.90. ∎

RELATIONSHIP AMONG LIFE ANNUITIES

In a manner similar to annuities certain, the relationship between annuities immediate and annuities due is simply the accumulation of the present value of the annuity immediate by one interest conversion period in order to obtain the present value of the annuity due. That is equivalent to adding the amount of one payment to the present value of an annuity immediate, and for an annuity of $P per period,

$$\ddot{A}_x = A_x + P. \tag{9.4-10}$$

If the interest rate remains constant over the life of the annuity, the present value of the combination of an n-payment temporary annuity and a whole life annuity deferred $n + 1$ payment periods is simply the present value of a whole life annuity at age x. That is,

$$A_x = P_1 \left[\frac{N_{x+t} - N_{x+t+n}}{D_x} \right] + P_2 \left[\frac{N_{x+t+n}}{D_x} \right], \tag{9.4-11}$$

which reduces to Equation (9.4-5) when $r_1 = r_2$ and $P_1 = P_2$.

EXAMPLE 9.4.9 A company's retirement plan will pay, for as long as the person is alive, $24,000 per year for the first 16 years of retirement and $18,000 per year for life thereafter. If the retirement age is 65, what amount of money (present value) is needed if the interest rate is 5%?

Solution Since the first payment is at age 65, $x = 65, t = 0, n = 16$, and $x + t + n = 81$. Then, from Equation (9.4-11),

$$A_{65} = 24,000 \left[\frac{N_{65} - N_{81}}{D_{65}} \right] + 18,000 \left[\frac{N_{81}}{D_{65}} \right]$$

$$= 24,000 \left[\frac{35,523.37 - 4,919.65}{3,234.37} \right] + 18,000 \left[\frac{4,919.65}{3,234.37} \right]$$

$$= 227,088.82 + 27,378.96$$

$$= \$254,467.78. \quad ∎$$

EXAMPLE 9.4.10 Suppose the company plan in Example 9.4.9 paid $24,000 per year for life. What amount would be needed at age 65?

Solution Since the payment is constant, Equation (9.4-11) reduces to

$$A_{65} = 24,000 \left[\frac{N_{65}}{D_{65}} \right]$$

$$= 24,000 \left[\frac{35,523.37}{3,234.37} \right]$$

$$= \$263,594.11.$$

Notice that the difference in the present values of Example 9.4.9 and 9.4.10 is only \$9,126. That is, an additional \$9,126 at age 65 will give the retiree an additional \$6,000 per year for life from age 81 forward. ∎

SETTLEMENT OPTIONS

When the value of a death benefit is to be settled upon the beneficiary of the benefit, the insuring body will usually offer several settlement options, Workman (1987). These could be four nonlife contingent options and four life contingent options as follows:

1. **Lump sum.** A single payment that represents the amount of the proceeds (or present value) of the policy.
2. **Interest only.** With this option, the company would retain the proceeds of the policy and the beneficiary would receive periodic payments of the interest credited to the fund account. With an agreed upon date or at the death of the beneficiary, the original amount of the policy would be distributed according to the stipulations of the agreement.
3. **Fixed period certain.** With this option, the proceeds of the policy would be distributed over the fixed period in the same manner as the amortization of a mortgage. The periodic payment would be comprised of interest and principal. At the end of the fixed period, the account will have been reduced to zero.
4. **Fixed payment certain.** With this option, the beneficiary receives a specified periodic payment until the fund is reduced to zero. Usually the final payment is smaller than the other payments.
5. **Contingent life income.** With this option, the beneficiary will receive equal periodic payments for as long as he or she is alive. Upon the death of the beneficiary, payments are terminated.
6. **Contingent life with period certain.** With this option, the beneficiary will receive equal payments for life with the payments guaranteed for period certain if death occurs. The periodic payment would be the amount determined by a life annuity. If the annuitant dies during the period certain, the payments will continue to a designated contingent beneficiary for the balance of period certain.
7. **Contingent life income with installment refund.** With this option, a fixed amount is paid periodically until the total amount paid equals the original present value. If the beneficiary survives beyond that period certain, the payment is continued until death. For example, if the present value of the

fund was $100,000 and a payment of $5,000 per year is to be made, that payment would continue for 20 $\left(\frac{100,000}{5,000}\right)$ years certain. If the insured dies before 20 years, the payment would be paid to a designated beneficiary.

8. **Contingent life income with cash refund.** This option is the same as Option 7 except that if death occurs during the period certain, a lump sum is paid to any contingent beneficiary or the estate of the original beneficiary.

The mathematics for options 2, 3, and 4 utilize the formulas for annuities certain. The mathematics for options 5 and 6 use the concepts of annuities certain in conjunction with the concepts that have been developed in this chapter; options 7 and 8 do not. The following example will illustrate options 2–6.

EXAMPLE 9.4.11   The beneficiary of a $500,000 life insurance policy is a female age 65. Determine the annual payments for the settlement options 2–6 using a 5% interest rate.

Solution   Normally, the periodic payment of the selected option would be a monthly payment commencing on the first of a month following death. However, for simplicity, we will treat the payments as annual annuities due.

Option 2. This is the interest only option and would generate an annual payment of $25,000 (5% of $500,000).

Option 3. This is the fixed payment certain option and would be the annual payment of the amortization of $500,000. The life expectancy of a person age 65 is shown to be 17.0 years in Table A-6. However, a more realistic life expectancy of a female at age 65 is 20 years. Using this life expectancy, the annual payment, beginning now, is determined as,

$$P = \frac{500,000}{\left[\dfrac{1-(1.05)^{-20}}{0.05}\right](1.05)}$$

$$= \left[\frac{500,000}{13.085321}\right]$$

$$= \$38,210.76.$$

Option 4. This is the fixed payment option, and let us assume a fixed payment of $40,000 per year beginning now. Utilizing Equation (4.4-4), the number of payments is determined as

$$n = -\frac{Ln\left[1-\dfrac{Ai}{P(1+i)}\right]}{Ln(1+i)}$$

$$= -\frac{Ln\left[1-\dfrac{(500,000)(0.05)}{(40,000)(1.05)}\right]}{Ln(1.05)}$$

$$= 18.537.$$

Then, this option will generate 18 payments of $40,000 and a final smaller payment. The balance of the original present value after 18 payments can be determined using Equation (4.4-7) as follows.

$$B_n = (V - P)(1 + i)^{n-1} - P \left[\frac{(1 + i)^{n-1} - 1}{i} \right]$$

$$= (500{,}000 - 40{,}000)(1.05)^{17} - (40{,}000) \left[\frac{(1.05)^{17} - 1}{0.05} \right]$$

$$= \$20{,}713.78.$$

This amount would accrue 5% interest for 1 additional year, and the final payment would be $21,749.47.

Option 5. This option is a whole life annuity due at age 65. From Equation (9.4-5), with $t = 0$,

$$500{,}000 = P \left[\frac{N_{65}}{D_{65}} \right]$$

$$= P \left[\frac{35{,}523.27}{3{,}234.37} \right].$$

Solving for P gives $45,524.67 per year for as long as the beneficiary is alive.

Option 6. The disadvantage of Option 5 is that should the beneficiary die shortly after selecting the option and signing an agreement for the whole life annuity, all payments cease, and the estate of the beneficiary could forfeit a considerable sum of money. This option guarantees that payments will continue to her estate for a specified number of years. If available, it would be reasonable to select the number of years of the life expectancy as the period certain guarantee. Thus, an annuity of life with 20 years certain combines the present value of a 20-year certain annuity due with a whole life annuity deferred 20 years. Then,

$$500{,}000 = P \left[\frac{1 - (1.05)^{-20}}{0.05} \right] (1.05) + P \left[\frac{N_{85}}{D_{65}} \right]$$

$$= P \left\{ \left[\frac{1 - (1.05)^{-20}}{0.05} \right] (1.05) + \left[\frac{2{,}357.66}{3{,}234.37} \right] \right\}.$$

Solving for P gives $36,194.48 per year until she dies. If death occurs before 20 years has elapsed, the same amount will be paid to her designated beneficiary for the remainder of the 20 years from the year of her death. ■

PROBLEM SET 9.4

1. Find the present value of a 10-year temporary life annuity immediate of $30,000 per year for an individual now aged 60 if the interest rate is 3%.

2. Find the present value of a 10-year temporary life annuity immediate of $30,000 per year for an individual now aged 60 if the interest rate is 5%.

3. Find the present value of a 10-year temporary life annuity due of $30,000 per year for an individual now aged 60 if the interest rate is 3%.

4. Find the present value of a 10-year temporary life annuity due of $30,000 per year for an individual now aged 60 if the interest rate is 5%.

5. Find the present value at age 64 of a whole life annuity immediate of $50,000 per year if the interest rate is 5%.

6. Find the present value at age 64 of a whole life annuity immediate of $50,000 per year if the interest rate is 3%.

7. Find the present value at age 65 of a whole life annuity due of $50,000 per year if the interest rate is 5%.

8. Find the present value at age 65 of a whole life annuity due of $50,000 per year if the interest rate is 3%.

9. Find the present value of a 10-year temporary life annuity due of $30,000 per year for an individual now aged 60 if the interest rate is 3% and the first payment is at age 65.

10. Find the present value of a 10-year temporary life annuity due of $30,000 per year for an individual now aged 60 if the interest rate is 5% and the first payment is at age 65.

11. Find the present value at age 55 of a whole life annuity of $50,000 per year if the interest rate is 5% and the first payment is at age 65.

12. Find the present value at age 55 of a whole life annuity of $50,000 per year if the interest rate is 3% and the first payment is at age 65.

13. Find the present value at age 62 of a whole life annuity of $100,000 per year if the interest rate is 5% and the first payment is at age 70.

14. Find the present value at age 55 of a whole life annuity of $50,000 per year if the interest rate is 3% and the first payment is at age 70.

15. Find the present value at age 64 of a 10-payment temporary life annuity immediate of $25,000 per year at 5% followed by a 10-payment temporary life annuity of $25,000 per year at 3%.

16. Find the present value at age 64 of a 10-payment temporary life annuity immediate of $25,000 per year at 3% followed by a 10-payment temporary life annuity of $25,000 per year at 5%.

17. Find the present value at age 64 of a 10-payment temporary life annuity immediate of $25,000 per year at 5% followed by a 10-payment temporary life annuity of $15,000 per year at 3%.

18. Find the present value at age 64 of a 10-payment temporary life annuity immediate of $25,000 per year at 3% followed by a 10-payment temporary life annuity of $15,000 per year at 5%.

19. Find the present value at age 64 of a 10-payment temporary life annuity immediate of $25,000 per year at 5% followed by a whole life annuity of $25,000 per year at 3%.

20. Find the present value at age 64 of a 10-payment temporary life annuity immediate of $25,000 per year at 3% followed by a whole life annuity of $25,000 per year at 5%.

21. Find the present value at age 64 of a 10-payment temporary life annuity immediate of $25,000 per year at 5% followed by a whole life annuity of $15,000 per year at 3%.

22. Find the present value at age 64 of a 10-payment temporary life annuity immediate of $25,000 per year at 3% followed by a whole life annuity of $15,000 per year at 5%.

23. Find the annual payment for a 10-year temporary life annuity immediate for an individual at age 60 if the present value is $250,000 and the interest rate is 3%.

24. Find the annual payment for a 10-year temporary life annuity immediate for an individual at age 60 if the present value is $250,000 and the interest rate is 5%.

25. Find the annual payment for a 10-year temporary life annuity due for an individual at age 60 if the present value is $250,000 and the interest rate is 3%.

26. Find the annual payment for a 10-year temporary life annuity due for an individual at age 60 if the present value is $250,000 and the interest rate is 5%.

27. Find the annual payment for a whole life annuity immediate for an individual at age 64 if the present value is $250,000 and the interest rate is 5%.

28. Find the annual payment for a whole life annuity immediate for an individual at age 64 if the present value is $250,000 and the interest rate is 3%.

29. Find the annual payment for a whole life annuity due for an individual at age 64 if the present value is $250,000 and the interest rate is 5%.

30. Find the annual payment for a whole life annuity due for an individual at age 64 if the present value is $250,000 and the interest rate is 3%.

For Problems 30–38, the present value is $500,000. Find the annual payment as indicated.

31. A 10-year temporary life annuity due for an individual now age 60 if the interest rate is 3% and the first payment is at age 65.

32. A 10-year temporary life annuity due for an individual now age 60 if the interest rate is 5% and the first payment is at age 65.

33. A whole life annuity for an individual now age 55 if the interest rate is 5% and the first payment is at age 65.

34. A whole life annuity for an individual now age 55 if the interest rate is 3% and the first payment is at age 65.

35. A whole life annuity for an individual now age 62 if the interest rate is 5% and the first payment is at age 70.

36. A whole life annuity for an individual now age 62 if the interest rate is 3% and the first payment is at age 70.

37. A 10-payment temporary life annuity immediate at 5% followed by a 10-payment temporary life annuity of at 3% for a person now age 64.

38. A 10-payment temporary life annuity immediate at 3% followed by a 10-payment temporary life annuity at 5% for a person now age 64.

For problems 39–50, the present value is $1,000,000. Find the indicated payments.

39. For an individual at age 55, a 10-payment temporary life annuity immediate of $P per year at 5% followed by a 10-payment temporary life annuity of 50% of P per year at 3%.

40. For an individual at age 55, a 10-payment temporary life annuity immediate of $P per year at 3% followed by a 10-payment temporary life annuity of 50% of P per year at 5%.

41. For an individual at age 62, a 10-payment temporary life annuity immediate of $P per year at 5% followed by a whole life annuity of $P per year at 3%.

42. For an individual at age 62, a 10-payment temporary life annuity immediate of $P per year at 3% followed by a whole life annuity of $P per year at 5%.

43. For an individual at age 70, a 10-payment temporary life annuity immediate of $P per year at 5% followed by a whole life annuity of $\left(\frac{2}{3}\right) P$ per year at 3%.

44. For an individual at age 64, a 10-payment temporary life annuity immediate of $P per year at 3% followed by a whole life annuity of $\left(\frac{2}{3}\right) P$ per year at 5%.

45. For a person age 65, a life annuity due with 20 years certain using 3% interest.

46. For a person age 65, a life annuity due with 20 years certain using 5% interest.

47. For a person age 55, a life annuity due with 20 years certain using 3% interest.

48. For a person age 55, a life annuity due with 20 years certain using 5% interest.

49. For a person age 70, a life annuity due with 10 years certain using 3% interest.

50. For a person age 70, a life annuity due with 10 years certain using 5% interest.

9.5 LIFE INSURANCE: NET SINGLE PREMIUM

In the previous section we discussed life annuity payments where the designated person has to be alive to receive the benefits. We now utilize the actuarial principles that are necessary in order to determine the cost (or premium) for the case where the designated person must die before the benefit is paid. The concept that death must occur before a payment is made is the fundamental concept in life insurance, and in this section we determine how to calculate the net single premium to pay for such insurance. The common practice is to calculate the premium per $1,000 of insurance. The actual premium will be greater than that indicated by the calculations since the net single premium does not take into account expenses, profits, and cancellation of policies. As with life annuities, the determination of a net single premium requires the use of a designated mortality table and a specified rate of interest.

NET SINGLE PREMIUM FOR ONE YEAR OF LIFE INSURANCE

The net single premium is the present value of the future benefit that is being offered by the insurance company. Suppose that everyone who is age x this year purchases a \$1,000 life insurance policy for one year. The net total payment would be the product of the net single premium per \$1,000 times the number of people alive at age x or,

$$\text{Total Paid In} = A_x l_x,$$

where,

A_x is the net single premium, and

l_x is the number alive at age x.

At the end of one year the future value of the total pay-in at an interest rate of $i\%$ would be

$$\text{Future Value} = A_x l_x (1 + i).$$

The amount of money that the insurance company would have to pay as benefits during the year would be 1,000 times the number of people who die during the year or,

$$\text{Total Paid Out} = 1,000 d_x,$$

where d_x is the number who will during the year of age x.

Now if everyone who was alive at the beginning of age x died by the end of age x, the future value of the total pay-in would just equal the total pay-out by the insurance company or,

$$A_x l_x (1 + i) = 1,000 d_x.$$

The net single premium for \$1,000 of life insurance for one year is determined by solving for A_x as

$$A_x = 1,000 \left(\frac{d_x v}{l_x} \right), \tag{9.5-1}$$

where v is the present value of \$1.00 and is defined as the reciprocal of $(1 + i)$.

Multiplying the numerator and denominator of Equation (9.5-1) by v^x enables the use of commutation functions, C and D, such that, for a benefit of \$$B$,

$$A_x = B \left(\frac{C_x}{D_x} \right). \tag{9.5-2}$$

EXAMPLE 9.5.1   Determine the net single premium for \$10,000 of life insurance for one year at ages 40, 60, and 80. Use a 3% table.

Solution   Using Equation (9.5-2) and the 3% table of Table A-6 for the commutation functions, the net single premium at age 40 is

$$A_{40} = 10,000 \left(\frac{65.48}{29,100.21} \right)$$

$$= \$22.50.$$

The net single premium at age 60 is

$$A_{60} = 10,000 \left(\frac{188.68}{14,211.07} \right)$$

$$= \$132.77.$$

The net single premium at age 80 is

$$A_{80} = 10,000 \left(\frac{271.16}{4,057.93} \right)$$

$$= \$668.22. \quad \blacksquare$$

It should be noted that the net single premium increases dramatically between ages 40 and 80. This is due to the fact that the number who will die at age 80 is significantly greater than at age 40, and the number alive at age 80 is significantly less than at age 40. Equation (9.5-1) shows that the net single premium varies directly as the number dying and inversely as the number alive at each age.

NET SINGLE PREMIUM FOR n-YEAR TERM INSURANCE

It may be wise to purchase insurance for a specified term in order to protect one's assets in the event of death during that period. The primary wage earner in a family might want to protect the family against economic hardship until the children graduate from college. A company might purchase insurance for an employee who might be undertaking a hazardous assignment or companies might purchase insurance for its employees as a fringe benefit of the employment.

The net single premium for n-year term insurance is the sum of the present values of the net single premiums for each year from one to n. Thus,

$$A_{x:\overline{n}|} = B \left[\frac{d_x v}{l_x} + \frac{d_{x+1} v^2}{l_x} + \frac{d_{x+2} v^3}{l_x} + \cdots + \frac{d_{x+n-1} v^n}{l_x} \right],$$

or in terms of commutation functions,

$$A_{x:\overline{n}|} = B \left[\frac{C_x}{D_x} + \frac{C_{x+1}}{D_x} + \frac{C_{x+2}}{D_x} + \cdots + \frac{C_{x+n-1}}{D_x} \right].$$

Combining the fractions over a single denominator gives

$$A_{x:\overline{n}|} = B \left[\frac{C_x + C_{x+1} + C_{x+2} + \cdots + C_{x+n-1}}{D_x} \right].$$

From the definitions of M_x and M_{x+n},

$$M_x = C_x + C_{x+1} + C_{x+2} + \cdots + C_{x+n-1} + C_{x+n} \cdots + \text{to the end of the table.}$$

$$M_{x+n} = \qquad\qquad\qquad\qquad\qquad\qquad C_{x+n} \cdots + \text{to the end of the table.}$$

Then, the net single premium can be determined from the commutation functions by subtracting the bottom function from the top function to get

$$A_{x:\overline{n}|} = B\left[\frac{M_x - M_{x+n}}{D_x}\right]. \qquad (9.5\text{-}3)$$

EXAMPLE 9.5.2 A person is 25 years of age. Find the net single premium for $1,000 of 3-year term insurance using a) 3% interest and b) 5% interest.

Solution Using Equation (9.5-3)

$$A_{25:\overline{3}|} = \left[\frac{M_{25} - M_{28}}{D_{25}}\right].$$

a) From the 3% table of Table A-6,

$$A_{25:\overline{3}|} = 1,000\left[\frac{11,260.03 - 11,088.88}{46,380.28}\right]$$

$$= \$3.69.$$

To illustrate that the $3.69 premium is sufficient, consider the following tabulation. We assume that at age 25, the 97,110 living people pay the $3.69 premium for a total of $358,335.90.

Year (1)	Fund Amount at Beginning of the Year (2)	Accumulated Amount at End of the Year (3) = 1.03(2)	Number Dying During the Year (4)	Amount of Claims Paid for Year (5)	Fund Amount at End of Year (6) = (3) − (5)
1	$358,335.90	$369,085.98	128	$128,000	$241,085.97
2	241,085.97	248,318.55	126	126,000	122,318.55
3	122,318.55	125,988.11	126	126,000	−11.89

The −$11.89 at the end of year 3 is due to the rounding of numbers to the nearest cent.

b) From the 5% table of Table A-6,

$$A_{25:\overline{3}|} = 1,000\left[\frac{3,150.11 - 3,048.23}{28,676.85}\right]$$

$$= \$3.55.$$

The premium at 3% is "only" 14 cents greater than the premium at 5%. However, when multiplied by thousands of insurance policies, that relatively small difference can be significant. ■

NET SINGLE PREMIUM FOR WHOLE LIFE INSURANCE

Whole life insurance may be considered to be an "infinite-year" term policy. Since the number of people alive at $x + \infty$ is 0, $M_{x+\infty} = 0$ and Equation (9.5-3) reduces to

$$A_x = B\left[\frac{M_x}{D_x}\right]. \qquad (9.5\text{-}4)$$

EXAMPLE 9.5.3   Find the net single premium for a $1,000 whole life policy for a) a person age 25, and b) a person age 65 using an interest rate of 3%.

Solution   Using the 3% table of Table A-6 and Equation (9.5-4),

a) For a person at age 25,

$$A_{25} = 1,000 \left[\frac{M_{25}}{D_{25}} \right]$$

$$= 1,000 \left[\frac{11,260.03}{46,380.28} \right]$$

$$= \$242.78.$$

b) For a person at age 65,

$$A_{65} = 1,000 \left[\frac{M_{65}}{D_{65}} \right]$$

$$= 1,000 \left[\frac{7,059.83}{11,289.40} \right]$$

$$= \$625.35.$$

Note that the premium at age 65 is nearly three times the premium at age 25. This is due to the greater probability that the 65-year old person will die sooner than the 25-year old person. ∎

ENDOWMENT INSURANCE

A pure endowment policy is one that pays a designated beneficiary a specified amount if the beneficiary is alive to receive it. The net present value for an endowment policy is the same as the present value with the benefit of survivorship as given by Equation (9.4-2). Pure endowment policies are generally illegal because the question of who is to receive the money if the designated beneficiary is not alive at the maturity date. However, a pure endowment may be combined with term life insurance with the insured designated as the beneficiary of the endowment and another person, or the estate of the insured person, designated as the beneficiary of the insurance payment. Thus, endowment insurance indicates that a benefit will be paid immediately if death occurs during a given term, or the benefit will be paid at the end of the given term.

The net single premium for endowment insurance is comprised of two parts: a pure endowment policy and a term insurance policy. As such,

$$A_x = B \left[\frac{D_{x+n}}{D_x} \right] + B \left[\frac{M_x - M_{x+n}}{D_x} \right]. \tag{9.5-5}$$

EXAMPLE 9.5.4   Find the net single premium for a 25-year, $10,000 endowment insurance for a person at age 40 if a 3% interest rate is used.

Solution   Using Equation (9.5-5),

$$A_{40} = 10,000 \left[\frac{D_{40+25}}{D_{40}} + \frac{M_{40} - M_{40+25}}{D_{40}} \right].$$

Substituting from the 3% table of Table A-6,

$$A_{40} = 10,000 \left[\frac{11,289.49}{29,100.21} + \frac{10,445.18 - 7,059.83}{29,100.21} \right]$$

$$= \$5,042.88. \quad \blacksquare$$

It can be seen that endowment insurance is relatively expensive; however, should the insured survive to age 65, he or she will receive a payment of $10,000. In terms of compound interest, an interest rate of 2.7762% compounded annually will have been received. However, should the insured die one day after the effective date of the policy, the $10,000 payment represents an enormous (i.e., "infinite") interest rate.

GENERAL FORMULA FOR NET SINGLE PREMIUM

The foregoing analyses for the net single premium for life insurance may be compiled into a single formula. The following equation is valid for an n-year term policy, a whole life policy, and an n-year endowment insurance policy.

$$A_x = B \left[\frac{M_x - M_{x+n} + D^E_{x+n}}{D_x} \right]. \tag{9.5-6}$$

The factor D^E_{x+n} is the factor for endowment insurance and equals 0 if a policy is not an endowment policy, and the term factor $M_{x+n} = 0$ for whole life insurance.

It might be wise to write Equation (9.5-6) in the solution for all problems. Simply set the endowment and term factors to 0 when appropriate.

PROBLEM SET 9.5

Use a 3% interest rate in Problems 1–10.

1. Calculate the net single premium for $10,000 of one-year term insurance at age 17.

2. Calculate the net single premium for $80,000 of 4-year term insurance at age 17.

3. Calculate the net single premium for $100,000 of 2-year term insurance at age 40.

4. Calculate the net single premium for $5,000 of whole life insurance at age 2.

5. Calculate the net single premium for $100,000 of whole life insurance at age 50.

6. Calculate the net single premium for $100,000 of whole life insurance at age 65.

7. Calculate the net single premium for $100,000 of whole life insurance at age 60.

8. Calculate the net single premium at age 2 for a $20,000 20-year endowment policy.

9. Calculate the net single premium at age 35 for a $100,000 20-year endowment policy.

10. Calculate the net single premium at age 35 for a $100,000 30-year endowment policy.

Use a 5% interest rate for the following problems.

11. Problem No. 1.

12. Problem No. 2.

13. Problem No. 3.

14. Problem No. 4.

15. Problem No. 5.

16. Problem No. 6.

17. Problem No. 7.

18. Problem No. 8.

19. Problem No. 9.

20. Problem No. 10.

9.6 LIFE INSURANCE: NET ANNUAL PREMIUM

We discussed the determination of the net single premium for term insurance and whole life insurance in Section 9.5. It may be unrealistic to pay the premium for a large amount of life insurance with a single payment. The more realistic approach is to pay premiums periodically, such as annually, semiannually, quarterly, or monthly. The usual method is to determine an annual premium and adjust it for more frequent payments during the year.

A general rule for determining the net annual premium is that the *present value of the annual premiums to be paid by the insured must equal the present value of the benefits to be paid by the insurer.* The net annual premium is determined based on interest and mortality only. In practice, the net annual premium must be increased for commissions to a sales force, company expenses, and company profits. The sum of the net annual premium and the increased amount gives the gross annual premium. In this section, we will limit our calculations to the net annual premium only.

The annual premiums are due on each policy anniversary date, only if the insured is alive to make the payment. Therefore, the annual premiums constitute contingent annuities due for the following types of insurance:

(a) Term life insurance with payments for a specified number of years.

(b) *n*-Payment life insurance with payments for a specified number of years.

(c) Whole life insurance with payments until death.

We shall discuss the three types.

NET ANNUAL PREMIUM FOR TERM LIFE INSURANCE

As the name implies, term life insurance is insurance for a specified number of years. Premiums are due for a fixed number of years, and a benefit will be paid if death occurs during the period. At the end of the term, the premiums and insurance terminate. This type of insurance is popular for younger people who wish to protect against a financial disaster during the early years of raising a family or paying a mortgage. Lending institutions require its purchase if a down payment for a house is less than 20% of the market value of the house.

We apply the general rule; the Present Value (Net Premiums) = Present Value (Future Benefits) for the term of a policy. Consider a 3-year term policy for a person age 30. Everyone of the cohort alive at age 30 would make three annual payments of P, and the insurance company would pay three annual benefits of B as shown in Figures 9-5a and 9-5b.

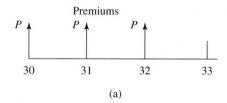

(a)

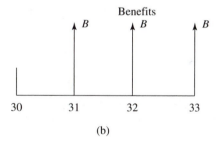

(b)

FIGURE 9-5 A 3-Payment Term Policy

The present value, $\ddot{A}_{30}$, of the premiums, P, is given by

$$\ddot{A}_{30} = Pl_{30} + Pl_{31}v + Pl_{32}v^2,$$

and the present value, A_{30}, of the benefits, B, is given by

$$A_{30} = Bd_{30}v + Bd_{31}v^2 + Bd_{32}v^3.$$

Since the present values must be equal,

$$_3P_{30} = B\left[\frac{d_{30}v + d_{31}v^2 + d_{32}v^3}{l_{30} + l_{31}v + l_{32}v^2}\right]. \qquad (9.6\text{-}1)$$

EXAMPLE 9.6.1   Find the net annual premium at age 30 for a 3-year term policy that will pay $1,000 upon death if a 3% interest rate is used.

Solution   The numbers of living and dying at each age are found in the 3% table of Table A-6. From Equation (9.6-1), the net annual premium is determined as

$$_3P_{30} = 1,000 \left[\frac{127(0.971) + 130(0.943) + 132(0.915)}{96,477 + 96,350(0.971) + 96,220(0.943)} \right]$$

$$= \$1.31. \quad \blacksquare$$

By using the definitions of the commutation functions M and N that were developed in Section 9.3, the Equation (9.6-1) may be written using the commutation functions. Then, the net annual premium, beginning at age x for n payments, may be determined by

$$_nP_x = B \left[\frac{M_x - M_{x+n}}{N_x - N_{x+n}} \right]. \tag{9.6-2}$$

EXAMPLE 9.6.2   Find the net annual premium, at age 30, for a 3-year term policy that will pay $1,000 upon death if a 3% interest rate is used.

Solution   Using the commutation functions of the 3% table of Table A-6 and Equation (9.6-2), the net annual premium is

$$_3P_{30} = 1,000 \left[\frac{M_{30} - M_{33}}{N_{30} - N_{33}} \right]$$

$$= 1,000 \left[\frac{10,983.09 - 10,832.04}{987,569.28 - 871,917.48} \right]$$

$$= \$1.31,$$

as above. $\blacksquare$

Term insurance remains in force for a specified number of years. At the end of the term both the premiums and insurance benefit terminate. However, life insurance that remains in force until the death of the insured is also available. It is available in two forms: whole life or n-payment life. The latter is also referred to as **temporary life**. Whole life insurance is also known as **ordinary life**, **continuous payment life**, or **straight life**.

NET ANNUAL PREMIUM FOR WHOLE LIFE INSURANCE

Whole life insurance is, essentially, "infinite-year" term insurance. M_{x+n} and N_{x+n} would be the values at the end of the mortality table and would be equal to 0. Then, Equation (9.6-2) reduces to

$$P = B \left[\frac{M_x}{N_x} \right]. \tag{9.6-3}$$

EXAMPLE 9.6.3   Determine the annual premium for $1,000 of whole life insurance at age 25 if the interest rate is 3%.

Solution   From Equation (9.6-3) and the 3% table of Table A-6, the net annual premium is

$$P_{25} = 1,000 \left[\frac{M_{25}}{N_{25}} \right]$$

$$= 1,000 \left[\frac{11,260.03}{1,205,795.10} \right]$$

$$= \$9.34. \quad \blacksquare$$

NET ANNUAL PREMIUM FOR LIMITED-PAYMENT LIFE INSURANCE

The numerator of Equation (9.6-2) represents the death benefit that must be paid each year from the year x to the year $x + t$, where t is to the end of the mortality table being used. The denominator represents the annual payments made from the year x to $x + n$. Since a death benefit must be paid whenever the insured dies, $M_{x+n} = 0$ in Equation (9.6-2); however, there will always be people living at $x + n$, therefore, $N_{x+n} \neq 0$. As such, Equation (9.6-2) reduces to the following for an n-payment life insurance policy:

$$_n P_x = B \left[\frac{M_x}{N_x - N_{x+n}} \right]. \tag{9.6-4}$$

EXAMPLE 9.6.4   Determine the net annual premium for a 20-payment life at age 25 using a 3% interest rate.

Solution   From Equation (9.6-4) and the 3% table of Table A-6, the net annual premium becomes

$$_{20} P_{25} = 1,000 \left[\frac{M_{25}}{N_{25} - N_{45}} \right]$$

$$= 1,000 \left[\frac{11,260.03}{1,205,950.10 - 503,899.09} \right]$$

$$= \$16.04. \quad \blacksquare$$

GENERAL FORMULA FOR THE NET ANNUAL PREMIUM

The foregoing can be compiled into a general formula to determine the net annual premium for term, whole life, temporary, and endowment insurance similar to that for the net single premium. This general formula is

$$_n P_x = B \left[\frac{M_x - M_{x+n} + D_{x+n}^E}{N_x - N_{x+n}} \right]. \tag{9.6-5}$$

For:

1. Term insurance. $D_{x+n}^E = 0$.
2. n-payment life insurance. $D_{x+n}^E = 0$ and $M_{x+n} = 0$.
3. Whole life insurance. $n = \infty$ and is omitted, $D_{x+n}^E = 0$, $M_{x+n} = 0$, and $N_{x+n} = 0$.

As indicated above, the net annual premium is based on an interest rate and mortality only. This premium must be increased to account for expenses and profits. Life insurance includes two broad categories, **participating** and **nonparticipating**. In **participating insurance policies**, the policy is credited with dividends by the insurance company. These dividends accrue over time to become a cash value of the policy, and the **cash value** is the amount that would be paid should the policy be canceled. Also, the insured may be able to borrow the cash value. In the event of death prior to the repayment of the loan, the amount due would be deducted from the death benefit. In addition to expenses and profits, the net annual premium would have to be adjusted to account for the projected dividends. **Nonparticipating policies** do not have a cash value, and the net annual premium is not affected by this factor.

PROBLEM SET 9.6

Use a 3% interest rate in Problems 1–13.

1. Calculate the net annual premium for $10,000 of 1-year term insurance at age 17.

2. Calculate the net annual premium for $80,000 of 4-year term insurance at age 65.

3. Calculate the net annual premium for $100,000 of 2-year term insurance at age 40.

4. Calculate the net annual premium at age 2 for a $20,000 20-payment temporary life policy.

5. Calculate the net annual premium at age 35 for a $100,000 20-payment temporary life policy.

6. Calculate the net annual premium at age 35 for a $100,000 30-payment temporary life policy.

7. Calculate the net annual premium for $5,000 of whole life insurance at age 2.

8. Calculate the net annual premium for $100,000 of whole life insurance at age 50.

9. Calculate the net annual premium for $100,000 of whole life insurance at age 65.

10. Calculate the net annual premium for $100,000 of whole life insurance at age 60.

11. Calculate the net annual premium at age 2 for a $20,000 20-year endowment policy.

12. Calculate the net annual premium at age 35 for a $100,000 20-year endowment policy.

13. Calculate the net annual premium at age 35 for a $100,000 30-year endowment policy.

Using a 5% interest rate in Problems 14–26.

14. Problem No. 1.

15. Problem No. 2.

16. Problem No. 3.

17. Problem No. 4.

18. Problem No. 5.

19. Problem No. 6.

20. Problem No. 7.

21. Problem No. 8.

22. Problem No. 9.

23. Problem No. 10.

24. Problem No. 11.

25. Problem No. 12.

26. Problem No. 13.

9.7 LIFE INSURANCE RESERVES

The premium for one year of life insurance at age x in shown by Equation (9.5-2) to be $B\left(\frac{C_x}{D_x}\right)$. As the age increases, C_x increases and D_x decreases due the greater probability of death. Therefore, the premium for each successive year will be greater than that of the preceding year, and that increasing premium could become prohibitive in the latter years of a term or whole life policy. The yearly premium is called the *natural* premium. In order to make the cost of insurance affordable, a level annual premium for the term of the insurance is determined and is shown in the general Equation (9.6-5) as

$$_n P_x = B\left[\frac{M_x - M_{x+n} + D_{x+n}^E}{N_x - N_{x+n}}\right].$$

At an insurance policy interest rate of 3%, using the 3% table of Table A-6, the natural premiums for a $1,000,000 10-year term policy at age 65 is $19,982.32 and will be $40,617.54 when the policyholder reaches age 74. Since this is a term, not endowment, policy, the level annual premium would be determined by

$$_{10} P_{65} = (1,000,000)\left[\frac{7,059.83 - 4,554.54}{145,218.21 - 56,081.30}\right]$$

$$= 28,106.08983 = \$28,106.09.$$

The natural premiums and the excess of the level premium above the natural premium for each policy year are shown in Table 9-1. Notice that the level premium between ages 65 and 69 is larger than the respective natural premium,

TABLE 9-1 NATURAL AND LEVEL PREMIUMS FOR A $1,000,000 10-YEAR TERM POLICY AT 3%

Age	D_X	C_X	$\left(\frac{C_x}{D_x}\right)$	Natural Premium	Level Premium	Excess at End of Age
65	11,289.49	225.59	0.01998232	19,982.32	28,106.09	8,123.77
66	10,735.08	231.03	0.02152069	21,520.69	28,106.09	6,585.40
67	10,191.38	236.36	0.02319180	23,191.80	28,106.09	4,914.29
68	9,658.19	242.48	0.02510625	25,106.25	28,106.09	2,999.84
69	9,134.40	248.81	0.02723833	27,238.33	28,106.09	867.76
70	8,619.54	255.41	0.02963208	29,632.08	28,106.09	−1,525.99
71	8,113.07	261.07	0.03217904	32,179.04	28,106.09	−4,072.95
72	7,615.70	265.72	0.03489087	34,890.87	28,106.09	−6,784.78
73	7,128.16	268.64	0.03768702	37,687.02	28,106.09	−9,580.93
74	6,651.91	270.18	0.04061754	40,617.54	28,106.09	−12,511.45

and it is smaller than the natural premium between ages 70 and 74. That is, the level premium will be insufficient to pay death benefits over the term of the policy. A fund must be accumulated during the years when the net annual premium exceeds the natural premium in order that sufficient funds will be available to pay the death benefits during the term of the policy. This fund is referred to as the *reserve*. The amount of the reserve is calculated at the end of each year, after all death benefits have been paid, and is known as the *terminal reserve.*

The excess at the end of each age is the basis for the terminal reserve for that year. The excess premium per person of $8,123.77, as shown in Table 9-1 opposite age 65, occurs at the beginning of the year. The addition of 3% interest, $243.71, for the year creates an excess of $8,367.48 for each person who paid the premium at age 65. The terminal reserve for those who survive to age 66 is determined as follows. Table A-6 indicates that $l_{65} = 77,107$ and that $l_{66} = 75,520$. Suppose the 77,107 alive at age 65 each paid the $28,106.09 for a $1,000,000 life insurance policy. The total premium would be $2,167,176,269. Adding 3% interest of $65,015,288 at the end of the year gives a total fund of $2,232,191,557. Based on our assumption that all death benefits are paid at the end of the year of death, then death benefits of $1,587,000,000 would be paid to the beneficiaries of the 1,587 people who will have died during the year. This leaves an excess fund of $645,191,557. Dividing this excess by the 77,107 people who paid the premium, the excess per person would be $8,367.48, the same as determined using the excess premium above. However, dividing the excess fund by the number of people alive at the beginning of the next year, 75,520 at age 66 in this case, determines the terminal reserve as $8,543.32 for each person alive at the beginning of age 66.

RETROSPECTIVE TERMINAL RESERVES

The terminal reserves using the process that is described above are referred to as the *retrospective terminal reserves* because they are determined each year by looking backwards from the end of the year. The combination of premiums, interest, and death benefits for a given year generates a terminal reserve for the next year. Letting:

$_tV_x$ be the terminal reserve at age $x + t$, and

i be the interest rate stated in the insurance policy,

the terminal reserve for year t may be determined by Equation (9.7-1) as

$$_tV_x = \frac{(Pl_t + V_{t-1})(1 + i) - Bd_t}{l_{t+1}}. \tag{9.7-1}$$

The 10 years of retrospective terminal reserves, for the illustration above, using Equation (9.7-1) are shown in Table 9-2.

The 49-cent terminal reserve at the end of age 74 should be 0. This apparent discrepancy is due to rounding of numbers to two decimal places in the equation for determining the net annual premium. The $12,511.90 terminal reserve at the end of age 73 that was available to pay the death benefits for those who reached age 74 and died prior to reaching age 75 exceeds the deficit in the excess at the end of age 74 in Table 9-1 by 45 cents. Again, this apparent discrepancy is due to rounding of numbers. A net annual premium of $28,106.06 would compensate for these rounding errors.

The difficulty is using Equation (9.7-1) in a particular year is the need to know the value of the fund at the end of the preceding year. Therefore, a more direct approach is necessary. Consider the accumulations, at time $x + t$, for each year of the net annual premium and the death benefit for an n-payment life policy as shown in Figures 9-6a and 9-6b.

The terminal reserve at time t would be the difference between the future values of the premiums and death benefits, at the interest rate of the policy, divided by the number of survivors at year $x + t$, as

$$_tV_x = \frac{P[l_x(1 + i)^t + \cdots + l_{x+t-1}(1 + i)] - B[d_x(1 + i)^{t-1} + \cdots + d_{x+t-1}]}{l_{x+t}}.$$

Substituting the definition $v = (1 + i)^{-1}$, this equation becomes

$$_tV_x = \frac{P[l_x v^{-t} + \cdots l_{l+t-1}v^{-1}] - B[d_x v^{-(t-1)} + \cdots + d_{x+t-1}]}{l_{x+t}}.$$

Multiplying the numerator and denominator by v^{x+t} gives

$$_tV_x = \frac{P[l_x v^x + \cdots + l_{x+t-1}v^{x+t-1}] - B[d_x v^{x+1} + \cdots + d_{x+t-1}v^{x+t}]}{l_{x+t}v^{x+t}}.$$

Using the definitions of D_x and C_x, Equation (9.3-1) and Equation (9.3-3) respectively, this reduces to

$$_tV_x = \frac{P[D_x + \cdots + D_{x+t-1}] - B[C_x + \cdots + C_{x+t-1}]}{D_{x+t}},$$

TABLE 9-2 RETROSPECTIVE TERMINAL RESERVES–10-YEAR TERM, $1,000,000 POLICY

(1)	(2)	(3)	Total Premiums Beginning of Year (4)	Fund Reserve Beginning of Year (5)	Interest Earned at End of Year 3.0% (6)	Death Benefits Paid at End of Year (7)	Fund Reserve at End of Year (8)	Terminal Reserve per Survivor (9)
Age	l_x	d_x	$(2) \times 28{,}106.09$	$(4) + \text{Previous }(8)$	$0.03*(5)$	$(3) \times 1{,}000{,}000$	$(5) + (6) - (7)$	$\dfrac{(8)}{\text{Next}(2)}$
64			0	0	0	0	0	0
65	77,107	1,587	2,167,176,269	2,167,176,269	65,015,288	1,587,000,000	645,191,557	8,543.32
66	75,520	1,674	2,122,571,904	2,767,763,461	83,032,904	1,674,000,000	1,176,796,365	15,935.82
67	73,846	1,764	2,075,522,310	3,252,318,675	97,569,560	1,764,000,000	1,585,888,235	22,001.17
68	72,082	1,864	2,025,943,167	3,611,831,403	108,354,942	1,864,000,000	1,856,186,345	26,434.62
69	70,218	1,970	1,973,553,416	3,829,739,761	114,892,193	1,970,000,000	1,974,631,953	28,933.18
70	68,248	2,083	1,918,184,419	3,892,816,372	116,784,491	2,083,000,000	1,926,600,864	29,118.13
71	66,165	2,193	1,859,639,434	3,786,240,298	113,587,209	2,193,000,000	1,706,827,506	26,680.85
72	63,972	2,299	1,798,002,779	3,504,830,285	105,144,909	2,299,000,000	1,310,975,194	21,256.87
73	61,673	2,394	1,733,386,878	3,044,362,072	91,330,862	2,394,000,000	741,692,934	12,511.90
74	59,279	2,480	1,666,100,899	2,407,793,834	72,233,815	2,480,000,000	27,649	0.49
75	56,799							

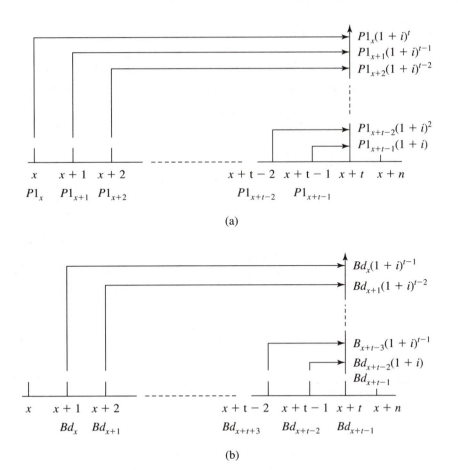

FIGURE 9-6 (a) Future Value of Annual Premiums on Date $x + t$; (b) Future Value of Annual Death Benefits on Date $x + t$

and from the definitions of N_x and M_x, Equation (9.3-2) and Equation (9.3-4) respectively, the terminal reserve for year t may be determined from

$$_tV_x = \frac{P[N_x - N_{x+t}] - B[M_x - M_{x+t}]}{D_{x+t}}. \quad t \le n \qquad (9.7\text{-}2)$$

In the case of limited payment life policies, the number of payments is less than the term of the policy. For the case of determining the reserve after payments have terminated, Equation (9.7-2) becomes

$$_tV_x = \frac{P[N_x - N_{x+n}] - B[M_x - M_{x+t}]}{D_{x+t}}. \quad t > n \qquad (9.7\text{-}3)$$

EXAMPLE 9.7.1   A 20-year term life insurance policy of $100,000 is issued at age 35. Find the 10th terminal reserve at 3%.

Solution   The net annual premium is determined using Equation (9.6-2), or
$$_{20}P_{35} = B\left[\frac{M_{35} - M_{55}}{N_{35} - N_{55}}\right].$$

From the 3% table of Table A-6, the net annual premium becomes
$$_{20}P_{35} = (100{,}000)\left[\frac{10{,}731.08 - 8{,}914.22}{800{,}567.41 - 290{,}795.18}\right] = 356.406$$
$$= \$356.41.$$

Since the year of the reserve is less than the number of payments ($t < n$), Equation (9.7-2) is used. Then, the 10th terminal reserve is determined as
$$_{10}V_{35} = \frac{(356.41)[N_{35} - N_{45}] - (100{,}000)[M_{35} - M_{45}]}{D_{45}}.$$

Substituting the commutation values from the 3% table of Table A-6 and solving gives the 10th terminal reserve as
$$_{10}V_{35} = \frac{(356.41)[800{,}567.41 - 503{,}899.09] - (100{,}000)[10{,}731.08 - 10{,}074.52]}{24{,}751.19}$$
$$= \$1{,}619.30. \quad\blacksquare$$

EXAMPLE 9.7.2   A 10-year term life insurance policy of $100,000 is issued at age 35 and is payable for 8 years. Find the 9th terminal reserve at 3%.

Solution   The net annual premium is determined using Equation (9.6-2) modified to accommodate a 10-year term with eight payments as follows.
$$_{10}P_{35} = B\left[\frac{M_{35} - M_{45}}{N_{35} - N_{43}}\right].$$

The net annual premium for this case becomes
$$_{10}P_{35} = (100{,}000)\left[\frac{10{,}731.08 - 10{,}074.52}{800{,}567.41 - 555{,}905.77}\right]$$
$$= \$268.35.$$

Since the terminal reserve is for the year following the final payment ($t > n$), Equation (9.7-3) is used as follows.
$$_9V_{35} = \frac{P[N_{35} - N_{35+8}] - B[M_{35} - M_{35+9}]}{D_{44}}.$$

Then,
$$_9V_{35} = \frac{(268.35)[800{,}567.41 - 555{,}905.77] - (100{,}000)[10{,}731.08 - 10{,}157.55]}{25{,}579.25}$$
$$= \$324.56. \quad\blacksquare$$

EXAMPLE 9.7.3   A 20-year term life insurance policy of $100,000 is issued at age 35. Find the 20th terminal reserve at 3%. This reserve should be zero, in theory.

Solution   The net annual premium was shown to be $356.41 in Example 9.7.1. Since the year of the reserve is the same as the number of payments ($t = n$),

Equation (9.7-2) is used. Then, the 20th terminal reserve is determined as

$$_{20}V_{35} = \frac{(356.41)[N_{35} - N_{55}] - (100{,}000)[M_{35} - M_{55}]}{D_{55}}.$$

Substituting the commutation values from the 3% table of Table A-6 and solving gives the 20th terminal reserve as

$$_{20}V_{35} = \frac{(356.41)[800{,}567.41 - 290{,}795.18] - (100{,}000)[10{,}731.08 - 8{,}914.22]}{17{,}383.99}$$

$$= \$0.11.$$

Note that the rounding of the commutation functions and the net annual premium of 356.4062364 to 356.41 causes a final excess terminal reserve of 11 cents, rather than the theoretical zero. ■

EXAMPLE 9.7.4 A 20-year endowment policy for $100,000 is issued at age 35. Find the 20th terminal reserve at 3%. If the insured is alive at the end of 20 years, $100,000 must be available to pay the endowment.

Solution The net annual premium for an endowment policy is shown by Equation (9.6-5) to be

$$_{20}P_{35} = B\left[\frac{M_{35} - M_{55} + D_{55}^{E}}{N_{35} - N_{55}}\right].$$

From the 3% table of Table A-6, the net annual premium becomes

$$_{20}P_{35} = (100{,}000)\left[\frac{10{,}731.08 - 8{,}914.22 + 17{,}383.99}{800{,}567.41 - 290{,}795.18}\right] = 3{,}766.554$$

$$= \$3{,}766.55.$$

Since the year of the reserve is the same as the number of payments ($t = n$), Equation (9.7-2) is used. Then, the 20th terminal reserve is determined as

$$_{20}V_{35} = \frac{(3{,}766.55)[N_{35} - N_{55}] - (100{,}000)[M_{35} - M_{55}]}{D_{55}}.$$

Substituting the commutation values from the 3% table of Table A-6 and solving gives the 20th terminal reserve as

$$_{20}V_{35} = \frac{(3{,}766.55)[800{,}567.41 - 290{,}795.18] - (100{,}000)[10{,}731.08 - 8{,}914.22]}{17{,}383.99}$$

$$= \$99{,}999.86.$$

The 14-cent deficit is due to rounding 3,766.55427219 to 3,766.55. ■

The retrospective terminal reserves are determined beginning with the age at issue. The terminal reserves are calculated based on the interest rate stated in the policy, as indicated above. However, if the investment rate of return of the company in a particular year is less that the policy's stated interest rate, sufficient reserves may not be available to satisfy legal requirements. At any time during the life of a policy, it might be necessary to look ahead at the benefits that the company will have to pay and determine that a sufficient terminal reserve is available at the *attained* age of the policyholder. The procedure for looking ahead generates a *prospective terminal reserve*.

PROSPECTIVE TERMINAL RESERVES

The fundamental basis for the prospective terminal reserve at a particular time is that, at the policy stated interest rate, the present value of the future death benefit obligations of the company must equal the then current terminal reserve plus the present value of the future net premiums, or

$$A_{x+t} = {}_tV_x + {}_tA_x, \tag{9.7-4}$$

where ${}_tV_x$ is the prospective terminal reserve at the attained age, $x+t$.

Now, the present value of the future death benefits is simply the net single premium that is determined at the issue age and is given by modifying the general formula Equation (9.5-6) to

$$A_{x+t} = B\left[\frac{M_{x+t} - M_{x+n} + D_{x+n}^E}{D_{x+t}}\right],$$

and, the present value, deferred t years, of annual premiums is given by Equation (9.4-4) modified to

$${}_tA_x = P_x\left[\frac{N_{x+t} - N_{x+n}}{D_{x+t}}\right],$$

where P_x is the net annual premium determined at the issue age. Now, the prospective terminal reserve at the attained age $x+t$ is determined by rearranging Equation (9.7-4) to

$${}_tV_x = A_{x+t} - {}_tA_x. \tag{9.7-5}$$

If the policy is **not** an endowment policy, this reduces to

$${}_tV_x = \frac{B[M_{x+t} - M_{x+n}] - P_x[N_{x+t} - N_{x+n}]}{D_{x+t}}. \tag{9.7-6}$$

If the policy is an endowment, Equation (9.7-5) reduces to

$${}_tV_x = \frac{B[M_{x+t} - M_{x+n} + D_{x+n}^E] - P_x[N_{x+t} - N_{x+n}]}{D_{x+t}}. \tag{9.7-7}$$

The terminal reserve at the attained age of $x + t$ determined prospectively must be the same as the tth terminal reserve determined retrospectively since both are values on the same date.

EXAMPLE 9.7.5 Using the prospective method confirm, within rounding errors, that the indicated terminal reserves of a) Example 9.7.1 to be $1,619.30, b) Example 9.7.2 to be $324.56, c) Example 9.7.3 to be $0.00, and d) Example 9.7.4 to be $100,000.00.

Solution For all the examples the death benefit equals $100,000, the interest rate is 10%, and the issue age is 35.

a) The policy is a 20-year term policy, and the 10th terminal reserve is the reserve at age 45. Then $x = 35$, $x + t = 35 + 10 = 45$, and $x + n = 35 + 20 = 55$. The net annual payment was determined as \$356.41. Then, from Equation (9.7-6)

$$_{10}V_{45} = \frac{B[M_{45} - M_{55}] - P_x[N_{45} - N_{55}]}{D_{45}}.$$

$$_{10}V_{45} = \frac{(100{,}000)[10{,}074.18 - 8{,}914.22] - (356.41)[503{,}899.09 - 290{,}795.18]}{24{,}751.19}$$

$$= \$1{,}617.85.$$

The \$1.45 difference between the prospective and retrospective terminal reserves is due to rounding.

b) The policy is a 10-year term policy, and the 9th terminal reserve is the reserve at age 43. Then $x = 35$ and $x + t = 35 + 9 = 44$. Since premiums were to be paid for eight years, the future premiums $= 0$. Then, from Equation (9.7-6)

$$_{9}V_{45} = \frac{B[M_{44} - M_{45}]}{D_{44}}.$$

$$_{9}V_{45} = \frac{(100{,}000)[10{,}157.55 - 10{,}074.52]}{25{,}579.25}$$

$$= \$324.60.$$

Again, the 4-cent difference between the prospective and retrospective reserves is due to rounding.

c) The policy is a 20-year term issued at age 35. The 20th terminal reserve is at age 55. Then $x = 35$, $x + t = 35 + 20 = 55$, $x + n = 35 + 20 = 55$, and $P = 356.41$. From Equation (9.7-6),

$$_{20}V_{35} = \frac{B[M_{55} - M_{55}] - P_x[N_{55} - N_{55}]}{D_{45}}$$

$$= \$0.00.$$

d) The policy is a 20-year endowment policy issued at age 35. The 20th terminal reserve is at age 55. Then $x = 35$, $x + t = 55$, and $x + n = 55$. From Eq (9.7-7),

$$_{20}V_{35} = \frac{B[M_{55} - M_{55} + D_{55}^{E}] - P_x[N_{55} - N_{55}]}{D_{55}},$$

which reduces to

$$_{20}V_{35} = B\left[\frac{D_{55}^{E}}{D_{55}}\right] = B = \$100{,}000.$$

This is the endowment amount if the policyholder survives to age 65. ■

USES OF THE TERMINAL RESERVE

The policyholder may make use of the terminal reserve, which is the value of an insurance policy at the end of any year, for three purposes.

1. The policyholder may terminate payments, surrender the policy to the insurance company, and receive the cash value of the reserve.

2. The policyholder may continue paying the premiums but borrow money from the reserve, up to the cash value of the policy. The reserve then becomes the security for the loan. This type of borrowing was very popular when the insurance company charged the policy interest rate as the interest rate on the loan. Since policy interest rates were generally low, the policyholder could invest the funds at market interest rates, which were generally higher than the policy rate. Insurance companies now charge market rates on loans from cash value reserves, and this type of borrowing by the insured is not as attractive as it once was.

3. The policyholder may stop making payments of the annual premium and use the reserve as the single premium payment for one of two choices.

   (a) Extend the term of the policy for a period that is supported by the reserve.

   (b) Purchase a new policy with a face value for which the reserve is the net single premium.

EXAMPLE 9.7.6   In order to protect a home mortgage, a 30-year term insurance policy of $100,000 was purchased at age 35. At the end of 15 years, the policyholder felt that the insurance would not be needed for the remaining 15 years. The policyholder informed the company that annual premiums were terminating and to limit the term of the policy to that which would be supported by the cash value of the reserve. Assume a 3% interest rate.

Solution   The net annual premium for a 30-year term policy of $100,000 at 3% issued at age 35 is determined by

$$30 P_{35} = B \left[\frac{M_{35} - M_{65}}{N_{35} - N_{65}} \right]$$

to be

$$30 P_{35} = (100,000) \left[\frac{10,731.08 - 7,059.83}{800,567.41 - 145,218.21} \right]$$

$$= \$560.20.$$

Since future premiums from age 50 are to be zero, the 15th terminal reserve is determined retrospectively from Equation (9.7-2) as

$$15 V_{35} = \frac{P(N_{35} - N_{50}) - B(M_{35} - M_{50})}{D_{50}},$$

and substituting the values gives

$$15 V_{35} = \frac{(560.20)[800,567.41 - 388,055.55] - (100,000)[10,731.08 - 9,575.14]}{20,877.73}$$

$$= \$5,531.98.$$

At the end of 15 years, the attained age of the insured is 50. The net single premium for an n-year term policy at age 50 is

$$A_{50} = {}_{15}V_{35} = B\left[\frac{M_{50} - M_{50+n}}{D_{50}}\right].$$

In order to extend the term, $x + n$ must be determined. Then, solving this equation for M_{50+n} gives

$$M_{50+n} = M_{50} - \frac{{}_{15}V_{35}D_{50}}{B}$$

$$= 9{,}575.14 - \frac{(5{,}531.98)(20{,}877.73)}{100{,}000}$$

$$= \$8{,}420.19,$$

and the 3% table of Table A-6 must be interpolated to determine n. Now, 8,494.74 is between $M_{59} = 8{,}265.17$ and $M_{58} = 8{,}437.72$. Linear interpolation would give $M_{50+n} = 58.102$. Then, the terminal reserve at age 50 would pay for a term of 8.102 years or 8 years and 37 days ∎

EXAMPLE 9.7.7   In order to protect a home mortgage, a 30-year term insurance policy of \$100,000 was purchased at age 35. At the end of 15 years, the policyholder felt that the face amount of the policy was no longer needed as protection because of the equity that had accrued in the property. The policyholder informed the company that annual premiums were terminating and to issue a whole life policy for a face amount that would be supported by the cash value of the reserve. Assume a 3% interest rate.

Solution   The net annual premium and the 15th terminal reserve are the same as in Example 9.7.11. The net single premium for a whole policy that is issued at age 50 is determined by

$$A_{50} = {}_{15}V_{35} = B\left(\frac{M_{50}}{D_{50}}\right),$$

which may be solved for the benefit B. Then,

$$B = {}_{15}V_{35}\left(\frac{D_{50}}{M_{50}}\right),$$

or

$$B = (5{,}531.98)\left(\frac{20{,}877.73}{9{,}575.14}\right)$$

$$= \$12{,}061.98.$$

Then, the 15th terminal reserve at age 50 would purchase a whole life policy of \$12,000. In general

$$M_{t+n} = M_t + \frac{{}_tV_xD_t}{B}, \tag{9.7-8}$$

and

$$B = {}_tV_x\left(\frac{D_t}{M_t}\right). \quad ∎ \tag{9.7-9}$$

PROBLEM SET 9.7

*Use a 3% interest rate for the following problems and solve Problems 1–20 to the next **higher dollar**.*

1. A 20-year term policy for $50,000 is issued to a person aged 25. Find the 10th retrospective terminal reserve.

2. A 20-year term policy for $50,000 is issued to a person aged 35. Find the 20th retrospective terminal reserve.

3. A 30-year term policy for $100,000 is issued to a person aged 40. Find the 20th retrospective terminal reserve.

4. A 30-year term policy for $100,000 is issued to a person aged 45. Find the 30th retrospective terminal reserve.

5. A whole life policy for $100,000 is issued to a person aged 35. Find the 30th retrospective terminal reserve.

6. A whole life policy for $100,000 is issued to a person aged 40. Find the 25th retrospective terminal reserve.

7. A 25-payment whole life policy for $100,000 is issued to a person aged 35. Find the 25th retrospective terminal reserve.

8. A 25-payment whole life policy for $100,000 is issued to a person aged 40. Find the 25th retrospective terminal reserve.

9. A 20-year endowment policy for $100,000 is issued for a child at birth. Find the 10th retrospective terminal reserve.

10. A 20-year endowment policy for $100,000 is issued for a child at birth. Find the 15th retrospective terminal reserve.

11. A 20-year term policy for $50,000 is issued to a person aged 25. Find the 10th prospective terminal reserve.

12. A 20-year term policy for $50,000 is issued to a person aged 35. Find the 20th prospective terminal reserve.

13. A 30-year term policy for $100,000 is issued to a person aged 40. Find the 20th prospective terminal reserve.

14. A 30-year term policy for $100,000 is issued to a person aged 45. Find the 30th prospective terminal reserve.

15. A straight life policy for $100,000 is issued to a person aged 35. Find the 30th prospective terminal reserve.

16. A whole life policy for $100,000 is issued to a person aged 40. Find the 25th prospective terminal reserve.

17. A 25-payment whole life policy for $100,000 is issued to a person aged 35. Find the 25th prospective terminal reserve.

18. A 25-payment whole life policy for $100,000 is issued to a person aged 40. Find the 25th prospective terminal reserve.

19. A 20-year endowment policy for $100,000 is issued for a child at birth. Find the 10th prospective terminal reserve.

20. A 20-year endowment policy for $100,000 is issued for a child at birth. Find the 15th prospective terminal reserve.

21. A whole life policy for $100,000 is issued to a person aged 35. Determine the term of a policy that may be purchased if payments are terminated and the 30th retrospective terminal reserve of Problem 5 is used as the net single premium.

22. A whole life policy for $100,000 is issued to a person aged 40. Determine the term of a policy that may be purchased if payments are terminated and the 25th retrospective terminal reserve of Problem 6 is used as the net single premium.

23. A whole life policy for $100,000 is issued to a person aged 35. Determine the face value of a whole life policy that may be purchased, using the 30th retrospective terminal reserve of Problem 5 as the net single premium, if payments are terminated.

24. A whole life policy for $100,000 is issued to a person aged 40. Determine the face value of a whole life policy that may be purchased, using the 25th retrospective terminal reserve of Problem 6 as the net single premium, if payments are terminated.

CHAPTER | 10

Social Security

10.1 A BRIEF HISTORY OF SOCIAL SECURITY

Note: Much of the following history has been extracted from the Social Security Administration document Social Security—A Brief History. The full text is available on the World Wide Web at *http://www.ssa.gov/history/history6.html.*

THE PROBLEM OF ECONOMIC SECURITY

Prior to the turn of the 20th century, the majority of people in the United States lived and worked on farms, and they depended upon their children, relatives, or charity to provide economic security. However, the Industrial Revolution changed this dependence. The extended family and the family farm as sources of economic security became less common.

The first American company to provide for its employees who left work due to old age was the American Express Company; it established the first private pension plan in 1875. The Baltimore & Ohio Railroad Company established the second plan in 1880. However, private plans such as these eased the economic problem of old age for only a small segment of workers. All people throughout human history have faced the uncertainties brought on by death, disability, and old age. Social Security addressed a universal human need, a measure of economic dignity during old age. The Great Depression of 1929 triggered a crisis in the nation's economic life. Factories closed, workers were deprived of "pensions," workers lost their dignity (families were evicted from their homes with their furniture piled on curbs outside), and many lived only because of welfare provided by the states, or were forced to work on "poor farms." It was against this backdrop that the Social Security Act emerged.

THE SOCIAL SECURITY ACT

President Franklin D. Roosevelt sent a message to the Congress of the United States on June 8, 1934 in which he announced his intention to provide a program for Social Security. Subsequently, the president, by Executive Order, created the Committee on Economic Security, which was comprised of Frances Perkins, secretary of labor, Henry Morgenthau, Jr., secretary of the treasury; Henry A. Wallace, secretary of agriculture; Homer S. Cummings, attorney general; and Harry L. Hopkins, Federal Emergency Relief administrator. The committee, which was chaired by Frances Perkins, was instructed to study the entire problem of economic insecurity and to make recommendations that would serve as the basis for legislative consideration by the Congress.

In early January 1935, the committee made its report to the president, and on January 17 the president introduced the report to both houses of Congress for simultaneous consideration. Each House passed its own version, but eventually the differences were resolved, and President Roosevelt signed the Social Security Act into law on August 14, 1935. In addition to several provisions for general welfare, the new Act created a social insurance program designed to pay retired workers, age 65, or older a continuing income after retirement.

MAJOR PROVISIONS OF THE ACT

The Social Security Act did not quite achieve all the aspirations its supporters had hoped by way of providing a "comprehensive package of protection" against the "hazards and vicissitudes of life." Certain features of that package, notably disability coverage and medical benefits, would have to await future developments. But it did provide a wide range of programs to meet the nation's needs. In addition to the program we now think of as Social Security, it included unemployment insurance, old age assistance, aid to dependent children, and grants to the states to provide various forms of medical care.

The two major provisions relating to the elderly were Title I–Grants to States for Old-Age Assistance, which supported state welfare programs for the aged, and Title II–Federal Old-Age Benefits. Title II was the new social insurance program that we now think of as Social Security. In the original Act benefits were to be paid only to the primary worker when he or she retired at age 65. Benefits were to be based on payroll tax contributions that the worker made during his or her working life. Taxes would first be collected in 1937 and monthly benefit payments would begin in 1942. The new social insurance program sought to address the long-range problem of economic security for the aged through a contributory system in which the workers themselves contributed to their own future retirement benefit by making regular payments into a joint fund. It was thus distinct from the welfare benefits provided under Title I of the Act and from the various states "old-age pensions." As President Roosevelt conceived of the Act, Title I was to be a temporary "relief" program that would eventually disappear as more people were able to obtain retirement income through the contributory system. The new social insurance system was a very moderate alternative to the radical calls to action that were so common in America in the 1930s.

A provision of the Act established a Social Security Board with a charge to create a protocol for implementing the program. One of its first tasks was to assign Social Security numbers (SSN) to all workers. (The lowest SSN was the number assigned to Grace Dorothy Owen, a New Hampshire resident. She received the number 001-01-0001.) The Social Security Board was abolished in 1946 and was replaced by the current Social Security Administration.

TRUST FUNDS

The concept of the Social Security Act was a pay-as-you-go concept. That is, those who are in the work force will pay the benefit of those who leave the work force due to old age. In order to accomplish this, the president signed the Federal Insurance Contributions Act (FICA); it became the law of the land. This Act mandated that employers withhold taxes from employees' wages, match the employees' contributions, and remit both to the Social Security Board.

After Social Security numbers were assigned, the first FICA taxes were collected, beginning in January 1937. Special trust funds were created for these dedicated revenues. Benefits were then paid from the monies in the Social Security Trust Funds. Over the years, more than $4.5 trillion has been paid into the Trust Funds, and more than $4.1 trillion has been paid out in benefits. The remainder is currently on reserve in the Trust Funds and will be used to pay future benefits.

AMENDMENTS TO THE ACT

1. 1939 Amendments

The original Act provided only retirement benefits to the worker. The 1939 Amendments made a fundamental change in the Social Security program. The Amendments added two new categories of benefits: payments to the spouse and minor children of a retired worker (so-called dependents' benefits) and survivors' benefits paid to the family in the event of the premature death of a covered worker. This change transformed Social Security from a retirement program for workers into a family-based economic security program. The 1939 Amendments also increased benefit amounts and accelerated the start of monthly benefit payments to 1940. Payments of monthly benefits began in January 1940, and were authorized not only for aged retired workers but also for their aged wives or widows, children under age 18, and surviving aged parents. On January 31, 1940, the first monthly retirement check was issued to Ida May Fuller of Ludlow, Vermont, in the amount of $22.54. Miss Fuller died in January 1975 at the age of 100. During her 35 years as a beneficiary, she received over $20,000 in benefits.

2. 1950 Amendments

From 1940 until 1950 virtually no changes were made in the Social Security program. Payment amounts were fixed and no major legislation was enacted. Because the program was still in its infancy, and because low levels of payroll taxation financed it, the absolute value of Social Security's retirement benefits was very low. Until 1951, the average value of the welfare benefits received under the old-age assistance provisions of the Act was higher than the retirement benefits received under Social Security. Also, there were more elderly Americans receiving old-age assistance than were receiving Social Security.

Because of these shortcomings in the program major amendments were enacted in 1950. These amendments increased benefits for existing beneficiaries for the first time, and they dramatically increased the value of the program to future beneficiaries. By February 1951 there were more Social Security retirees than welfare pensioners, and by August of that year, the average Social Security retirement benefit exceeded the average old-age assistance grant for the first time. However, it became necessary to increase benefits in order to offset the corrosive effects of inflation. These increases in benefits are known as Cost of Living Allowances (COLAs). The 1950 Amendments that were enacted by Congress were the first legislated increase in benefits. These recomputations were effective for September 1950 and appeared for the first time in the October 1950 checks. A second increase was legislated for September 1952. Together these two increases almost doubled the value of Social Security benefits for existing beneficiaries. From that point on, benefits were increased only when Congress enacted special legislation for that purpose.

3. The 1954 and 1956 Amendments

The Social Security Amendments of 1954 initiated a disability insurance program that provided the public with additional coverage against economic insecurity. At first, there was a disability "freeze," (signed by President Eisenhower) of workers' Social Security records during the years when they were unable to work. While this measure offered no cash benefits, it did prevent such periods of disability from reducing or wiping out retirement and survivor benefits. On August 1, 1956, the Social Security Act was amended to provide benefits to disabled workers aged 50 to 65 and disabled adult children. Over the next two years, Congress broadened the scope of the program, permitting disabled workers under age 50 and their dependents to qualify for benefits. By 1960, 559,000 people were receiving disability benefits, with the average benefit amount being around $80 per month. Today, the disability benefit is determined as if the worker was age 65 at the beginning of the waiting period for the disability determination.

4. The 1961 Amendments

The decade of the 1960s brought major changes to the Social Security program. Under the Amendments of 1961, the age at which men are first eligible for old-age insurance was lowered to 62, with benefits actuarially reduced (women previously were given this option in 1956). Later in the decade (1965) Medicare coverage was enacted to help protect older Americans from being economically devastated due to illness. In order to be qualified for benefits, 39 quarters of covered employment was necessary for those who turned 62 in 1989 or 1990, and 40 quarters were needed for those who turned 62 in 1991 and later. Before 1978, a quarter of coverage was earned for each calendar quarter that a worker had earned $50, up to a maximum of four quarters in one year. In 1978, the requirement was changed to $250 of covered earnings for each calendar quarter. This requirement has been indexed with inflation until in 2001 the covered earnings had increased to $830 in a calendar quarter. However, a subsequent change in the determination of credited quarters was that the annual amount could be earned in one calendar quarter. Thus, if a worker earned $3,320, or more, in any one quarter he or she would be given credit for four quarters for that year. After 2001, covered earnings will be determined by multiplying $250 by the ratio of the National Average Wage Index two years earlier to $9246.48 rounded to the nearest multiple of $10. Application of this formula requires $870 per quarter or $3,480 for the year to earn four quarters of coverage in 2002.

5. The 1972 Amendments

In 1972 legislation by Congress, the law was changed to provide, beginning in 1975, for automatic annual cost of living allowances (i.e., COLAs) based on the annual increase in the Consumer Price Index. Between 1950 and 1978 inclusive, the monthly benefit was increased 636%. The increases between 1979 and this year are included in Table 10-2 at the end of this section. Annual increases are determined in October and take effect with the following January's check.

6. The 1983 Amendments

In the early 1980s the Social Security program faced a serious long-term financing crisis. President Reagan appointed a blue-ribbon panel, known as the Greenspan Commission, to study the financing issues and make recommendations for legislative changes. The final bill, signed into law in 1983, made numerous changes in the Social Security and Medicare programs, including the taxation of Social Security benefits, the first coverage of Federal employees under Social Security, and an increase in the retirement age in the next century. The following table shows the Social Security Normal Retirement ages for the future.

    Benefits to workers may begin at age 62; however, benefits will be permanently reduced according to the following formulas. Let:

R_w be the worker's benefit reduction,

R_s be the spouse's benefit reduction,

M_n be the normal retirement age of the worker or spouse in months, and

M_a be the number of months to the month of the actual retirement age. However, the first eligibility month is the entire month in which a worker was the retirement age (62 to 70), and therefore, the first benefit month will be one month later. Thus, $M_a = 12(\text{Retirement Age}) + 1$.

    For workers whose birth year is ≤ 1937,

$$R_w = \frac{5}{9}(780 - M_a)\%. \tag{10.1-1}$$

For workers whose birth year is > 1937,

$$R_w = \frac{5}{9}(M_n - M_a)\%. \quad M_n - M_a \leq 36 \tag{10.1-2a}$$

$$R_w = 20\% + \frac{5}{12}(M_n - M_a - 36)\%. \quad M_n - M_a > 36 \tag{10.1-2b}$$

Spousal benefits are also reduced, upon the retirement of the worker, depending on the spouse's age relative to his or her own normal retirement age. The maximum spousal benefit is 50% of the worker's benefit if the spouse's age is his or her normal retirement age at the retirement of the worker. The formulas for the reduction from this maximum benefit are as follows:

For spouses whose birth year is ≤ 1937,

$$R_s = \frac{25}{36}(780 - M_a)\%. \tag{10.1-3}$$

For spouses whose birth year is > 1937,

$$R_s = \frac{25}{36}(M_n - M_a)\%. \quad M_n - M_a \leq 36 \tag{10.1-4a}$$

$$R_s = 25\% + \frac{5}{12}(M_n - M_a - 36)\%. \quad M_n - M_a > 36 \tag{10.1-4b}$$

EXAMPLE 10.1.1 A worker was born in October 1940 and will retire in 2002 at the age of 62. The spouse was also born in October 1940 and will also be age 62. Determine the reduction of benefits for the worker and spouse.

Solution Since both the worker and spouse were born in 1940, their normal retirement ages are 65 years, 6 months. The worker's normal retirement age, M_n, in months is 786, and the actual benefit age, M_a, in months is 745 ($12 \times 62 + 1$) and is 41 months early. For the worker, the reduction in benefits may be determined from Equation (10.1-2b) as

$$R_w = 20\% + \frac{5}{12}(786 - 745 - 36) = 22.1\%.$$

Since the spouse is the same age as the worker, the basic spousal benefit is 50% of the worker's benefit. Because the spouse's age at retirement is less than his or her normal retirement age, the spousal reduction, *from the 50% basic benefit*, may be determined from Equation (10.1-4b) as

$$R_s = 25\% + \frac{5}{12}(786 - 745 - 36) = 27.1\%.$$

Thus, the spouse's benefit will be 36.45% ($50\% - 0.271 \times 50\%$) of the worker's primary insurance amount (PIA), a concept that will be defined in Section 10.2. Letting the subscript s represent the spouse and the subscript w represent the worker, in general,

$$PIA_s = \left(1 - \frac{R_s}{100}\right)(0.50)PIA_w. \quad \blacksquare \tag{10.1-5}$$

The reduction formulas for both the worker and spouse will be revisited in Section 10.2.

7. The 1990s

From its modest beginnings, Social Security has grown to become an essential facet of modern life. One in seven Americans receive a Social Security benefit, and more than 90% of all workers are in jobs covered by

Social Security. From 1940, when slightly more than 222,000 people received monthly Social Security benefits, until today, when over 42 million people receive such benefits, Social Security has grown steadily. However, the viability of the current system is being questioned. The accumulated trust fund, due to the 1983 changes, and the number of workers who will be contributing to the system during the first half of the next century, will be sufficient to provide approximately 75% of the benefits to which those retirees will be entitled. Therefore, President George W. Bush formed the *Social Security Reform Commission* to review alternatives that will ensure the continuing viability of a Social Security system in the United States. The Commission has submitted its report, but as of this writing, Congress has not acted.

8. The 2000 Amendment

Because the country was in an economic depression during the 1930s one of the provisions of the Act was to discourage retirees from working in order to have jobs available for younger workers. Each year retirees between ages 62 and 65 and between 65 and 70 had respective earnings limits imposed whereby the monthly Social Security benefit would be reduced for each dollar that was earned above the limits. In April 2000, then President Clinton signed the "Senior Citizens Right to Work Act of 2000" that removed the limits for retirees between ages 65 and 70. Table 10-3 shows the history of the FICA wage limits, tax rates, and National Average Wage Indices since 1951.

PROBLEM SET 10.1

1. Verify the maximum reduction in benefits in Table 10-1 for a worker born in 1937.

2. Verify the maximum reduction in benefits in Table 10-1 for a worker born in 1938.

3. Verify the maximum reduction in benefits in Table 10-1 for a worker born in 1939.

4. Verify the maximum reduction in benefits in Table 10-1 for a worker born in 1940.

5. Verify the maximum reduction in benefits in Table 10-1 for a worker born in 1941.

6. Verify the maximum reduction in benefits in Table 10-1 for a worker born in 1942.

7. Verify the maximum reduction in benefits in Table 10-1 for a worker born in 1950.

8. Verify the maximum reduction in benefits in Table 10-1 for a worker born in 1955.

Year of Birth of Individual Receiving Benefit (Workers and Spouses)	Social Security Normal Retirement Age (SSNRA)	Maximum Reduction of Worker's Benefit Old Age Benefit (Retirement at Age 62)
1937 and before	65 years	20.00%
1938	65 years, 2 months	20.83%
1939	65 years, 4 months	21.67%
1940	65 years, 6 months	22.50%
1941	65 years, 8 months	23.33%
1942	65 years, 10 months	24.17%
1943–1954	66 years	25.00%
1955	66 years, 2 months	25.83%
1956	66 years, 4 months	26.67%
1957	66 years, 6 months	27.50%
1958	66 years, 8 months	28.33%
1959	66 years, 10 months	29.17%
1960 and later	67 years	30.00%

TABLE 10-1 SOCIAL SECURITY NORMAL RETIREMENT AGES

9. Verify the maximum reduction in benefits in Table 10-1 for a worker born in 1956.

10. Verify the maximum reduction in benefits in Table 10-1 for a worker born in 1957.

11. Verify the maximum reduction in benefits in Table 10-1 for a worker born in 1958.

12. Verify the maximum reduction in benefits in Table 10-1 for a worker born in 1959.

13. Verify the maximum reduction in benefits in Table 10-1 for a worker born in 1960.

Determine the reduction in old age benefits for Problems 14–20.

14. A worker was born in 1939 and will retire at age 62 years and six months.

15. A worker was born in 1940 and will retire at age 63 years.

16. A worker was born in 1940 and will retire at age 63 years and nine months.

17. A worker was born in 1950 and will retire at age 64 years.

18. A worker was born in 1950 and will retire at age 64 years and three months.

19. A worker was born in 1960 and will retire at age 63.

20. A worker was born in 1960 and will retire at age 63 years and six months.

21. A spouse who was born in 1937 will be 62 on the worker's retirement date.

    a) What will be the reduction in the spousal benefit?

    b) What percent of the worker's benefit will the spouse receive?

22. A spouse who was born in 1940 will be 63 on the worker's retirement date.

    a) What will be the reduction in the spousal benefit?

    b) What percent of the worker's benefit will the spouse receive?

23. A spouse who was born in 1940 will be 64 on the worker's retirement date.

    a) What will be the reduction in the spousal benefit?

    b) What percent of the worker's benefit will the spouse receive?

24. A spouse who was born in 1952 will be 62 on the worker's retirement date.

    a) What will be the reduction in the spousal benefit?

    b) What percent of the worker's benefit will the spouse receive?

25. A spouse who was born in 1961 will be 62 on the worker's retirement date.

    a) What will be the reduction in the spousal benefit?

    b) What percent of the worker's benefit will the spouse receive?

10.2 DETERMINATION OF BENEFITS

The first calculation in the determination of benefits is the *Primary Insurance Amount*, or PIA. For most current and future retirees, the PIA is determined using the 35 highest inflation-adjusted wages for earnings from 1951 or age 22, whichever is later, through the year before the retirement year. It is a three-step calculation.

1. The Average Indexed Monthly Earnings (AIME)

The Average Indexed Monthly Earnings are based on actual earnings from the latter of 1951 or the year of age 22 to the year prior to the beginning of Social Security benefits. The actual earnings, up to the FICA wage limit, are multiplied by an index factor that is determined by dividing the National Average Wage Index for the age-60 year by the National Average Wage Index for each year from 1951 or the year of age 22. Indexing stops at the age-60 year and actual earnings from age 60 through one year before the retirement year are used. *The wages earned during the retirement year are not used.*

For a worker who turned 65 in 2002, the age-60 year would be 1997. The National Average Wage Index for 1996 was $27,426.00. As an example for an arbitrary particular year, the National Average Wage Index for 1980 was $12,513.46. Then, the index factor for 1980 would be $\frac{27,426.00}{12,513.46}$, or 2.19 when rounded to the second decimal place. The indexed wage for 1980 would be the *actual* wage amount, up to the FICA wage limit for that year, multiplied by 2.19. That product indicates that, for a person who earned the wage limit of $25,900 in 1980, the indexed earnings would be $56,721. The same procedure is used for all years from 1951, or age 22, whichever is later, through the year prior to the start of Social Security benefits. Table 10-4 shows the index factors and indexed earnings for all workers whose wages were equal to or greater than the FICA wage limit and will have retired in 2002 at age 65, and that the age-60 year is 1997. Table 10-5 shows the index factors and indexed earnings for all workers whose wages were equal to or greater than the FICA wage limit and will have retired in 2002 at age 62, and that the age-60 year is 2000. Table 10-6 shows the index factors and indexed earnings for all workers whose wages were equal to or greater than the FICA wage limit and will have retired in 2002 at age 67, and that the age-60 year, is 1995. The base year for the benefit determination is the age-62 year and the terminating year for retirement in 2002 is 2001. The number of years from 1951 or age 22 to age 62 is 40, or more. The 35 highest indexed earnings years, as shown in Tables 10-4, 10-5, and 10-6, are added and then divided by 420 (35 years × 12 months/year) rounded to the next lower $1.00 to arrive at the number which is referred to as the Average Indexed Monthly Earnings (AIME).

2. The Primary Insurance Amount (PIA)

The AIME does not translate, dollar for dollar, into the Social Security benefit. The Primary Insurance Amount is the base for the determination of the benefit. In order to insure that more of the earnings of lower wage earners are included in the Social Security benefit, the Social Security Administration uses a concept known as "bend points," that are adjusted each year for inflation. The bend points for the years 1979 through 2002 are shown in Table 10-2. The Primary Insurance Amount is determined using the bend points for the age-62 year. Therefore, for a worker who retires in 2002 at age 65, the age-62 year would be 1999, and for the worker who retires in 2002 at age 62, the age-62 year is 2002. The formula for the PIA is as follows:
Using the bend points of the age-62 year,

A = 90% of the Lower Bend Point, and

B = 32% of the (AIME − Lower Bend Point) if the AIME ≤ Upper Bend Point,
or

B = 32% of the Middle Range if the AIME > Upper Bend Point, and

C = 15% of the (AIME − Upper Bend Point).

Then,

PIA = $A + B$ if the AIME ≤ Upper Bend Point, or

PIA = $A + B + C$ if the AIME > Upper Bend Point.

The applicable sum is *rounded down* to the next lower 10 cents to obtain the final base PIA.

3. The Initial Monthly Benefit

The result of Step 2 gives the worker's Primary Insurance Amount. As indicated in Section 10.1, the PIA is reduced for workers who retire before their normal retirement date. There is no reduction for workers who retire during their normal retirement year. After the appropriate reduction, if any, the PIA is increased by cost of living adjustments (COLAs) to the retirement year. Thus, for retirement

TABLE 10-2 SOCIAL SECURITY BENEFIT PARAMETERS								
Bend Points			COLAs		Delayed Credits			
Year of Age 62	Lower Point	Middle Range	Upper Point	Year	Rate	Year of Birth	Year of Age 62	Delayed Credit
	90%	32%	15%					
1979	180	905	1,085	1979	9.90%	1917	1979	3.00%
1980	194	977	1,171	1980	14.30%	1918	1980	3.00%
1981	211	1,063	1,274	1981	11.20%	1919	1981	3.00%
1982	230	1,158	1,388	1982	7.40%	1920	1982	3.00%
1983	254	1,274	1,528	1983	3.50%	1921	1983	3.00%
1984	267	1,345	1,612	1984	3.50%	1922	1984	3.00%
1985	280	1,411	1,691	1985	3.10%	1923	1985	3.00%
1986	297	1,493	1,790	1986	1.30%	1924	1986	3.00%
1987	310	1,556	1,866	1987	4.20%	1925	1987	3.50%
1988	319	1,603	1,922	1988	4.00%	1926	1988	3.50%
1989	339	1,705	2,044	1989	4.70%	1927	1989	4.00%
1990	356	1,789	2,145	1990	5.40%	1928	1990	4.00%
1991	370	1,860	2,230	1991	3.70%	1929	1991	4.50%
1992	387	1,946	2,333	1992	3.00%	1930	1992	4.50%
1993	401	2,019	2,420	1993	2.60%	1931	1993	5.00%
1994	422	2,123	2,545	1994	2.80%	1932	1994	5.00%
1995	426	2,141	2,567	1995	2.60%	1933	1995	5.50%
1996	437	2,198	2,635	1996	2.90%	1934	1996	5.50%
1997	455	2,286	2,741	1997	2.10%	1935	1997	6.00%
1998	477	2,398	2,875	1998	1.30%	1936	1998	6.00%
1999	505	2,538	3,043	1999	2.50%	1937	1999	6.50%
2000	531	2,671	3,202	2000	3.50%	1938	2000	6.50%
2001	561	2,820	3,381	2001	2.60%	1939	2001	7.00%
2002	592	2,975	3,567	2002		1940	2002	7.00%
2003				2003		1941	2003	7.50%
2004				2004		1942	2004	7.50%
2005				2005		1943	2005	8.00%
2006				2006		1944	2006	8.00%

TABLE 10-3 FICA HISTORY

Year	FICA Wage Limit	Social Security Tax Rate	Social Security Tax	National Average Wage Index	Year	FICA Wage Limit	Social Security Tax Rate	Social Security Tax	National Average Wage Index
1951	3,600	1.50%	54.00	2,799.16	1987	43,800	5.70%	2,496.60	18,426.51
1952	3,600	1.50%	54.00	2,973.32	1988	45,000	5.70%	2,565.00	19,334.04
1953	3,600	1.50%	54.00	3,139.44	1989	48,000	6.06%	2,908.80	20,099.55
1954	3,600	2.00%	72.00	3,155.64	1990	51,300	6.06%	3,108.78	21,027.98
1955	4,200	2.00%	84.00	3,301.44	1991	53,400	6.20%	3,310.80	21,811.60
1956	4,200	2.00%	84.00	3,532.36	1992	55,500	6.20%	3,441.00	22,935.42
1957	4,200	2.25%	94.50	3,641.72	1993	57,600	6.20%	3,571.20	23,132.67
1958	4,200	2.25%	94.50	3,673.80	1994	60,600	6.20%	3,757.20	23,753.53
1959	4,800	2.50%	120.00	3,855.80	1995	61,200	6.20%	3,794.40	24,705.66
1960	4,800	3.00%	144.00	4,007.12	1996	62,700	6.20%	3,887.40	25,913.90
1961	4,800	3.00%	144.00	4,086.76	1997	65,400	6.20%	4,054.80	27,426.00
1962	4,800	3.13%	150.00	4,291.40	1998	68,400	6.20%	4,240.80	28,861.44
1963	4,800	3.63%	174.00	4,396.64	1999	72,600	6.20%	4,501.20	30,469.84
1964	4,800	3.63%	174.00	4,576.32	2000	76,200	6.20%	4,502.20	32,154.82
1965	4,800	3.63%	174.00	4,658.72	2001	80,400	6.20%	4,984.80	N/A
1966	6,600	3.90%	257.40	4,938.36	2002	84,900	6.20%	5,263.80	N/A
1967	6,600	3.90%	257.40	5,213.44					
1968	7,800	3.90%	304.20	5,571.76					
1969	7,800	4.20%	327.60	5,893.76					
1970	7,800	4.20%	327.60	6,186.24					
1971	7,800	4.60%	358.80	6,497.08					
1972	9,000	4.60%	414.00	7,133.80					
1973	10,800	4.85%	523.80	7,580.16					
1974	13,200	4.95%	653.40	8,030.76					
1975	14,100	4.95%	697.95	8,630.92					
1976	15,300	4.95%	757.35	9,226.48					
1977	16,500	4.95%	816.75	9,779.44					
1978	17,700	4.95%	876.15	10,556.03					

TABLE 10-3 *continued*

Year	FICA Wage Limit	Social Security Tax Rate	Social Security Tax	National Average Wage Index	Year	FICA Wage Limit	Social Security Tax Rate	Social Security Tax	National Average Wage Index
1979	22,900	5.05%	1,156.45	11,479.46					
1980	25,900	5.08%	1,315.72	12,513.46					
1981	29,700	5.08%	1,508.76	13,773.10					
1982	32,400	5.35%	1,733.40	14,531.34					
1983	35,700	5.40%	1,927.80	15,239.24					
1984	37,800	5.40%	2,041.20	16,135.07					
1985	39,600	5.40%	2,138.40	16,822.51					
1986	42,000	5.70%	2,394.00	17,321.82					

in 2002 at age 65, the COLAs are for the years of age 63 (2000), age 64 (2001), and age 65 (2002). Since the fiscal year of the U.S. Government is from October 1 though September 30, the COLAs would have been announced in October of the years 1999, 2000, and 2001, respectively, and each would take effect on January 1 of 2000, 2001, and 2002 respectively. Cost of living adjustment rates are given in Table 10-2 for the years in which they are announced. Each cost of living rate is applicable for the following year. The accumulated amount for each year is also *rounded down* to the next lower 10 cents and is used as the basis for the following year. The final accumulation is *rounded down* to the next lower $1.00. This final amount is the retiree's basic Social Security benefit. The final rounding takes place after the deduction of the premium, if any, for Medicare Part-B. We will not consider that aspect in this writing.

EXAMPLE 10.2.1   Normal Retirement in 2002 at Age 65

A worker will retire in 2002 at age 65, and the worker's wages have been at least the FICA wage limit from 1958 through 2001. The total of the highest 35 years of indexed earnings through 2001 is $2,003,775 as shown in Table 10-4. Determine the initial monthly benefit.

Solution   The AIME for all workers whose wages have been equal to at least the FICA wage limit and who retire in 2002 will be $4,770 $\left(\frac{2,003,775}{420}\right)$. For age 65 in 2002, the age-62 year is 1999, and the bend points for that year are obtained from Table 10-2 as 505 for the lower bend point, 2,538 for the middle range, and 3,043 for the upper bend point. Since the AIME is greater than the upper bend point, the PIA is determined using as follows:

$A = 90\%$ of $505 = 454.50$,

$B = 32\%$ of $2,538 = 812.16$, and

$C = 15\%$ of $(4,770 - 3,043) = 259.05$.

		FICA Wage Limit	Assumed Actual Wages	National Average Wages	Index Factor	Indexed Wages	Increasing Indexed Wages	
TABLE 10-4 INDEXED WAGES: RETIREMENT AT AGE 65								
Year	Age							
1959	22	4,800	4,800.00	3,855.80	7.11	34,143.00	28,258.00	
1960	23	4,800	4,800.00	4,007.12	6.84	32,852.00	28,767.00	
1961	24	4,800	4,800.00	4,086.76	6.71	32,213.00	29,943.00	
1962	25	4,800	4,800.00	4,291.40	6.39	30,677.00	30,677.00	
1963	26	4,800	4,800.00	4,396.64	6.24	29,943.00	32,213.00	
1964	27	4,800	4,800.00	4,576.32	5.99	28,767.00	32,852.00	
1965	28	4,800	4,800.00	4,658.72	5.89	28,258.00	32,924.00	
1966	29	6,600	6,600.00	4,938.36	5.55	36,657.00	34,143.00	
1967	30	6,600	6,600.00	5,213.44	5.26	34,723.00	34,578.00	1
1968	31	7,800	7,800.00	5,571.76	4.92	38,392.00	34,605.00	2
1969	32	7,800	7,800.00	5,893.76	4.65	36,294.00	34,723.00	3
1970	33	7,800	7,800.00	6,186.24	4.43	34,578.00	36,294.00	4
1971	34	7,800	7,800.00	6,497.08	4.22	32,924.00	36,657.00	5
1972	35	9,000	9,000.00	7,133.80	3.85	34,605.00	38,392.00	6
1973	36	10,800	10,800.00	7,580.16	3.62	39,075.00	39,075.00	7
1974	37	13,200	13,200.00	8,030.76	3.42	45,078.00	44,810.00	8
1975	38	14,100	14,100.00	8,630.92	3.18	44,810.00	45,078.00	9
1976	39	15,300	15,300.00	9,226.48	2.97	45,487.00	45,487.00	10
1977	40	16,500	16,500.00	9,779.44	2.80	46,266.00	45,985.00	11
1978	41	17,700	17,700.00	10,556.03	2.60	45,985.00	46,266.00	12
1979	42	22,900	22,900.00	11,479.46	2.39	54,709.00	54,709.00	13
1980	43	25,900	25,900.00	12,513.46	2.19	56,773.00	56,773.00	14

TABLE 10-4 *continued*

Year	Age	FICA Wage Limit	Assumed Actual Wages	National Average Wages	Index Factor	Indexed Wages	Increasing Indexed Wages	
1981	44	29,700	29,700.00	13,773.10	1.99	59,133.00	59,133.00	15
1982	45	32,400	32,400.00	14,531.34	1.89	61,139.00	61,139.00	16
1983	46	35,700	35,700.00	15,239.24	1.80	64,260.00	63,855.00	17
1984	47	37,800	37,800.00	16,135.07	1.70	64,260.00	64,260.00	18
1985	48	39,600	39,600.00	16,822.51	1.63	64,548.00	64,260.00	19
1986	49	42,000	42,000.00	17,321.82	1.58	66,486.00	64,548.00	20
1987	50	43,800	43,800.00	18,426.51	1.49	65,175.00	65,175.00	21
1988	51	45,000	45,000.00	19,334.04	1.42	63,855.00	65,400.00	22
1989	52	48,000	48,000.00	20,099.55	1.37	65,520.00	65,520.00	23
1990	53	51,300	51,300.00	21,027.98	1.30	66,896.00	66,337.00	24
1991	54	53,400	53,400.00	21,811.60	1.26	67,124.00	66,378.00	25
1992	55	55,500	55,500.00	22,935.42	1.20	66,378.00	66,486.00	26
1993	56	57,600	57,600.00	23,132.67	1.19	68,314.00	66,896.00	27
1994	57	60,600	60,600.00	23,753.53	1.16	69,993.00	67,124.00	28
1995	58	61,200	61,200.00	24,705.66	1.11	67,932.00	67,932.00	29
1996	59	62,700	62,700.00	25,913.90	1.06	66,337.00	68,314.00	30
1997	60	65,400	65,400.00	27,426.00	1.00	65,400.00	68,400.00	31
1998	61	68,400	68,400.00	28,861.44	1.00	68,400.00	69,993.00	32
1999	62	72,600	72,600.00	30,469.84	1.00	72,600.00	72,600.00	33
2000	63	76,200	76,200.00	32,154.82	1.00	76,200.00	76,200.00	34
2001	64	80,400	80,400.00	N/A	1.00	80,400.00	80,400.00	35

Sum of 35 Greatest Indexed Earnings = 2,003,782

AIME =   4,770

Year	Age	FICA Wage Limit	Assumed Actual Wages	National Average Wages	Index Factor	Indexed Wages	Increasing Indexed Wages	
1962	22	4,800	4,800	4,291.40	7.49	35,965.68	33,129.94	
1963	23	4,800	4,800	4,396.64	7.31	35,104.79	33,726.47	
1964	24	4,800	4,800	4,576.32	7.03	33,726.47	35,104.79	
1965	25	4,800	4,800	4,658.72	6.90	33,129.94	35,965.68	
1966	26	6,600	6,600	4,938.36	6.51	42,974.15	38,603.13	
1967	27	6,600	6,600	5,213.44	6.17	40,706.68	40,542.82	1
1968	28	7,800	7,800	5,571.76	5.77	45,014.07	40,566.51	2
1969	29	7,800	7,800	5,893.76	5.46	42,554.77	40,706.68	3
1970	30	7,800	7,800	6,186.24	5.20	40,542.82	42,554.77	4
1971	31	7,800	7,800	6,497.08	4.95	38,603.13	42,974.15	5
1972	32	9,000	9,000	7,133.80	4.51	40,566.51	45,014.07	6
1973	33	10,800	10,800	7,580.16	4.24	45,813.29	45,813.29	7
1974	34	13,200	13,200	8,030.76	4.00	52,852.24	52,530.09	8
1975	35	14,100	14,100	8,630.92	3.73	52,530.09	52,852.24	9
1976	36	15,300	15,300	9,226.48	3.49	53,321.39	53,321.39	10
1977	37	16,500	16,500	9,779.44	3.29	54,252.04	53,916.13	11
1978	38	17,700	17,700	10,556.03	3.05	53,916.13	54,252.04	12
1979	39	22,900	22,900	11,479.46	2.80	64,144.60	64,144.60	13
1980	40	25,900	25,900	12,513.46	2.57	66,553.12	66,553.12	14
1981	41	29,700	29,700	13,773.10	2.33	69,337.92	69,337.92	15
1982	42	32,400	32,400	14,531.34	2.21	71,694.43	71,694.43	16
1983	43	35,700	35,700	15,239.24	2.11	75,327.06	74,840.38	17

TABLE 10-5 INDEXED WAGES: RETIREMENT AT AGE 62

TABLE 10-5 *continued*

Year	Age	FICA Wage Limit	Assumed Actual Wages	National Average Wages	Index Factor	Indexed Wages	Increasing Indexed Wages	
1984	44	37,800	37,800	16,135.07	1.99	75,329.84	75,327.06	18
1985	45	39,600	39,600	16,822.51	1.91	75,692.09	75,329.84	19
1986	46	42,000	42,000	17,321.82	1.86	77,965.39	75,692.09	20
1987	47	43,800	43,800	18,426.51	1.75	76,432.33	76,200.00	21
1988	48	45,000	45,000	19,334.04	1.66	74,840.38	76,205.13	22
1989	49	48,000	48,000	20,099.55	1.60	76,789.35	76,432.33	23
1990	50	51,300	51,300	21,027.98	1.53	78,445.11	76,614.77	24
1991	51	53,400	53,400	21,811.60	1.47	78,722.67	76,676.34	25
1992	52	55,500	55,500	22,935.42	1.40	77,809.45	76,789.35	26
1993	53	57,600	57,600	23,132.67	1.39	80,065.02	77,800.22	27
1994	54	60,600	60,600	23,753.53	1.35	82,033.37	77,809.45	28
1995	55	61,200	61,200	24,705.66	1.30	79,652.80	77,965.39	29
1996	56	62,700	62,700	25,913.90	1.24	77,800.22	78,445.11	30
1997	57	65,400	65,400	27,426.00	1.17	76,676.34	78,722.67	31
1998	58	68,400	68,400	28,861.44	1.11	76,205.13	79,652.80	32
1999	59	72,600	72,600	30,469.84	1.06	76,614.77	80,065.02	33
2000	60	76,200	76,200	32,154.82	1.00	76,200.00	80,400.00	34
2001	61	80,400	80,400	N/A	1.00	80,400.00	82,033.37	35

Sum of 35 Greatest Indexed Earnings = 2,309,775
AIME =   5,499

TABLE 10-6 INDEXED WAGES: RETIREMENT AT AGE 67

Year	Age	FICA Wage Limit	Assumed Actual Wages	National Average Wages	Index Factor	Indexed Wages	Increasing Indexed Wages	
1957	22	4,200	4,200	3,641.72	6.78	28,493.00	25,455.00	
1958	23	4,200	4,200	3,673.80	6.73	28,245.00	25,916.00	
1959	24	4,800	4,800	3,855.80	6.41	30,754.00	26,972.00	
1960	25	4,800	4,800	4,007.12	6.17	29,592.00	27,634.00	
1961	26	4,800	4,800	4,086.76	6.05	29,016.00	28,245.00	
1962	27	4,800	4,800	4,291.40	5.76	27,634.00	28,493.00	
1963	28	4,800	4,800	4,396.64	5.62	26,972.00	29,016.00	
1964	29	4,800	4,800	4,576.32	5.40	25,916.00	29,592.00	
1965	30	4,800	4,800	4,658.72	5.30	25,455.00	29,664.00	
1966	31	6,600	6,600	4,938.36	5.00	33,020.00	30,754.00	
1967	32	6,600	6,600	5,213.44	4.74	31,278.00	31,154.00	1
1968	33	7,800	7,800	5,571.76	4.43	34,586.00	31,167.00	2
1969	34	7,800	7,800	5,893.76	4.19	32,698.00	31,278.00	3
1970	35	7,800	7,800	6,186.24	3.99	31,154.00	32,698.00	4
1971	36	7,800	7,800	6,497.08	3.80	29,664.00	33,020.00	5
1972	37	9,000	9,000	7,133.80	3.46	31,167.00	34,586.00	6
1973	38	10,800	10,800	7,580.16	3.26	35,198.00	35,198.00	7
1974	39	13,200	13,200	8,030.76	3.08	40,604.00	40,355.00	8
1975	40	14,100	14,100	8,630.92	2.86	40,355.00	40,604.00	9
1976	41	15,300	15,300	9,226.48	2.68	40,974.00	40,974.00	10
1977	42	16,500	16,500	9,779.44	2.53	41,679.00	41,418.00	11
1978	43	17,700	17,700	10,556.03	2.34	41,418.00	41,679.00	12

TABLE 10-6 *continued*

Year	Age	FICA Wage Limit	Assumed Actual Wages	National Average Wages	Index Factor	Indexed Wages	Increasing Indexed Wages	
1979	44	22,900	22,900	11,479.46	2.15	49,281.00	49,281.00	13
1980	45	25,900	25,900	12,513.46	1.97	51,127.00	51,127.00	14
1981	46	29,700	29,700	13,773.10	1.79	53,282.00	53,282.00	15
1982	47	32,400	32,400	14,531.34	1.70	55,080.00	55,080.00	16
1983	48	35,700	35,700	15,239.24	1.62	57,870.00	57,510.00	17
1984	49	37,800	37,800	16,135.07	1.53	57,872.00	57,870.00	18
1985	50	39,600	39,600	16,822.51	1.47	58,173.00	57,872.00	19
1986	51	42,000	42,000	17,321.82	1.43	59,892.00	58,173.00	20
1987	52	43,800	43,800	18,426.51	1.34	58,736.00	58,736.00	21
1988	53	45,000	45,000	19,334.04	1.28	57,510.00	58,992.00	22
1989	54	48,000	48,000	20,099.55	1.23	58,992.00	59,774.00	23
1990	55	51,300	51,300	21,027.98	1.18	60,278.00	59,892.00	24
1991	56	53,400	53,400	21,811.60	1.13	60,503.00	60,278.00	25
1992	57	55,500	55,500	22,935.42	1.08	59,774.00	60,503.00	26
1993	58	57,600	57,600	23,132.67	1.07	61,517.00	61,200.00	27
1994	59	60,600	60,600	23,753.53	1.04	63,024.00	61,517.00	28
1995	60	61,200	61,200	24,705.66	1.00	61,200.00	62,700.00	29
1996	61	62,700	62,700	25,913.90	1.00	62,700.00	63,024.00	30
1997	62	65,400	65,400	27,426.00	1.00	65,400.00	65,400.00	31
1998	63	68,400	68,400	28,861.44	1.00	68,400.00	68,400.00	32
1999	64	72,600	72,600	30,469.84	1.00	72,600.00	72,600.00	33
2000	65	76,200	76,200	32,154.82	1.00	76,200.00	76,200.00	34
2001	66	80,400	80,400	N/A	1.00	80,400.00	80,400.00	35

Sum of 35 Greatest Indexed Earnings = 1,843,942

AIME =   4,390

The sum $A + B + C = 1,525.71$ and is rounded down to the next lower 10 cents, or $1,525.70 for the PIA. The applicable COLA rates may be obtained from Table 10-2 as 2.5% for the age-63 year (2000), 3.5% for the age-64 year (2001), and 2.6% for the age-65 year (2002). The accumulation to arrive at the monthly benefit is as follows:

$(1,525.70)(1.025) = 1,563.84$ rounded down to 1,563.80.

$(1,563.80)(1.035) = 1,618.58$ rounded down to 1,618.50.

$(1,618.50)(1.026) = 1,660.66$ rounded down to 1,660.00.

Then, for normal retirement in 2002 at age 65, a worker who has earned at least the FICA wage limit from 1958 through 2001 will receive an initial monthly benefit of $1,660 per month. Also, this is the maximum possible benefit available for retirement in 2002 at the normal retirement age of 65. ■

EXAMPLE 10.2.2   Early Retirement in 2002 at Age 62
    A worker will retire in 2002 at age 62, and the worker's wages have been at least the FICA wage limit from 1962 through 2001. The total of the 35 years of highest indexed earnings through 2001 is $2,309,776 as shown in Table 10-5. Determine the initial monthly benefit.

Solution   Age 62 is the first eligible year for Social Security retirement. The AIME for all workers whose wages have been equal to at least the FICA wage limit and who retire at age 62 in 2002 will be $5,499 $\left(\frac{2,309,775}{420}\right)$. The PIA would be determined using the bend points of 2002, the age-62 year, and are given in Table 10-2 as 592 for the lower bend point, 2,975 for the middle range, and 3,567 for the upper bend point. Since the AIME is greater than the upper bend point, the PIA is determined as follows:

$A = 90\%$ of $592 = 532.80$,

$B = 32\%$ of $2,975 = 952.00$, and

$C = 15\%$ of $(5,499 - 3,567) = 289.80$.

The sum of $A + B + C = 1,774.60$ and is "rounded down to $1,774.60" for the base PIA. Since this worker will retire before the normal retirement year, a permanent reduction in the base PIA is imposed. For age 62 in 2002, the birth year was 1940, and the normal retirement age is 65 years, six months. The reduction formula for the normal retirement age is given in Section 10.1 as Equation (10.1-1). The number of months to the normal retirement age is 786 ($12 \times 65 + 6$), and the number of months to the beginning of benefits is 745 ($12 \times 62 + 1$). The difference is 41 months. Thus, if a worker retires in 2002 at age 62, Equation (10.1-2b) shows that the base *PIA* will be reduced by

$$R_w = 20\% + \frac{5}{12}(786 - 745 - 36) = 22.1\%.$$

and the final PIA will be 77.9% of the base PIA. Then, this worker's PIA will be reduced to

$$PIA = (1 - 0.221)(1,774.60) = 1,382.41,$$

which is rounded to 1,382.40. Since this worker is 62 in 2002, the first COLA would be implemented in January 2003; therefore, there will not be a COLA adjustment. The initial monthly benefit for retirement in 2002 at age 62 will be $1,382, after dropping the cents. ■

DELAYED RETIREMENT

If a worker delays retirement benefit beyond the normal retirement date, he or she will have foregone earned benefits. For the normal retirement age of 65 in 2002, the loss of benefits can be as much as $1,660 per month (the maximum possible benefit for retirement in 2002 at age 65). In order to compensate for this loss of benefit, the law mandates a delayed retirement credit (DRC) to be included in the calculation of benefits when retirement does occur. The rates of the delayed credits are shown in Table 10-2, and the worker's PIA would be increased by a simple interest calculation for the number years and months that the retirement was delayed as follows:

$$B_d = \left[1 + \left(\frac{DRC}{12} \right) (M_d - M_n) \right] PIA_n \qquad (10.2\text{-}1)$$

where,

PIA_n is the PIA for the normal retirement age,

DRC is the Delayed Retirement Credit,

M_d is the delayed retirement age in the number of whole months,

M_n is the normal retirement age in the number of whole months, and

B_d is the delayed benefit.

Social Security benefits will begin automatically at age 70; therefore, the maximum delay in retirement is five years for workers born in 1937 who would be 65 in 2002. Then, for a person who is age 65 in 2002 and does not collect benefits until age 70, Table 10-2 shows that the DRC for age 62 in 1999 is 6.5%. The maximum delayed credit will be 32.5% ($\frac{6.5\%}{12}$ for each month of delay times 60 months). After applying COLAs to the PIA for the years from age 62 through the delayed retirement year in the same manner as in the determination of early and normal retirement benefits the resulting benefit is increased by the appropriate DRC. Recall that each accumulation is rounded down to the next lower 10 cents, and this value is used as the base for the next accumulation. As before, the final accumulation is rounded down to the next lower $1.00. Further, the worker's PIA may also increase in that for each year of delayed benefits a high earnings year may replace a low earnings year in the 35 indexed earnings years that are used to determine the AIME.

EXAMPLE 10.2.3   Delayed Benefits to 2002 at Age 67

A worker who was born in January 1935 began to collect benefits in January 2002 at age 67. Since age 22, this worker has always earned wages that were at least the FICA wage limit each year. Determine the initial monthly benefit.

Solution   Because the worker did not begin to collect benefits until after the normal retirement age of 65 for workers born before 1938, the AIME will be determined by using the 35 highest indexed wages from age 22 through age 66. Referring to Table 10-6, the sum of 35 highest indexed wages is $1,843,942 giving an AIME of $4,390. The age-62 year was 1997 and from Table 10-2, the bend points are 455 for the lower, 2,286 for the middle range, and 2,741 for the upper. Since the AIME is greater than the upper bend point, the base PIA is determined as follows.

$A = 90\%$ of $455 = 409.50$,

$B = 32\%$ of $2,286 = 731.52$, and

$C = 15\%$ of $(4,390 - 2,741) = 247.35$.

The base PIA is the sum of $A + B + C$ and equals 1,388.37 and is rounded down to 1,388.30. The COLAs from age 62 will be used to increase this base PIA as follows. Table 10-2 show these to be 2.1% for the age-63 year (1998), 1.3% for the age-64 year (1999), and 2.5% for the age-65 year (2000), 3.5% for the age-66 year (2001), and 2.6% for the age-67 year, 2002. The accumulation to arrive at the monthly benefit is as follows:

$(1,388.30)(1.021) = 1,471.45$ rounded down to 1,417.40.

$(1,417.40)(1.013) = 1,435.88$ rounded down to 1,435.80.

$(1,435.80)(1.025) = 1,471.70$ rounded down to 1,471.70.

$(1,471.70)(1.035) = 1,523.21$ rounded down to 1,523.20.

$(1,523.20)(1.026) = 1,562.80$.

For a worker whose age-62 year was 1997, the delayed retirement credit is 6.0%. The number of months to the delayed benefit age 67 is 804, and the number of months at the normal retirement age 65 is 780. Since the delay is 24 months, Equation (10.2-1) gives the delayed benefit as

$$B_d = \left[1 + \left(\frac{0.06}{12}\right)(24)\right](1,562.80) = 1,750.34.$$

As always, the final value is rounded down to the next lower $1.00, and the worker's monthly benefit will be $1,750. ■

A note of caution is necessary. The Delayed Retirement Credit calculation is based on calendar years and is applied through December of the year preceding the year of the delayed retirement. Had the worker of Example 10.2.3 been born in August, the number of months of delayed credit would have been 17 (five for the year of age 65 and 12 for the year of age 66). The delayed benefit, beginning in August 2001 would be $1,695. However, the delayed credit of the seven months of the age-67 year (2002) would be credited at the end of 2002.

SPOUSAL BENEFITS

Every worker who is covered by Social Security is entitled to his or her own benefit based on indexed earnings as shown above. However, for married couples, the spouse of an eligible worker is entitled to a "spousal benefit" whether or not the spouse had worked under covered employment. If the spouse had worked in employment that is covered by Social Security and that retirement benefit is greater than the spousal benefit, the spouse will receive the greater benefit. The formulas for spousal benefits are given in Section 10.1 as Equations (10.1-3), (10.1-4a), and (10.1-4b). The PIA for the spousal benefit is based on the age of the spouse when the worker retires. The maximum spousal benefit will be received if the "nonworking" spouse is at his or her normal retirement age when the worker retires, regardless of the worker's age. Reduced benefits are given to spouses from age 62 through 64. The maximum spousal PIA is 50% of the workers' PIA when the worker retires as indicated in Section 10.1. That 50% maximum is multiplied by the complement of the results of the reduction formulas, Equations (10.1-3), (10.1-4a), or (10.1-4b).

EXAMPLE 10.2.4   A worker has a PIA of $1,525.70 for retirement at age 65 in 2002. Determine the spousal benefit if the spouse is a) age 65 and b) age 62.

Solution

a) For a spouse who is age 65 at the retirement of the worker in 2002, the birth year would be 1936, and the normal retirement age is obtained from Table 10-1 as 65. Then, the benefit reduction is 0%, and the spousal PIA equals 50% of the worker's PIA or $762.80 after rounding down to the next lower 10 cents. The spouse's age-62 year is 1998, the same as the worker's. As with the worker, the applicable COLA rates may be obtained from Table 10-2 as 2.5%, for the age-63 year (2000), 3.5%, for the age-64 year (2001), and 2.6% for the age-65 year (2002). The accumulation to arrive at the monthly benefit is as follows:

$(762.80)(1.025) = 781.87$ rounded down to 781.80.
$(781.80)(1.035) = 809.16$ rounded down to 809.10.
$(809.10)(1.026) = 830.14$ rounded down to 830.00.

The spouse's initial monthly benefit, then, is $830.

b) If the spouse is age 62 at the retirement of the worker in 2002, the birth year would be 1940, and the normal retirement age will be 65 years plus 6 months. The number of months at the normal retirement age is 786 and the number of months to the beginning of benefits 745 ($12 \times 62 + 1$). Since the difference in months is 41 and is greater than 36, Equation (10.1-4b) is used to determine the PIA reduction. Then,

$$R_S = 25\% + \frac{5}{12}(786 - 745 - 36) = 27.1\%.$$

From Equation (10.1-5),
$$PIA_S = (1 - 0.271)(0.50)PIA_w = 0.3645 PIA_w$$

or 556.12, which is reduced to 556.10. Since the spouse's age-62 year is 2002, there will not be any cost of living adjustment until January 2003 and the spousal benefit is the PIA rounded down to the next lower dollar, $556. ∎

EARNINGS LIMITS AFTER RETIREMENT

Many people continue to work in other environments after retiring from a primary employer. For such workers the annual amounts that they can earn from FICA wage sources are limited. A reduction of the monthly benefit amount during the following year is imposed when earnings exceed the annual limits in a current year. The employer apprises the Social Security Administration of the worker's earnings when the worker's W-2 is filed with the IRS at the end of the year, or by the contractor when a Form 1099 is filed with the IRS at the end of the year. Thus, if a worker expects to earn an amount that is significantly greater than the earnings limit for a given year, it would be prudent to notify the Social Security Administration. The appropriate reductions could be taken in the given year.

For workers who retire between ages 62 and less than 65 the limit is $11,280 ($940 per month) in 2002 and will be indexed with inflation. The reduction that is imposed is $1.00 for every $2.00 for any month in which FICA earnings are in excess of that limit. In general, the penalty may be determined as follows. Let:

P = the penalty,
E = the earnings in the given year,
L = the earnings limit for the given year, and
B = the annual benefit.

Then,

$$P = \frac{1}{2}(E - L). \quad 62 \leq Age < 65 \tag{10.2-2}$$

The maximum earnings is the amount at which the penalty would be equal to the benefit and may be determined from

$$E_{\text{Max}} = 2B + L. \quad 62 \leq Age < 65 \tag{10.2-3}$$

where, E_{Max} is the earnings for a given year that results in $0.00 in benefits, the maximum reduction.

EXAMPLE 10.2.5 A worker retired in 2002 at age 62 and will receive a Social Security benefit, B, of $1,382 per month. The worker continued to work and earned $12,000 in year 2002.

(a) Determine the penalty reduction for this worker in 2003.

(b) Determine the amount of earned income that this worker must earn in 2002 such that the entire Social Security benefit would be forfeited in 2003 if the Social Security Administration were not notified in 2002.

Solution

a) Since the penalty-free amount of earnings in year 2002 is $11,280 or $940 per month, the reduction in the Social Security benefit can be determined from Equation (10.2-2) as

$$P = \frac{1}{2}(12,000 - 11,280) = 360.$$

Thus, if this worker earns $12,000 in 2002, the Social Security payments in 2003 will be reduced by $360 or $30 per month.

b) Equation (10.2-3) indicates that the maximum penalty will occur if the earnings are $44,448. Thus, if a worker will earn $44,448 in year 2002, the benefit for 2003 will be reduced by $16,584. ∎

The Social Security Administration will arrange for a reasonable reduction of the monthly benefit during the following year in order to avoid imposing a hardship on the retiree, if the earnings limit is exceeded in a given year.

The "Senior Citizens Right to Work Act of 2000" removed earnings limits for those workers who receive benefits between ages 65 and 70. After age 65 there is no penalty reduction. Stated positively, at ages 65 and older, a retiree may earn as much as he or she is able without a reduction in the Social Security benefit. However, in the year in which the worker will reach the normal retirement age, a limit of $30,000 ($2,500 per month) for 2002 is imposed from the beginning of the year to the month prior to age 65, if the individual had already been receiving Social Security benefits due to early retirement. Should monthly income exceed this amount, a reduction of $1.00 for every $3.00 above this monthly amount is imposed. Then, in that year the reduction may be determined by

$$P = \frac{1}{3}(E - L). \tag{10.2-4}$$

The maximum amount that may be earned in the year of the normal retirement age without forfeiting the entire benefits that are being received for that year may be determined by

$$E_{\text{Max}} = 3B + L. \quad 64 < \text{Age} < 65 \tag{10.2-5}$$

Consider a retired worker that is to receive $1,500 per month in 2002 due to retirement prior to the normal retirement age. Suppose that person will reach the normal retirement age in October 2002, but returns to work in January 2002 and earns $42,000 between January and September. Then, the benefits for January through September of 2002 would be reduced according to Equation (10.2-4) by,

$$P = \frac{1}{3}(42,000 - 30,000) = \$4,000.$$

This aspect of the law places those individuals who turn 65 late in the year at a disadvantage relative to those reaching it early in the year. The person who turns 65 in January could begin collecting benefits without the effect of the earnings limit. The person who turns 65 in December could lose prior benefits through November. Perhaps the law will be amended to eliminate this bias. Between ages 65 and 70, the individual may forego his or her benefit, and for each month that a benefit is not received, the delayed retirement credit will be applied.

OTHER BENEFITS

Although not essential for the purposes of this text, there are other benefits in the Social Security System of the United States. These benefits include payments to disabled persons, to widows and/or widowers, to divorced spouses, to dependent children under 18, and to disabled dependent children at any age. The Social Security Administration provides several booklets that explain these benefits free of charge. Some of these are SSA publication numbers. 05-10024, 05-10035, 05, 10069, and 05-10077.

LIMITATIONS ON SOCIAL SECURITY BENEFITS

One of the limitations on Social Security benefits is that if a worker is entitled to both Social Security and a public pension, such as Federal or State Civil Service, the Social Security benefit is reduced. That aspect is beyond the needs of this text. However, private pensions do not affect Social Security benefits, and elements of private pensions are discussed in Chapter 11.

PROBLEM SET 10.2

1. A person, born in 1940, retires in 2002 at age 62. The AIME through 2001 is 4,145. Determine the PIA and initial monthly benefit. These would be "high" wages.

2. A person, born in 1939, retires in 2002 at age 62. The AIME through 2001 is 2,683. Determine the PIA and initial monthly benefit. These would be "average" wages.

3. A person, born in 1939, retires in 2002 at age 62. The AIME through 2001 is 1,207. Determine the PIA and initial monthly benefit. These would be "low" wages.

4. A person, born in 1938, retires in 2002 at age 63. The AIME through 2001 is 5,230. Determine the PIA and initial monthly benefit.

5. A person born in 1938, retires in 2002 at age 64. The AIME through 2001 is 4,983. Determine the PIA and initial monthly benefit.

6. A person, born in 1937, retires in January 2002 at age 65. The AIME through 2001 is 2,322. Determine the PIA and initial monthly benefit. These would be "average" wages.

7. A person, born in 1936, retires in January 2002 at age 65. The AIME through 2001 is 3,590. Determine the PIA and initial monthly benefit. These would be "high" wages.

8. A person, born in 1935, retires in January 2002 at age 66. The AIME through 2001 is 4,555. Determine the PIA and initial monthly benefit. These are "maximum" wages.

9. A person, born in 1934, retires in January 2002 at age 67. The AIME through 2001 is 4,390. Determine the PIA and initial monthly benefit. These are "maximum" wages.

10. A person, born in 1933, retires in January 2002 at age 68. The AIME through 2001 is 4,265. Determine the PIA and initial monthly benefit. These are "maximum" wages.

11. A person, born in 1932, retires in January 2002 at age 69. The AIME through 2001 is 4,189. Determine the PIA and initial monthly benefit. These are "maximum" wages.

12. A person, born in 1932, retires in January 2002 at age 70. The AIME through 2001 is 4,165. Determine the PIA and initial monthly benefit. These are "maximum" wages.

In Problems 13–24 determine the initial monthly benefit of the spouse.

13. The spouse of the worker in Problem 1 was born in 1940 and is 62.

14. The spouse of the worker in Problem 2 was born in 1940 and is 62.

15. The spouse of the worker in Problem 3 was born in 1940 and is 62.

16. The spouse of the worker in Problem 4 was born in 1939 and is 63.

17. The spouse of the worker in Problem 5 was born in 1938 and is 64.

18. The spouse of the worker in Problem 6 was born in 1937 and is 65.

19. The spouse of the worker in Problem 7 was born in 1937 and is 65.

20. The spouse of the worker in Problem 8 was born in 1936 and is 66.

21. The spouse of the worker in Problem 9 was born in 1935 and is 67.

22. The spouse of the worker in Problem 9 was born in 1934 and is 68.

23. The spouse of the worker in Problem 9 was born in 1933 and is 69.

24. The spouse of the worker in Problem 9 was born in 1935 and is 70.

11

Private Pensions

11.1 A BRIEF HISTORY OF PRIVATE PENSIONS

Note: Much of the following was extracted from the U.S. Department of Labor at its Web site address, *http://www.dol.gov/dol/pwba/public/aboutpwba/ history4.htm.*

THE EARLY YEARS

The first private pension plans in the United States were developed in the late 1800s and early 1900s. Initially, the Internal Revenue Service (IRS) was the primary regulator of those plans. The Revenue Acts of 1921 and 1926 allowed employers to deduct pension plan contributions from corporate income and allowed for the income of the pension fund's portfolio to accumulate tax-free. The participants in the plan did not realize an income until monies were distributed to them, provided the plan was "tax qualified." To qualify for such favorable tax treatment, the plans had to meet certain minimum employee coverage and employer contribution requirements. The Revenue Act of 1942 provided stricter participation requirements and, for the first time, disclosure requirements. The Department of Labor became involved in the regulation of employee benefit plans upon passage of the Welfare and Pension Plans Disclosure Act in 1959 (WPPDA). Plan sponsors (e.g., employers and labor unions) were required to file plan descriptions and annual financial reports with the government; these materials were also to be made available to plan participants and beneficiaries. This legislation was intended to provide employees with enough information regarding plans so that they could monitor their plans to prevent mismanagement and abuse of plan funds. The WPPDA was amended in 1962, at which time the secretary of labor was given enforcement, interpretative, and investigative powers over employee benefit plans to prevent mismanagement and abuse of plan funds. In general, the WPPDA had a very limited scope, and many workers were left without pensions upon retirement, despite the precautions of the WPPDA.

EMPLOYEE RETIREMENT INCOME SECURITY ACT (ERISA)

Because of widespread losses of pensions, Congress enacted the Employee Retirement Income Security Act in 1974. The administration of ERISA is divided among the Department of Labor (DOL), the Internal Revenue Service of the Department of the Treasury, and the Pension Benefit Guaranty Corporation (PBGC) under four Titles (Chapters).

Title I: Protection of Employee Benefits

Title II: Types of Asset Accumulations

Title III: Jurisdictional Coordination between the DOL and IRS

Title IV: Pension Plan Benefit Insurance for Defined Benefit Plans

ERISA was the culmination of a long line of legislation concerned with the labor and tax aspects of employee benefit plans. Since its enactment in 1974, ERISA has been amended to meet the changing retirement and health care needs of employees and their families. The role of the Pension and Welfare Benefit Administration (PWBA) has also evolved to meet these challenges. The provisions of Title I of ERISA, which are administered by the Department of Labor, were enacted to address public concern that funds of private pension plans were being mismanaged and abused.

Prior to reorganization in 1978, there was overlapping responsibility for administration of the parallel provisions of Title I of ERISA and the tax code by the labor department and the IRS, respectively. As a result of this reorganization, the labor department has primary responsibility for reporting, disclosure, and fiduciary requirements; and the IRS has primary responsibility for participation, vesting, and funding issues. However, the labor department may intervene in any matters that materially affect the rights of participants, regardless of primary responsibility.

The goal of Title I of ERISA is to protect the interests of participants and their beneficiaries in employee benefit plans. Among other things, ERISA requires that sponsors of private employee benefit plans provide participants and beneficiaries with adequate information regarding their plans. Also, those individuals who manage plans (and other fiduciaries) must meet certain standards of conduct, derived from the common law of trusts and made applicable (with certain modifications) to all fiduciaries. The law also contains detailed provisions for reporting to the government and disclosure to participants. Furthermore, there are civil enforcement provisions aimed at assuring that plan funds are protected and that participants who qualify receive their benefits.

ERISA covers pension plans and welfare benefit plans (e.g., employment-based medical and hospitalization benefits, apprenticeship plans, and other plans described in section 3(1) of Title I). Plan sponsors must design and administer their plans in accordance with ERISA. Title II of ERISA contains standards that must be met by employee pension benefit plans in order to qualify for favorable tax treatment. Noncompliance with these tax qualification requirements of ERISA may result in disqualification of a plan and/or other penalties.

ERISA has been amended many times since 1974, and the responsibilities of the PWBA have been increased. Changes to ERISA were contained in:

1. The Tax Reform Act of 1976.
2. The Social Security Amendments of 1977.
3. The Revenue Act of 1978.
4. The Amendments to the Age Discrimination Act of 1978.
5. The Multiemployer Pension Plans Act of 1980.
6. The Economic Recovery Act (ERTA) of 1981.
7. The Tax Equity and Fiscal Responsibility Act (TEFRA) of 1982.
8. The Tax Reform Act of 1984 (TRA '84).
9. The Retirement Equity Act (REA '84) of 1984.
10. The Consolidated Omnibus Budget Reconciliation Act of 1985 (COBRA).
11. The Tax Reform Act of 1986 (TRA '86).
12. The Omnibus Budget Reconciliation Act of 1987 (OBRA '87).
13. The Technical and Miscellaneous Revenue Act of 1988 (TAMRA '88).
14. OBRA '89.
15. OBRA '90.
16. The Deficit Reduction Act of 1992 (DRA '92).
17. Unemployed Compensation Act of 1991.
18. The Retirement Protection Act of 1994.
19. Uniformed Services Employment and Reemployment Act (USERRA), 1994.
20. The Pension Simplification Act of 1995.
21. Uruguay Round of General Tariffs and Trade (GATT), 1995.
22. The Small Business Job Protection Act of 1996.
23. The Taxpayer Relief Act of 1997.
24. The Transportation Revenue Act of 1998.
25. The IRS Restructuring and Reform Act of 1998.
26. The Economic Growth and Tax Relief Act of 2001.

Each of the Acts above includes many changes to the original ERISA enactment, and there is no intent to give a thorough presentation, only some of the more significant changes. Through item 6 above, private pension plans were extremely beneficial to small businesses, such as physicians, lawyers, accountants, and in general small professional corporations. Pensions in excess of $136,000 could be funded or contributions as high as $45,000 per year toward pensions could be deducted from income for income tax purposes. Further, there were very favorable loan provisions in that participants could borrow from the plan and then repay the loan at market interest rates. This did not cause a tax consequence to the qualified plan, and the borrower could deduct the interest part of the periodic payment from income for income tax purposes. Normally, the greatest amount of the plan assets would be for the principal (physician, lawyer, etc.). Thus, the principal could borrow from the plan to purchase a home, for example, repay himself or herself in the plan and deduct the interest from income taxes. The Economic Recovery Act of 1981 also enabled all workers to save for retirement by way of a tax-deductible contribution to an individual retirement account (IRA) of up to $2,000 ($2,250 spousal) per year of earned income. For sole-proprietorships, contributions to Keogh (HR-10) and Simplified Employee Pension (SEP) plans were increased to the lesser of $15,000 or 15% of compensation.

TEFRA (1982) was the beginning of changes in private plans in a negative direction. Some of the monetary changes were as follows. The maximum pension that could be funded was reduced to $90,000 to be indexed with inflation, and the maximum contributions to contributory plans were reduced to $30,000. TEFRA also restricted contributions to dual pension plans—those plans where a benefit was defined and a contribution was specified. Loans made after August 13, 1982 could be treated as taxable as distributions from plans. REA '84 delayed an increase in the $90,000 pension until 1988. It also reduced the maximum age that an employer may require for participation in a pension plan, and it lengthened the period of time a participant could be absent from work without losing pension credits. Until this Act, the rights of spouse's access to plan assets were limited; however, REA '84 created spousal rights to pension benefits through qualified domestic relations orders (QDROs) in the event of divorce and through preretirement survivor annuities.

The Consolidated Omnibus Budget Reconciliation Act of 1985 (COBRA) added a new part 6 to Title I of ERISA that provides for the continuation of health care coverage for employees and their beneficiaries (for a limited period of time) if certain events would otherwise result in a reduction in benefits. TRA '86 imposed a host of changes that further protected employees' retirement benefits. However, it placed limits on tax-deductible IRAs, and allowed for nondeductible contributions to an IRA with tax-deferred treatment of the growth of those investments. The Omnibus Budget Reconciliation Act of 1989 requires the secretary of labor to assess a civil penalty equal to 20% of any amount recovered for violations of fiduciary responsibility. The Department of Labor Regulations stipulates a "prudent man rule" in exercising investment authority of plans. More recently, the Health Insurance Portability and Accountability Act of 1996 (HIPAA) added a new Part 7 to Title I of ERISA aimed at making health care coverage more portable and secure for employees, and gave the department broad additional responsibilities with respect to private health plans.

The next major change was OBRA '90 wherein an excise tax was imposed on excess plan assets that were greater than necessary to meet the "current" obligations of the plan. If a participant of a plan died and the assets that were designated for the deceased exceeded the present value of a life annuity under specified conditions, the amount of assets in excess of this present value would be subject to a nondeductible 15% excise tax. The most recent changes of significance are in the Taxpayer Relief Act of 1997 wherein the excise tax on excessive assets at death was repealed. It also repealed the limitations on dual plans as of January 1, 2000.

There are four reasons, decrements, for a worker to leave the pension plan of an employer. These are death, disability, termination, and retirement. We will be concerned with the retirement decrement only, and there are only two broad categories of pension plans by which retirement can be funded. These are a **defined benefit pension plan** and a **defined contribution pension plan**. As the names imply, the defined benefit plan specifies the benefit that the worker can expect on retirement. It becomes the obligation of the employer to fund the plan with sufficient assets in order to meet the benefit. The defined contribution plan specifies a fixed contribution, usually a percent of compensation. When the worker retires, the magnitude of the pension fund will determine the annual benefit based on the actuarial life expectancy of the worker, and, if there is a spouse, the joint life expectancy of the worker and his or her spouse. The Tax Reform Act of 1986 mandated a life and 50% survivor benefit for the distribution from all pension plans unless the spouse signs a nonrevocable waiver of that provision.

PROBLEM SET 11.1

1. A medical group is comprised of a two physicians with annual compensation of $160,000 each, two nurses with compensations of $40,000 each, and a receptionist with a compensation of $25,000. The annual revenue of the medical practice is $1,000,000. Discuss why you feel that it was necessary for the government to have enacted such a myriad of legislation for any private pension plan that this group may establish.

11.2 DEFINED BENEFIT PENSION PLANS

As the name implies, the defined benefit pension plan "defines" the benefit that is to be received upon retirement. There are two basic types of these plans, flat benefit plans and unit benefit plans. In flat benefit plans, the benefit is usually specified as a percent of compensation, and that could be the average of the highest three salaries, the average of the highest five salaries, or of the highest compensation received, usually that of the final year. In unit benefit plans, the benefit could be defined as some percent of compensation multiplied by the number of years of employment up to a limit that would be no greater than 100% of compensation. Therefore, given a current compensation, a salary scale may be included in the pension plan in order to give a final compensation upon which the benefit is to be based. However, it is common practice to "define the benefit" as a percent of current compensation. If the compensation changes each year, as with salary increases, the "defined benefit" changes according to the formula in the plan. In order to increase private pension plan coverage for employers with less than 100 employees, the Transportation Revenue Act of 1998 included the *Secure Money Annuity or Retirement Trust* (SMART) plans.

CONTRIBUTIONS TO CORPORATE PLANS

The contribution to a defined benefit plan for a given year depends upon several factors. These include:

- The benefit formula.
- The interest rate that is assumed for the growth of the fund.
- The amount necessary to provide the benefit upon retirement.
- The amount then accumulated in the pension plan fund.
- Actuarial assumptions on mortality, disability, termination, and retirement.

We will consider the retirement assumption only. For the year 2002, Section 415(b)(1)(A) of the Internal Revenue Code limits the maximum defined benefit that can be funded to provide an annual benefit equal to 100% of compensation up to $160,000; however, a minimum defined benefit of $10,000 per year may be funded regardless of the compensation. Once the benefit formula has been established the problem is to determine the amount of money that will be necessary on the date of the first payment. Since the benefit is to be paid for life, that amount of money is the present value of an annuity due *with the benefit of survivorship*, usually paid on a monthly basis. Such life annuities are discussed in Chapter 9. In those life annuities, it was assumed that anyone who died in a given year died at the end of the year, but that is not a rigorous assumption. It would be more reasonable to assume that the same number of people died in each month, and for advanced ages, more of the age group died toward the end of the year than at the beginning of the year.

Table A-7 of the Appendix uses the assumption that the number of deaths during the year is spread equally each month. Then, if a person is to retire at age 65, Table A-7 indicates the present value of an annuity due that is necessary to provide a defined benefit of $1.00 per month is $122.66 if the interest rate is 5%. It would be $113.20 at an interest rate of 6%. The interest rate that may be used for funding defined benefit plans is specified by the Internal Revenue Code, Section 412(c)(7) and will be based on Moody's Aa Corporate Bond Rate until 2004. We will assume 5% and 6% interest rates.

EXAMPLE 11.2.1 What amount of money will be necessary in order to provide for a defined benefit of $160,000 per year, payable on the first day of each month, beginning at age 65 if the interest rate is 5%?

Solution   A benefit of $160,000 per year translates to a monthly benefit of $11,666.67. From the 5% table of Table A-7, the present value of that annuity due is $122.66 for $1.00 per month. Then, the amount of the fund that will be necessary at age 65 is the present value of an annuity due and is determined as

$$\ddot{A}_{65} = 122.66 \left(\frac{160,000}{12} \right) = \$1,635,467. \quad \blacksquare$$

The law requires that contributions to pension plans be made at least quarterly in an amount equal to one-fourth the annual cost. That expectation might not be practical since the total contribution liability will not be known until the year-end of the plan. Since the total contribution need not be made until seven months after the plan-year-end, contributions are usually made between the plan-year-end and that final date. However, if the contributions are not made on a quarterly basis, interest is charged to the unpaid amount. We shall not be concerned with this quarterly requirement but will be concerned with the annual amount of a contribution only. The amount that is to be accumulated for a plan participant is referred to as the full funding limit, the liability of the plan to be able to provide the benefit. Then, the contribution stream is an annuity of payments such that the full funding limit will be attained for each participant in the pension plan. Let:

i be the effective annual interest rate assumption of the plan,

AA be the attained age (age at last birthday) at the end of a plan-year,

RA be the normal retirement age,

s_x be the accumulation annuity factor for an annuity that begins at age x for a number of years based on AA and RA,

S be the annual compensation (salary),

a_x be the present value factor from Table A-7 (in actuarial notation it is $12\ddot{a}_x^{(12)}$),

$\ddot{A}_x$ the full funding limit (the present value that is necessary at the retirement age), and

C be the periodic contribution.

The full funding limit is determined as

$$\ddot{A}_x = a_x \left(\frac{S}{12} \right). \tag{11.2-1}$$

Since contributions are made the end of a plan-year, they represent the payments to an immediate annuity, and therefore, the annuity accumulation factor would be determined by

$$s_x = \frac{(1+i)^{RA-AA+1} - 1}{i}. \tag{11.2-2}$$

The exponent in Equation (11.2-2) is determined as follows. Payments to an immediate annuity begin one payment period after the start of the annuity. Since the contributions to the pension plan are to be made at each subsequent attained age, AA, the "annuity" of the contributions is considered to have begun at each subsequent attained age $AA - 1$. This is shown in Figure 11-1.

Then the normal cost, periodic contribution, would be determined by

$$C = \frac{\ddot{A}_x}{s_x}. \tag{11.2-3}$$

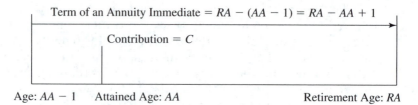

Age: $AA - 1$ Attained Age: AA Retirement Age: RA

FIGURE 11-1 Term of an Annuity Immediate

Except for annuity factors, all monetary amounts are rounded to the nearest dollar.

EXAMPLE 11.2.2   A defined benefit pension plan for 100% of compensation has been newly qualified. The key participant of the pension plan is 45 years old at the beginning of the plan-year and has an annual compensation of $126,000. The normal retirement age of the plan is 65, and the interest rate for the plan is 6%. Determine the annual contributions if the investment rate of return of the plan always will be 6% per year.

Solution   The 6% interest rate table of Table A-7 shows that at age 65 the value of $a_x = \$113.20$ per $1.00 per month of benefit. From Equation (11.2-1)

$$\ddot{A}_x = 113.20 \left(\frac{126,000}{12} \right) = \$1,188,600.$$

The contribution is made at the end of the first year at the attained age of 46 and, as such, represents an immediate annuity that began at age 45. The contribution is the amount that is necessary in order to accumulate the present value, $\ddot{A}_x$, of the pension fund. Then the annuity factor from the attained age to the normal retirement age would be determined from Equation (11.2-2) as

$$s_{45} = \frac{(1 + 0.06)^{20} - 1}{0.06} = 36.785591.$$

From Equation (11.2-3), the normal annual cost (the payment for the immediate annuity) is determined as

$$C = \frac{1,188,600}{36.785591} = \$32,312.$$

If the compensation and interest rate remain constant for the next 20 years, the cost to provide an annual benefit of $126,000 to this employee would be $32,312 per year. ■

EXAMPLE 11.2.3   For the newly qualified plan of Example 11.2.2, one of participants is 30 years old at the beginning of the plan-year. The annual compensation for this employee also is $126,000. Determine the annual contributions if the investment rate of return of the plan always will be 6% per year.

Solution   At a 6% interest rate, $a_x = \$113.20$ per $1.00 per month of benefit. For a $126,000 annual benefit, the monthly benefit will be $10,500, and the full funding limit, $\ddot{A}_x$, will be $1,188,600 as before. Since the contribution will be made at the end of the first year, the participant will have an attained age of 31. From Equation (11.2-2), the accumulation annuity factor is determined as

$$s_{31} = \frac{(1 + 0.06)^{65-31+1} - 1}{0.06} = 111.43478.$$

From Equation (11.2-3), the contribution is determined as

$$C = \frac{1,188,600}{111.43478} = \$10,666.$$

If the compensation and interest rate remain constant for the next 35 years, the cost to provide an annual benefit of $126,000 to this employee would be $10,666 per year. ■

A comparison of Examples 11.2.2 and 11.2.3 shows that the defined benefit plan is more costly for older participants than it is for relatively young participants. Thus, if the key employees of a small corporation are the primary stockholders and are the older employees, the greater cost for the defined benefit plans translates to a greater income tax deduction for the corporation. Under these circumstances, a defined benefit plan is desirable.

UNFUNDED LIABILITY AND SUBSEQUENT CONTRIBUTIONS

The Internal Revenue Code Section 412(c)(7) stipulates that, for full funding calculations, an interest rate be used that is within a specified range of 90% to 110% of a *weighted* average of the interest rate using Moody's Aa Corporate Bond Rate through 2004.

In Examples 11.2.2 and 11.2.3, the assumption was that the investment rate of return would be 6% per year from the inception of the plan to the respective retirement dates. If the assumption proved to be true, the annual, deductible cost to the corporation for the two employees would be $42,978 and both defined benefits would be available when each participant retired at age 65. However, suppose the pension plan fund actually earned an effective rate of return of 10% during the second year. The first-year contribution of $32,312 for the first participant would have been made at the end of that year (at which time the participant's attained age would be 46). The contribution for the second year also would be made at the end of that year (at which time the participant's attained age would be 47). The contribution from the first year would have accumulated to $(1.1)(32,312) = \$35,543$ at the attained age of 47 and would have 18 additional years (age 65–age 47) to grow. The contribution for the second year would be for an immediate annuity that began at the attained age of 46, and the exponent in the annuity factor would be 19 (age 65–age 46). The following diagram illustrates the time frame for the second and subsequent contributions.

Using the assumed interest rate of the plan for the remaining 18 years to the normal retirement age, the asset amount at the end of the second year would accumulate to $101,452 $(\$35,543(1.06)^{18})$, at age 65 of the key employee of Example 11.2.2. Since $1,188,600 will be necessary to fully fund the defined benefit, the fund is deficient by $1,087,148 (1,188,600–101,452). The difference between the full funding limit and the projected funding amount is referred to

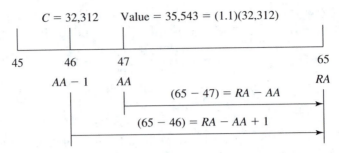

as the unfunded liability. Since the key employee would be 47, 18 years remain in which to accumulate this unfunded liability as an immediate annuity. Then, with the plan assumption of a 6% interest rate and Equation (11.2-2),

$$s_{47} = \frac{(1+0.06)^{19} - 1}{0.06} = 33.759992,$$

and the contribution for the second year would be determined using Equation (11.2-3) as

$$C = \frac{1,087,148}{33.759992} = \$32,202.$$

Note that the cost (contribution) for this employee for the second year is less than that of the first year because the pension plan fund earned an interest rate greater than that assumed in the plan. Had the return been less, the cost for the second year would have been greater than that for the first year. This analysis would be necessary for each employee in the plan.

In general, let:

AA be the attained age at the end of the current plan-year,

RA be the normal retirement age of the plan,

B be the balance of the fund (assets) at the end of the current year,

L be the unfunded liability at the end of the current year, and

i be the effective annual interest rate.

Then, the contribution to the defined benefit pension plan for a given current year would be determined as follows. Repeating Equation (11.2-1) here for convenience,

$$\ddot{A}_x = a_x \left(\frac{S}{12} \right),$$

the unfunded liability is determined by

$$L = \ddot{A}_x - B(1+i)^{RA-AA}. \tag{11.2-4}$$

The annuity factor becomes

$$s_x = \frac{(1+i)^{RA-AA+1} - 1}{i}. \tag{11.2-5}$$

Then, the contribution for plan-year just ended is determined by

$$C = \frac{L}{s_x}. \tag{11.2-6}$$

Equation (11.2-4) may be used for the first year of a pension plan by setting $B = 0$ since there will not be any plan asset at that time. The unfunded liability is a function of the investment performance of the plan and any change in compensation from the preceding year. This will be shown in Example 11.2.4.

EXAMPLE 11.2.4   The employee of Example 11.2.2 receives an increase in compensation to \$130,000 the following year. The investment performance of the fund was 10%, giving a value at the end of the first year of \$35,543, as indicated above. The employee's attained age is now 47, and the plan formula is 100% of compensation and a 6% interest rate. Determine the contribution for the second year of the plan.

Solution   Using the 6% table of Table A-7, $a_x = 113.20$. Then,

$$\ddot{A}_x = 113.20 \left(\frac{130,000}{12} \right) = \$1,226,333.$$

The unfunded liability may be determined from Equation (11.2-4) as

$$L = 1,226,333 - 35,543(1 + 0.06)^{65-47} = \$1,124,881.$$

The annuity factor may be determined from Equation (11.2-5) as

$$s_x = \frac{(1 + 0.06)^{65-47+1} - 1}{0.06} = 33.75992.$$

Then, the contribution for the second year from Equation (11.2-6) is

$$C_2 = \frac{1,124,881}{33.75992} = \$33,320. \quad \blacksquare$$

A similar analysis is necessary for subsequent years, and therefore, in every defined benefit plan, the census of participants and the assets of the plan (for each participant) must be determined at the end of each plan-year. With this information, the unfunded liability of a plan can be determined in order to establish the cost for the current year.

CONTRIBUTIONS TO NONCORPORATE PLANS

Defined benefit plans for businesses other than corporations are the same as that for corporations with the exception that the contribution to the plan must be taken into account in order to determine the compensation of the employee. Using a sole proprietorship as an example, the expenses of the business must include the contribution to the pension plan. Therefore, the net income, after all other expenses, except the contribution to the pension plan, must be reduced in order to determine the earned income. It is this earned income that becomes the base for the defined benefit. Let:

E be the earned income for a given year,

N be the net income for a given year,

a_x be the present value factor from Table A-7,

L be the unfunded liability,

C be the contribution to a defined benefit pension plan for a given year,

B be the asset balance at the end of the current year,

P be the percent of earned income that is the defined benefit,

AA be the attained age at the end of the current plan year,

RA be the normal retirement age,

C' be an annual contribution that will provide a monthly benefit of $1.00 at the normal retirement date,

s_x be an annuity factor at the beginning of the annuity that will accumulate to $1.00 at the plan retirement date, and

i be the effective annual interest rate.

In order to determine the contribution for the first year of a qualified plan, the following procedure is necessary. The contribution that is necessary to provide a benefit of $1.00 per month is determined as

$$C' = \frac{\ddot{A}_x}{12s_x}, \tag{11.2-7}$$

where s_x is determined from Equation (11.2-5), repeated here for convenience as

$$s_x = \frac{(1+i)^{RA-AA+1} - 1}{i}.$$

The total annual contribution to the plan for the current year is determined by

$$C = C'PE. \tag{11.2-8}$$

However, E is unknown, but equals the net income minus the net contribution or $N - C$. Therefore,

$$N = E + C. \tag{11.2-9}$$

Substituting Equation (11.2-8) into Equation (11.2-9) gives

$$N = E + C'PE = (1 + C'P)E,$$

and solving for the earnings, E, gives

$$E = \frac{N}{1 + PC'}. \tag{11.2-10}$$

The contribution to the pension plan for the first year is the difference between net income and earnings, or

$$C = N - E. \tag{11.2-11}$$

Then, the annual contribution for subsequent years takes into account the unfunded liability as with corporate plans except that earnings replaces compensation or,

$$\ddot{A}_x = a_x \left(\frac{E}{12} \right), \tag{11.2-12}$$

$$L = \ddot{A}_x - B(1+i)^{RA-AA}, \tag{11.2-13}$$

and

$$C = \frac{L}{s_x}, \tag{11.2-14}$$

the same as for corporate plans. L in Equation (11.2-13) may be used the first year of a noncorporate pension plan by setting $B = 0$ since there will not be any plan asset at that time.

EXAMPLE 11.2.5 A sole proprietor, who has an attained age of 56, establishes a qualified defined benefit plan at 100% of earnings. The interest rate in the plan is 5%, and the normal retirement age is 65. If the net income, exclusive of the contribution to the pension plan, is $100,000, determine the contribution for the first year.

Solution From the 5% table of Table A-7, $a_x = 122.66$. From Equation (11.2-5),

$$s = \frac{(1 + 0.05)^{65-56+1} - 1}{0.05} = 12.577893.$$

From Equation (11.2-7),

$$C' = \frac{122.66}{12(12.577893)} = 0.812669.$$

Since the plan formula is 100% of earnings, $P = 1$ and from Equation (11.2-10)

$$E = \frac{100,000}{1 + (1)(0.812669)} = \$55,167.$$

Then, the earnings for the first year are $55,167, and from Equation (11.2-11), the contribution for the first year will be

$$C = 100,000 - 55,167 = \$44,833. \quad \blacksquare$$

EXAMPLE 11.2.6 The sole proprietor of Example 11.2.5 makes the contribution of $44,833 to the plan on the date of the plan year-end. During the following year, the investment performance of the contribution was such that at the end of the second plan-year, it had accumulated to $50,000. The net income of the sole proprietorship for the second year is $120,000. All other elements of the plan are the same. Determine the contribution cost for the second year.

Solution Since the sole proprietor had an attained age of 56 for the first year, he or she has an attained age of 57 for the 2nd year. From its definition,

$$s_x = \frac{(1 + 0.05)^{65-57+1} - 1}{0.05} = 11.026564.$$

In order to fund a benefit of $1.00 per month, its definition gives

$$C' = \frac{122.66}{12(11.026564)} = 0.927004.$$

Then, the earnings for 2nd plan-year are determined from Equation (11.2-10) as

$$E = \frac{120,000}{1 + (1)(0.927004)} = \$62,273.$$

In order to have sufficient assets to fund a benefit of $62,273, the amount necessary on the retirement date is determined from Equation (11.2-12) as

$$\ddot{A}_x = 122.66 \left(\frac{62,273}{12} \right) = \$636,534.$$

The unfunded liability for second year is determined from Equation (11.2-4) as

$$L = 636,534 - 50,000(1 + 0.05)^{65-57}$$

$$= \$562,661.$$

The contribution for the 2nd year is determined from

$$C = \frac{L}{s_x} = \frac{562,661}{11.026564},$$

and is $51,028. A similar analysis must be performed every year. ■

PROBLEM SET 11.2

1. The Internal Revenue Service has just qualified a new, corporate defined benefit plan. The plan formula is for 100% of compensation and a 6% interest rate. The normal retirement age is 65. A participant had an attained age of 40 and a compensation for $160,000 for the plan-year just ended. Determine the contribution for the first plan-year.

2. A corporate defined benefit plan formula is for 100% of compensation. The interest rate in the plan is 6%, and the normal retirement age is 65. A participant has an attained age of 36 with assets in the plan of $25,000. The compensation for the plan-year just ended was $40,000. Determine the contribution.

3. A corporate defined benefit plan formula is for 100% of compensation. The interest rate in the plan is 6%, and the normal retirement age is 65. A participant has an attained age of 50 with assets in the plan of $100,000. The compensation for the plan-year just ended was $100,000. Determine the contribution.

4. A corporate defined benefit plan formula is for 100% of compensation. The interest rate in the plan is 5%, and the normal retirement age is 65. A participant has an attained age of 50 with assets in the plan of $65,000. The compensation for the plan-year just ended was $200,000. Determine the contribution.

5. A corporate defined benefit plan formula is for 100% of compensation. The interest rate in the plan is 6%, and the normal retirement age is 65. A participant has an attained age of 36 with assets in the plan of $15,000. The compensation for the plan-year just ended was $40,000. Determine the contribution.

6. The Internal Revenue Service has just qualified a new defined benefit plan. The plan formula is for 100% of compensation and a 5% interest rate. The normal retirement age is 67. A participant had an attained age of 40 and had compensation of $160,000 for the plan-year just ended. Determine the contribution for the first plan-year.

7. A corporate defined benefit plan formula is for 100% of compensation. The interest rate in the plan is 6%, and the normal retirement age is 67. A participant has an attained age of 36 with assets in the plan of $25,000. The compensation for the plan-year just ended was $40,000. Determine the contribution.

8. A corporate defined benefit plan formula is for 100% of compensation. The interest rate in the plan is 5%, and the normal retirement age is 67. A participant has an attained age of 50 with assets in the plan of $100,000. The compensation for the plan-year just ended was $100,000. Determine the contribution.

9. A corporate defined benefit plan formula is for 100% of compensation. The interest rate in the plan is 5%, and the normal retirement age is 66. A participant has an attained age of 50 with assets in the plan of $65,000. The compensation for the plan-year just ended was $200,000. Determine the contribution.

10. A corporate defined benefit plan formula is for 100% of compensation. The interest rate in the plan is 5%, and the normal retirement age is 66. A participant has an attained age of 36 with assets in the plan of $15,000. The compensation for the plan-year just ended was $40,000. Determine the contribution.

11. The Internal Revenue Service has just qualified a new, noncorporate defined benefit plan. The plan formula is for 100% of earnings and a 5% interest rate. The normal retirement age is 65. A participant had an attained age of 40 and a net income of $150,000 for the plan-year just ended. Determine the contribution for the first plan-year.

12. A noncorporate defined benefit plan formula is for 100% of earnings. The interest rate in the plan is 5%, and the normal retirement age is 65. A participant has an attained age of 50 with assets in the plan of $150,000. The net income for the plan-year just ended was $200,000. Determine the contribution.

13. A noncorporate defined benefit plan formula is for 100% of earnings. The interest rate in the plan is 5%, and the normal retirement age is 66. A participant has an attained age of 45 with assets in the plan of $125,000. The net income for the plan-year just ended was $250,000. Determine the contribution.

14. A noncorporate defined benefit plan formula is for 50% of earnings. The interest rate in the plan is 5%, and the normal retirement age is 65. A participant has an attained age of 50 with assets in the plan of $150,000. The net income for the plan-year just ended was $200,000. Determine the contribution.

15. A noncorporate defined benefit plan formula is for 75% of earnings. The interest rate in the plan is 6%, and the normal retirement age is 67. A participant has an attained age of 45 with assets in the plan of $110,000. The net income for the plan-year just ended was $150,000. Determine the contribution.

11.3 DEFINED CONTRIBUTION PENSION PLANS

As the name implies, the annual contribution to these types of plans is specified in the plan, and each participant has a segregated account. The benefit at retirement depends solely upon the value of the account, postretirement interest rates, and life expectancy. A common practice is to remove the participant from a defined contribution plan upon retirement by a lump sum distribution of the accumulation either directly to the individual (with income tax implications) or a direct rollover to an individual retirement account. The amount of the accumulation on the retirement date, in conjunction with an interest rate and a term based on a life expectancy, may be used to determine an annual benefit. The danger in life expectancy terms is that one could outlive the fund, and to avoid that, the accumulation may be used to purchase a life annuity from an insurance company. Then, the annual benefit would be determined by using tables for life annuities.

Section 415 of the Internal Revenue Code stipulates that the total contribution by the employer be the lesser of 100% of compensation or $40,000. Also, the limit on compensation that may be considered in calculations is $200,000 in 2002. There are several forms of defined contribution plans and include the following with the respective corporate income tax-deductible contribution limits for 2002.

- Money Purchase Plans. The contribution is fixed as a percent of salary. The maximum contribution rate is 25% of compensation up to $40,000.

- Profit Sharing Plans and Age-Based Profit Sharing Plans. No contribution is required in any year where there are no profits. The basic maximum contribution is 25% of salary up to $40,000. In age-based plans, greater contributions, up to the $40,000 maximum, can be made for older participants as long as the total of the contributions is no more than 25% of the total compensation or if benefit accrual rates for younger employees are at least the benefit accrual rate for older, or highly compensated employees. A highly compensated employee is one whose compensation is at least $90,000 in 2002.

- 401(k) Plans. These plans are contributory plans with or without a matching contribution formula for the employer. The employees' contributions are tax deductible to the employees. The combined contribution must be no more than 25% of compensation up to $40,000. This includes individual elective deferrals of up to $11,000, and if the individual is over 50, he or she may contribute an additional $1,000 in 2002.

- Simplified Employee Pensions (SEP). These types of plans allow for contributions to be made to Individual Retirement Accounts (IRAs) of employees and are limited to 25% of compensation up to $40,000. These plans also are useful to small employers and sole proprietors, and the contribution may be omitted in any given year.

- Savings Incentive Match Plans for Employees (SIMPLE). Employers with 100 or fewer employees may use these plans. They became effective after January 1, 1997 and replaced new SEPs. These plans may be established as Simple IRA or Simple 401(k) type plans. Contributions are mandatory in both types of plans. However, in the Simple IRA the contribution may be 100% of compensation with a $7,000 employee elective deferral limit, and in the Simple 401(k), contributions may be 25% of compensation up to $40,000 with an $11,000 employee elective deferral limit. Both types of plans are also appropriate for noncorporate plans, but for the 401(k) plan, the contribution is limited to 25% of compensation up to $40,000.

- Target Benefit Plans. These plans are money purchase pensions plans in which the periodic contribution in each participant's account is determined by the benefit that is to be provided at retirement. If the older, key employee establishes a level of contribution to his or her account, the ensuing accumulation will provide a "target" benefit. If all employees are given this same benefit, the plan will qualify; however, if the other employees are significantly younger than the key employee, the respective contributions could be less than in a conventional money purchase plan. In effect, a target benefit plan is an age-based money purchase plan.

- Comparability Plans. Comparability Plans are plans where employees are grouped into groups where the estimated benefit at retirement as a percent of current compensation for each employee in the group is equal to or greater than the estimated benefit of a highly compensated employee that must head each group. The salary level definition of a highly compensated employee is $90,000 for 2002 as indicated above. Therefore, Comparability Plans could be more beneficial than money purchase plans. Where there is more than one highly compensated, older employee the contribution to fund a benefit at retirement for an older employee would normally be greater than the contributions for younger employees. However, except for profit sharing plans, contribution rates are specified in plan documents, and those plans would have to be funded at those rates regardless of circumstances at the end of the plan-year. The allocation rates in profit sharing plans are variable; therefore, Comparability Plans are most suitable for such and may be considered extensions of age-based profit sharing plans. Because the contribution rate in profit sharing plans is equal to that in money purchase pension plans, the latter is effectively obsolete in that contributions are mandatory in the money purchase plan and are not in the profit sharing plans. Since both enable the same contribution rate, the mandatory requirement becomes a negative attribute. Further, the contributions in all plans are limited by the Section 415 limit of $40,000. Therefore, if an employee's compensation were $100,000, a profit sharing plan for 25% of compensation would enable a contribution of $25,000 in the plan. Further, with an additional 401(k) plan, if the employer matched the employee contribution up to 3%, for example, the employee could defer a total of $11,000 (possibly $12,000) for a total addition to his or her pension fund of $39,000 ($25,000 from the profit sharing plan + a $3,000 employee contribution + a $3,000 employer match + an additional $8,000 to reach the $11,000 deferral limit) in 2002.

- Individual Retirement Account (IRA).

There are two basic IRAs—preincome tax and postincome tax contributions. The classic, preincome tax IRA enables an individual to defer up to $3,000, in

2002, of earned income and the future growth of the investment from income taxes. Upon retirement, the annual withdrawals would be subject to income taxes. The postincome tax contribution would have been made after income taxes have been paid on earned income. However, all future growth and distribution would be free of income taxes. This latter type of IRA is referred to as a Roth IRA after the congressman who introduced the legislation for it. For both types, there are earned income limits if the individual is covered by a retirement plan at work. The $3,000 amount is phased down to 0 for single persons with earned incomes between $34,000 and $44,000 and for married persons with earned incomes between $54,000 and $64,000. For the Roth IRA, the phase-out limits are between $95,000 to $110,000 for singles and $150,000 to $160,000 for married couples. In addition, the IRA becomes a powerful device into which to transfer (rollover) pension plan assets upon termination of employment or retirement. Many employers would prefer that the rollover be utilized in that there would be no further responsibility for including the employee in the company's records.

As with defined benefit plans, the contributions to corporate plans are based on the employee's compensation, but in noncorporate plans the employee earnings are established after the contribution. In the following, let:

P be the percent of compensation,

S be the annual compensation (salary) for corporate plans,

N be the net income, exclusive of the contribution, in noncorporate plans,

C be the annual contribution,

a_x be the present value factor from Table A-7, and

s_x be the annuity factor at the beginning of the annuity for an annuity immediate.

CORPORATE PLANS

The annual contribution for corporate defined contributions plans is

$$C = PS. \tag{11.3-1}$$

Thus, if a plan called for a contribution of 25% of compensation and the compensation for a particular year was $100,000, the contribution would be $25,000 (25% of 100,000).

NONCORPORATE PLANS

The contribution for noncorporate defined contribution plans is determined after the contribution is subtracted from the net income or,

$$C = P(N - C),$$

and solving for C gives

$$C = \left(\frac{P}{1 + P}\right) N. \tag{11.3-2}$$

Thus, in noncorporate plans that have a 25% of compensation as a limit, the contribution for a net income of $100,000 would be

$$C = \left(\frac{0.25}{1.25}\right)(100,000) = \$20,000.$$

EXAMPLE 11.3.1 The president of a 3-employee corporation wishes to install a qualified money purchase pension plan with a contribution level of 25% of compensation. Determine the contribution for each employee and the total contribution for the first year. The employees and their compensations are as follows:

Employee	Compensation
President	$200,000
Employee A	40,000
Employee B	30,000

Solution Since the business is a corporation Equation (11.3-1) would be used to determine the contribution of each employee. Then

Employee	Formula	Contribution
President	$0.25(200,000) = 50,000$	$40,000 limit due to Section 415
Employee A	$0.25(40,000)$	10,000
Employee B	$0.25(30,000)$	7,500

The total contribution for the first year would be the sum of the contributions or $57,500. ■

TARGET BENEFIT PLANS

In this type of plan, the key employee could establish the level of contribution, up to the $40,000 maximum for 2002, and that would establish the formula as a percent of compensation. Let:

P be the percent of compensation,

C be the annual contribution,

S be the compensation,

F be the future value (accumulation) per dollar of contribution,

i be the effective annual interest rate of the plan,

s_x be the accumulation value for an annuity immediate,

RA be the normal retirement age,

AA be the attained age at the end of the plan year,

k represents the key employee, and

e represents an employee other than the key employee.

Then, from the plan formula,

$$P = \frac{C}{S},$$

and establishes the contribution rate, or dollar of contribution per dollar of compensation for the key employee. The future amount per dollar of contribution for the key employee is

$$F_k = P s_x, \tag{11.3-3}$$

where s_x is the accumulation of an immediate annuity for a periodic payment of $1.00.

$$s_x = \frac{(1+i)^{RA-AA+1} - 1}{i}.$$

Once the value of F is determined for the key employee, it is used for all employees, and the contribution per dollar of compensation for the remaining employees is determined by solving Equation (11.3-3) for P, or

$$P_e = \frac{F_k}{(s_x)_e}, \tag{11.3-4}$$

where, as indicated the subscript k refers to the key employee, and the subscript e refers to an employee other than the key employee. The contribution for each of the other employees is determined by

$$C_e = P_e S_e. \tag{11.3-5}$$

EXAMPLE 11.3.2 Compare the contributions of Example 11.3.1 with the contribution of a target benefit plan if a $40,000 contribution is to be made to the president's account. The interest rate of the plan is to be 6%, and the normal retirement age is 65. The following table summarizes the necessary information.

Employee	Compensation	Age
President	$200,000	50
Employee A	40,000	35
Employee B	30,000	25

Solution The dollars of contribution per dollar of compensation is determined by the ratio of the desired contribution for the key employee to the compensation of the key employee. From the president's information,

$$P = \frac{40,000}{200,000} = 0.20.$$

The value of s_x for the president is determined as

$$s_x = \frac{(1 + 0.06)^{65-50+1} - 1}{0.06} = 25.672528.$$

The dollars of accumulation per dollar of salary is determined from Equation (11.3-3) as

$$F_k = (0.20)(25.672528) = 5.134505.$$

This value of F_k is used for each of the other employees. The contribution for employee A is determined as follows.

$$s_x = \frac{(1 + 0.06)^{65-35+1} - 1}{0.06} = 84.801677.$$

Then, the contribution per dollar of compensation is determined by Equation (11.3-4) as

$$P_A = \frac{5.134505}{84.801677} = 0.060547.$$

The contribution for employee A is determined by Equation (11.3-5) as

$$C = 0.060547(40,000) = \$2,422.$$

The contribution for Employee B is determined in the same manner.

$$s_x = \frac{(1 + 0.06)^{65-25+1} - 1}{0.06} = 165.047684.$$

$$P_B = \frac{5.134505}{165.047684} = 0.031109.$$

$$C = 0.031109(30,000) = \$933.$$

The total contribution would be $40,000 + 2,422 + 933$, or $43,355. This represents a cost savings to the corporation of $14,145 over a conventional money purchase plan that is illustrated in Example 11.3.1.

The validity of a target benefit plan is that the accumulation, F_k, of the key employee, in conjunction with the compensation, will be the $\ddot{A}_x$ of a defined benefit plan. Since the same value of F_k is used for all employees, the respective $\ddot{A}_x$ is appropriate as the defined benefit for each employee. In Example 11.3.2, $\ddot{A}_x$ for the president would be $1,026,901 (5.134505 \times 200,000)$, $\ddot{A}_x$ for Employee A would be $205,380 (5.134505 \times 40,000)$, and $\ddot{A}_x$ for Employee B would be $154,035 (5.134505 \times 30,000)$ if the compensations and investment rate of return remained constant. Since that would not normally be the situation, the actual amount of the contribution each year that will be available at retirement will depend on the actual investment rate of return from the date of the contribution to the retirement date. The final values of $\ddot{A}_x$ may not be the values indicated by the calculations, and therefore the benefits to be derived are "Targets." ∎

AGE-BASED PROFIT SHARING PLANS

In effect, an age-based profit sharing plan is a target benefit plan utilizing profit sharing regulations rather than the regulations of money purchase plans. However, as with target benefit plans, the age differential between the key employees and the remaining employees of the business is crucial to the effectiveness of these plans. There are three concepts that could be explored.

A. EQUAL PERCENT OF COMPENSATION BENEFIT AND SPECIFYING THE TOTAL CONTRIBUTION.

This would direct the greater contribution amounts to the older employees and lesser contribution amounts for the younger employees and yet maintain the total contribution to no more than 25% of the total compensation. However, the indicated contribution to the older employee could be in excess of the Section 415 limit of $40,000 for 2002 and would need to be reduced to the limit. We will investigate this type of age-based profit sharing plan below.

B. EQUAL BENEFITS FOR ALL EMPLOYEES.

Since the total contribution cannot be greater than 25% of total compensation, this could serve to diminish the contribution amounts for the key employees and increase the contribution amounts for the remaining employees. We will not investigate this option because "our goal" is to maximize contributions to key employees with minimum contributions to non-key employees.

C. EQUAL PERCENT OF COMPENSATION BENEFIT AND SPECIFYING THE CONTRIBUTIONS FOR THE KEY EMPLOYEES.

This approach could cause the total contribution to exceed 25% of the total compensation. In such cases, the contributions for the key employees would have to be reduced from the specified amounts, and the entire process recalculated. Obviously, this would be an iterative process and either a spreadsheet or specialized software would be necessary. We will illustrate this method below. Comparability Plans utilize this option but must be cross-tested under IRS Regulation Section 1.401(a)(4) for discrimination against non-highly compensated employees. That testing is beyond the level of this book.

AGE-BASED PROFIT SHARING: EQUAL PERCENT OF COMPENSATION BENEFIT—SPECIFY TOTAL CONTRIBUTION

This type of profit share plan is that of A above. Let:

a_x be the present value of the annuity due of $1.00 per month from Table A-7,

A_x be the present value of a_x at the end of the plan-year,

A be the present value of the future accumulation of the compensation at the end of the plan-year,

T be the total contribution for all employees,

S be the compensation,

C be the contribution,

B be the benefit as a percent of compensation,

M be the monthly pension benefit,

i be the interest rate,

RA be the normal retirement age,

AA be the attained age at the end of the plan-year, and

F be the future amount of the contribution.

Internal Revenue Regulation Section 1.401(a)(4)-12 allows a "standard interest rate" between 7.5% and 8.5% to be used in the calculation. However, we will use 6% for illustrative purposes. Then, for each employee,

$$A_x = a_x(1 + i)^{-(RA-AA)}, \qquad (11.3\text{-}6)$$

and

$$A = A_x S. \qquad (11.3\text{-}7)$$

B represents the benefit as a percent of compensation. Since the benefit is a function of the contribution, B may be determined as a function of the present value of the necessary future accumulation that is to provide the benefit. It would be the ratio of the total contribution to the total present value of the future necessary accumulations, or

$$B = \frac{T}{\sum A}. \qquad (11.3\text{-}8)$$

Then, the contribution is determined by

$$C = BA. \tag{11.3-9}$$

The projected accumulation for the contribution is determined by

$$F = C(1 + i)^{RA-AA}, \tag{11.3-10}$$

and the monthly benefit to be derived by the current contribution is determined by

$$M = \frac{F}{a_x}. \tag{11.3-11}$$

EXAMPLE 11.3.3   Determine the contribution to an age-based profit sharing plan for the corporation of Example 11.3.2. The pre- and postretirement interest rate is to be 6%, and the normal retirement age is 65. The total contribution to the plan is to be 25% of the total compensation. The following table summarizes the necessary information. Determine the contribution for each employee and the monthly (and annual) pension benefit to be derived by the contribution.

Employee	Compensation	Age
President	$200,000	50
Employee A	40,000	35
Employee B	30,000	25
Total	$270,000	

Solution   From Table A-7, the value of the benefit factor for a normal retirement age of 65 is 113.20. The total contribution is to be 25% of 270,000 or $67,500.

	$RA - AA$	A_x = Equation (11.3-6)	A = Equation (11.3-7)
President	15	$\left(\frac{113.20}{12}\right)(1.06)^{-15} = 3.93620$	$3.93620(200,000) = 787,240.08$
Employee A	30	$\left(\frac{113.20}{12}\right)(1.06)^{-30} = 1.64244$	$1.64244(40,000) = 65,697.56$
Employee B	40	$\left(\frac{113.20}{12}\right)(1.06)^{-40} = 0.91713$	$0.91713(30,000) = 27,513.88$

$$\sum A = 880,451.52$$

The results may vary depending upon the decimal precision use in A_x. Since $\sum A$ represents the present value of total benefits at the end of the current plan year and T represents the total contribution at the same time, the contribution as a percent of current compensation is given by Equation (11.3-8) as

$$B = \frac{67,500}{880,451.52} = 0.07667,$$

or 7.667% of current compensation. Equation (11.3-9) gives the contributions as follows:

President $C = 0.07667(787,240.08) = \$60,357.70.$
Employee A $C = 0.07667(65,697.56) = \$5,037.03.$
Employee B $C = 0.07667(27,513.88) = \$2,109.49.$

The total of these contributions is $67,504. The $4.00 discrepancy from the expected $67,500 is due to rounding. Since the president's contribution exceeds Section 415 limit, the A_x must be reduced to a value that would give a $40,000 contribution. Then, from the calculated values,

$$A_x = \frac{40,000}{787,240.08} = 0.05081.$$

The future accumulations and monthly pension benefit are given by Equations (11.3-10) and (11.3-11) as follows:

President $F = 40,000(1.06)^{15} = \$95,862.33.$ $M = \frac{95,862.33}{113.20} = \$846.84.$

Employee A $F = 5,037.03(1.06)^{30} = \$28,930.14.$ $M = \frac{28,930.14}{113.20} = \$255.57.$

Employee B $F = 2,109.49(1.06)^{40} = \$21,697.62.$ $M = \frac{21,697.62}{113.20} = \$191.68.$

The respective annual amounts and the percent of current compensation represented by those annual amounts are as follows:

President $12(846.84) = \$10,162$, and $B = 5.081\%.$
Employee A $12(255.57) = \$3,067$, and $B = 7.667\%.$
Employee B $12(191.68) = \$2,300$, and $B = 7.667\%.$

On occasion, there will be slight discrepancies in the Bs due to rounding to the nearest dollar. The contributions of a conventional profit sharing plan at 25% of compensation, compared to an age-based plan, would be as follows:

President $0.25(200,000) = \$50,000$ reduced to $40,000.
Employee A $0.25(40,000) = \$10,000$ compared to $5,037.
Employee B $0.25(30,000) = \$7,500$ compared to $2,109.

This age-based approach to profit sharing generates a savings of $20,353. It can be readily seen that the age-based profit sharing plan is advantageous to the older employee and detrimental to younger employees for the specified 25% of total compensation. Further savings can ensue if Option C above, **equal percent of compensation benefit and specifying the contributions for the key employees**, is used as the following example will illustrate. ∎

EXAMPLE 11.3.4   Determine the contribution to an age-based profit sharing plan for the corporation of Example 11.3.2. The pre- and postretirement interest rate is to be 6%, and the normal retirement age is 65. It is desired to contribute the Section 415 limit for the president. The following table summarizes the necessary information. Determine the contribution for each employee and the monthly (and annual) pension benefit to be derived by the contribution.

Employee	Compensation	Age	
President	$200,000	50	Contribution = $40,000
Employee A	40,000	35	
Employee B	30,000	25	
Total	$270,000		

Solution   From Table A-7, the value of the benefit factor for a normal retirement age of 65 is 113.20. The total contribution is to be 25% of 210,000 or $67,500.

	$RA - AA$	$A_x =$ Equation (11.3-6)	$A =$ Equation (11.3-7)
President	15	$\left(\frac{113.20}{12}\right)(1.06)^{-15} = 3.93620$	$3.93620(200,000) = 787,240.08$
Employee A	30	$\left(\frac{113.20}{12}\right)(1.06)^{-30} = 1.64244$	$1.64244(40,000) = 65,697.56$
Employee B	40	$\left(\frac{113.20}{12}\right)(1.06)^{-40} = 0.91713$	$0.91713(30,000) = 27,513.88$
			$\sum A = 880,451.52$

Again, $\sum A$ represents the present value of total benefits at the end of the current plan-year. The desired $40,000 contribution is used to determine the percent of current compensation to be the benefit. From Equation (11.3-8)

$$B = \frac{40,000}{787,240.08} = 0.050810421,$$

or 5.081%, and this percent is applied to all employees current compensation as follows (Results depend upon the decimal precision used in B):

	Annual Benefit	Monthly Benefit
President	$0.05081(200,000) = \$10,162.08.$	$M = \$846.84.$
Employee A	$0.05081(40,000) = \$2,032.42.$	$M = \$169.33.$
Employee B	$0.05081(30,000) = \$1,524.31.$	$M = \$127.03.$

These monthly benefits require future amounts at retirement and current contributions as follows (Results depend upon the decimal precision used in B):

	$RA - AA$	$A =$ Equation (11.3-10)	$C =$ Equation (11.3-7)
President	15	$(846.84)(113.2) = 95,862.33.$	$95,862.33(1.06)^{-15} = \$40,000.$
Employee A	30	$(169.37)(113.2) = 19,172.47.$	$19,172.47(1.06)^{-30} = \$3,338.12.$
Employee B	40	$(127.03)(113.2) = 14,379.35.$	$14,379.35(1.06)^{-40} = \$1,397.99.$

The total contribution is $44,502, a savings of $2,645 over Option A, **equal percent of compensation benefit and specifying the total contribution**, above. ∎

The justification for qualification by the IRS, however, is that all employees will be receiving the same percent of compensation as a retirement benefit for either option. Depending upon the actual ages and compensations, one of these two options will direct a specified contribution for the key employee with minimal contributions to the other employees.

PROBLEM SET 11.3

1. A money purchase pension plan of a corporation specifies a contribution of 25% of compensation. An employee has an attained age of 25, and the normal retirement age is 65. What will be the least amount of the accumulation for a current compensation of $25,000 if the investment rate of return is never less than 7% annually?

2. A money purchase pension plan of a corporation specifies a contribution of 25% of compensation. An employee has an attained age of 35, and the normal retirement age is 65. What will be the least amount of the accumulation for a current compensation of $50,000 if the investment rate of return is never less than 7% annually?

3. A money purchase pension plan of a corporation specifies a contribution of 25% of compensation. An employee has an attained age of 25, and the normal retirement age is 65. What will be the least amount of the accumulation for a current compensation of $25,000 if the investment rate of return is never less than 10% annually?

4. A money purchase pension plan of a corporation specifies a contribution of 25% of compensation. An employee has an attained age of 35, and the normal retirement age is 65. What will be the least amount of the accumulation for a current compensation of $50,000 if the investment rate of return is never less than 10% annually?

5. A profit sharing plan of a corporation specifies a contribution of 25% of compensation. An employee has an attained age of 25, and the normal retirement age is 65. What will be the least amount of the accumulation for a current compensation of $75,000 if the investment rate of return is never less than 8% annually?

6. A profit sharing plan of a corporation specifies a contribution of 25% of compensation. An employee has an attained age of 35, and the normal retirement age is 65. What will be the least amount of the accumulation for a current compensation of $125,000 if the investment rate of return is never less than 10% annually?

7. A profit sharing plan of a sole proprietorship specifies a contribution of 25% of compensation. The sole proprietor has an attained age of 55, and the normal retirement age is 65. What will be the least amount of the accumulation for a net income of $250,000 if the investment rate of return is never less than 7% annually?

8. A profit sharing plan of a sole proprietorship specifies a contribution of 25% of compensation. The sole proprietor has an attained age of 55, and the normal retirement age is 65. What will be the least amount of the accumulation for a net income of $160,000 if the investment rate of return is never less than 10% annually?

9. A profit sharing plan of a sole proprietorship specifies a contribution of 25% of compensation. The sole proprietor has an attained age of 55, and the normal retirement age is 65. What will be the least amount of the accumulation for a net income of $300,000 if the investment rate of return is never less than 10% annually?

10. A money purchase pension plan of a corporation specifies a contribution of 25% of compensation and a normal retirement age of 65. The attained age of an employee is 35 at the end of the first year being eligible to participate in the plan. The compensation for the year was $160,000. If the compensation increases at 3% per year and the investment rate of return is never less than 7%, what will be the accumulation at the normal retirement date?

11. A money purchase pension plan of a corporation specifies a contribution of 25% of compensation and a normal retirement age of 65. The attained age of an employee is 35 at the end of the first year being eligible to participate in the plan. The compensation for the year was $35,000. If the compensation increases at 4% per year and the investment rate of return is never less than 8%, what will be the accumulation at the normal retirement date?

12. A money purchase pension plan of a corporation specifies a contribution of 25% of compensation and a normal retirement age of 65. The attained age of an employee is 35 at the end of the first year being eligible to participate in the plan. The compensation for the year was $60,000. If the compensation increases at 4% per year and the investment rate of return is never less than 8%, what will be the least accumulation at the normal retirement date?

13. A money purchase pension plan of a corporation specifies a contribution of 25% of compensation and a normal retirement age of 65. The attained age of an employee is 35 at the end of the first year being eligible to participate in the plan. The compensation for the year was $60,000. If the compensation increases at 4% per year until it reaches $40,000 and remains constant thereafter, what will be the least accumulation at the normal retirement date if the investment rate of return is never less than 8%?

14. A two-employee corporation has established a target benefit plan, and the owner wishes to make a $25,000 contribution to her account. The interest rate of the plan is to be 6%, and the normal retirement age is 65. At the end of the first year, the owner has an attained age of 40 with compensation of $100,000. The employee has an attained age of 20 with a compensation of $25,000. Determine the contribution that must be made for the employee.

15. A three-employee corporation has established a target benefit plan, and the owner wishes to make a $25,000 contribution to his account. The interest rate of the plan is to be 6%, and the normal retirement age is 65. At the end of the first year, the owner has an attained age of 40 with compensation of $100,000. One employee has an attained age of 25 with a compensation of $25,000, and the second employee had an attained age of 20 with a compensation of $20,000. Determine the contribution that must be made for each of the employees.

16. A two-employee corporation has established an age-based profit sharing plan. The owner's compensation is $100,000 and the attained age is 40. The employee's compensation is $25,000 and the attained age is 20. The owner wishes to make a contribution that is 15% of the total compensation. The interest rate of the plan is 6%, and the normal retirement age is 65. Determine the contribution for each employee for the current year.

17. A three-employee corporation has established an age-based profit sharing plan. The owner's compensation is $100,000 and the attained age is 40. One employee's compensation is $25,000 and the attained age is 25 and the second employee's compensation is $20,000 and the attained age is 20. The owner wishes to make a contribution that is 15% of the total compensation. The interest rate of the plan is 6%, and the normal retirement age is 65. Determine the contribution for each employee for the current year.

18. You begin making $2,000 annual contributions to a regular IRA at the attained age of 22. If the investment rate of return never falls below 10% annually, how much will you have accumulated by age 60?

19. You begin making $2,000 annual contributions to a regular IRA at the attained age of 22. If the investment rate of return never falls below 10% annually, how much will you have accumulated by age 65?

20. You begin making $2,000 annual contributions to a regular IRA at the attained age of 22. If the investment rate of return never falls below 10% annually, how much will you have accumulated by age 70?

11.4 DISTRIBUTIONS FROM PENSION PLANS

Normally, distributions from tax-deferred accumulations must begin no later than age $70\frac{1}{2}$, or by April 15 of the year following the age $70\frac{1}{2}$. This age is referred to as the minimum distribution age. In the latter case, two distributions would be required in one year, and therefore, delaying a distribution into that year could present significant income tax considerations. However, the Employee Pension Portability and Accountability Act of 1998 allowed for a delay in the start of pension benefits as long as the employee remained employed in a full time capacity with the employer. This delay is not available for individuals who own 5% or more of a company, Individual Retirement Accounts, Supplemental Retirement Accounts, or tax-deferred annuities that were purchased from accumulations from prior employers.

Until age $70\frac{1}{2}$ distributions that are controlled by the employee may be in any amount, unless the distribution is to be a life annuity. Distributions are mandatory at age $70\frac{1}{2}$. The following table summaries the nature of distributions.

DEFINED BENEFIT PLANS

Upon retirement with a defined benefit plan, the plan usually will state that the pension will begin on the first of the month that follows the month in which the normal retirement age is attained. If the corporation remains as an ongoing entity, it may choose to remain the payer of the benefit. However, the corporation may decide to forego the "bookkeeping" aspect of the pension, or the corporation will be dissolved upon the retirement of the key individual (physician, lawyer). For those situations, the plan will have been designed to allow the retiree to rollover the appropriate amount to an Individual Retirement Account or to purchase a tax-deferred or an immediate annuity from an insurance company.

If the corporation remains the payer and a defined benefit of $135,000 had been funded, a single life pension for the retiree will be a monthly benefit of $11,250. However, the law specifies that for married participants, the primary form of pension benefit will be life with a 50% benefit to the survivor. This means that the pension will be paid for the life of the participant and spouse, and upon the death of the participant the benefit will be reduced by 50%. Except for the single life only benefit, this form of benefit produces the maximum monthly benefit while the participant is alive but the least survivor benefit for the spouse upon the death of the participant. Any option that will provide the spouse with a benefit that is less than the 50% amount would have to be waived by the spouse. For example, a joint and 50% to the survivor with a period certain would generate a lesser benefit to the participant than the same election without the period certain. As such, upon the death of the participant, the benefit to the spouse would be less than the option without the period certain and such an election by the participant would have to be waived by the spouse. The determination of joint and survivor benefits other than 100% to the survivor is beyond the level of this text. For joint and 100% to the survivor annuities, we will use the joint life expectancies as published in the Federal Register, and are included as Table A-8, Table A-9, and Table A-10 in the Appendix.

If the employee is allowed to or is mandated to rollover funds to an IRA and self-manage the distribution, an amount equal to $\ddot{A}_x$, the present value of an annuity due, of $\$M$ per month must be transferred out of the pension plan assets. Letting a_x be the accumulation factor from Table A-7,

$$\ddot{A}_x = M a_x. \tag{11.4-1}$$

EXAMPLE 11.4.1   A defined benefit pension plan mandates the rollover of pension funds to individual control upon retirement. A person is to retire at age 65 with a benefit of $135,000 per year. If the interest rate is 6%, what amount of money must be transferred to an IRA?

Solution   From the 6% table of Table A-7, $a_x = 113.20$, and M equals the annual benefit divided by 12. Then, the amount of money that must be transferred is determined from Equation (11.4-1) as

$$\ddot{A}_x = (113.20) \left(\frac{135,000}{12} \right) = \$1,273,500.$$

The monthly benefit will be $11,250 $\left(\frac{135,000}{12} \right)$. ∎

DEFINED CONTRIBUTION PLAN

The monthly benefit for defined contribution plans is based on the amount of money available in the participant's account on the date the distribution is to begin. The maximum benefit that will be available is for a single life annuity and may be determined by solving Equation (11.4-1) for M.

EXAMPLE 11.4.2   A defined contribution pension plan mandates the rollover of pension funds to individual control upon retirement. A person is to retire at age 65 with an accumulation of $929,938, and this amount is transferred to the employee's IRA. If the interest rate of the IRA never falls less than 6%, what will be the annual amount for a single life annuity?

Solution   From the 6% table of Table A-7, $a_x = 113.20$. Rearranging Equation (11.4-1) to solve for M gives

$$M = \frac{\ddot{A}_x}{a_x}$$

$$= \frac{929,938}{113.20} = \$8,215.00.$$

The annual benefit will be $12M$ or $98,580. ∎

MINIMUM DISTRIBUTIONS

Distributions from tax-deferred funds must begin age $70\frac{1}{2}$ under the conditions indicated above. The annual distribution is based on the value of the account on December 31 of the previous year. The first annual amount would be the fund amount divided by the single or joint life expectancy as published in the Federal Register, dated April 17, 2002. The regulations of the IRS stipulate the joint life expectancy of spouses and nonspouses will be based on an assumed 10-year age differential and will be recalculated each year using the life expectancy of that age, unless the spouse is more than 10 years younger. In the latter case, the actual age difference may be used. Table A-8 shows the minimum distribution factors for a single life and Table A-9 shows the minimum distribution factors for joint lives based on a 10-year age difference. Upon the death of the participant the benefit would continue to a spouse based on the remaining life expectancies shown in Table A-8 or a one-year-less life expectancy for a beneficiary other than a spouse. Table A-10 is used if the age difference between spouses is greater than 10 years. For example, if the participant is age 70 and the spouse is age 55, the initial minimum distribution factor is 31.1 and may be found at the intersection of ages 70 and 55. Table A-9 would also be used if the spouse could rollover a lump sum into his or her own IRA and name his or her own beneficiary. However, if the beneficiary is a not a spouse and is more than 10 years younger than the participant, the maximum age differential that may be used is 10 years. Upon the death of the participant after the beginning of distributions, the benefit will continue as indicated in Table 11-1. Let:

a_x be the single or joint life expectancy from Tables A-8, A-9, or A-10,

R be the annual amount of the annual distribution for each year, and

$\ddot{A}_x$ be the fund amount on December 31 of the previous year.

Then,

$$R = \frac{\ddot{A}_x}{a_x}. \tag{11.4-2}$$

TABLE 11-1 DISTRIBUTIONS FROM PENSION PLANS		
Designated Beneficiary	**Death Prior to Distributions**	**Death after Distributions Begin**
Spouse	1. May take distributions over his/her life expectancy. 2. May delay taking distributions until the deceased spouse would have been $70\frac{1}{2}$. 3. May rollover into his/her own IRA and name his/her own beneficiary.	1. May take distributions over the remainder of his or her life single expectancy. 2. May rollover into his/her own IRA and name his/her own beneficiary with a new joint life expectancy.
Nonspouse or Trust	1. May take distributions over a term certain (one-year-less) that is his/her life expectancy. 2. The life expectancy, when a trust is the designated beneficiary, is based on the oldest beneficiary of the trust and uses the term certain (one-year-less) method.	1. May take distributions over the remainder of his or her single life expectancy using a term certain (one-year-less) method. 2. The original joint life expectancy, when a trust is the designated beneficiary, is based on the oldest beneficiary of the trust and uses the term certain (one-year-less) method.
Estate	The entire amount must be distributed over the following five years.	The entire amount must be distributed by December 31 of the year following the year of death.

EXAMPLE 11.4.3 Upon retirement this year, the value of an individual's retirement plan was $1,000,000 on December 31 of last year. The participant and his wife, who is the beneficiary, will be 70 on the retirement date. If the fund is invested at 10% effective, determine the annual distribution for the first five years.

Solution The initial minimum distribution factor from Table A-9 is 27.4 years. The distribution for the first year will be

$$R = \frac{1,000,000}{27.4} = \$36,496.$$

The simplifying assumption that will be made is the total amount is removed from the fund on the first day of the year, and the balance will accumulate at the rate of return of the fund. Thus, the prewithdrawal balance for age 71 is the postwithdrawal balance for age 70 times 1.10. The first five years for recalculated life expectancies are shown in the following table.

Age	Pre-withdrawal Balance	Joint Life Expectancy	Distribution	Post-withdrawal Balance
70	1,000,000	27.4	36,496	963,504
71	1,059,854	26.5	39,994	1,019,860
72	1,121,846	25.6	43,822	1,078,024
73	1,185,827	24.7	48,009	1,137,818
74	1,251,600	23.8	52,588	1,199,012 ■

EXAMPLE 11.4.4 Upon retirement this year, the value of an individual's retirement plan was $1,000,000 on December 31 of last year. The participant will be $70\frac{1}{2}$ on the retirement date and designates a 40-year-old daughter as the beneficiary. The participant dies during the 1st year. If the fund is invested at 10% effective, determine the annual distribution to the daughter for the next five years.

Solution The prewithdrawal balance at the beginning of the 2nd year is shown in Example 11.4.3 as $1,059,854. The minimum distribution to the daughter for the 2nd year would be based on the remaining life expectancy of 42.7 years for the daughter who would be 41. The distribution for the 1st year to the daughter is

$$R = \frac{1,059,854}{42.7} = \$24,821.$$

The simplifying assumption that will be made again is the total amount is removed from the fund on the 1st day of the year, and the balance will accumulate at the rate of return of the fund. Thus, the prewithdrawal balance for age 42, the daughter's age "next" year, is the postwithdrawal balance for age a 41 times 1.10. The 1st five years for one-year-less life expectancies are shown in the following table.

Age	Pre-withdrawal Balance	One-Year-Less Life Expectancy	Distribution	Post-withdrawal Balance
41	1,059,854	42.7	24,821	1,035,033
42	1,138,536	41.7	27,303	1,111,233
43	1,222,357	40.7	30,033	1,192,323
44	1,311,556	39.7	33,037	1,278,519
45	1,406,371	38.7	36,340	1,370,031 ■

Note that the postwithdrawal balances in both examples are increasing. This is due to the relative high rate of return on the fund. A maximum will be reached at some point in the future and the fund will decrease annually thereafter due to the shorter remaining life expectancies and the subsequent larger minimum required withdrawals.

Since estates may have other significant assets, it becomes imperative to become informed about the alternatives for the distribution of money from pension plans, especially spousal rollover options and the utilization of trusts to extend the distributions.

PROBLEM SET 11.4

1. A defined benefit pension plan calls for the rollover of assets to individual retirement accounts upon retirement. The interest rate of the plan is 6%. For a benefit of $50,000 per year, what amount would be the rollover for a person retiring at age 65?

2. A defined benefit pension plan calls for the rollover of assets to individual retirement accounts upon retirement. The interest rate of the plan is 6%. For a benefit of $100,000 per year, what amount would be the rollover for a person retiring at age 65?

3. A defined benefit pension plan calls for the rollover of assets to individual retirement accounts upon retirement. The interest rate of the plan is 5%. For a benefit of $125,000 per year, what amount would be the rollover for a person retiring at age 65?

4. A defined benefit pension plan calls for the rollover of assets to individual retirement accounts upon retirement. The interest rate of the plan is 6%. For a benefit of $50,000 per year, what amount would be the rollover for a person retiring at age 66?

5. A defined benefit pension plan calls for the rollover of assets to individual retirement accounts upon retirement. The interest rate of the plan is 5%. For a benefit of $75,000 per year, what amount would be the rollover for a person retiring at age 66?

6. A defined benefit pension plan calls for the rollover of assets to individual retirement accounts upon retirement. The interest rate of the plan is 5%. For a benefit of $125,000 per year, what amount would be the rollover for a person retiring at age 66?

7. A defined benefit pension plan calls for the rollover of assets to individual retirement accounts upon retirement. The interest rate of the plan is 6%. For a benefit of $50,000 per year, what amount would be the rollover for a person retiring at age 67?

8. A defined benefit pension plan calls for the rollover of assets to individual retirement accounts upon retirement. The interest rate of the plan is 6%. For a benefit of $100,000 per year, what amount would be the rollover for a person retiring at age 67?

9. A defined benefit pension plan calls for the rollover of assets to individual retirement accounts upon retirement. The interest rate of the plan is 5%. For a benefit of $75,000 per year, what amount would be the rollover for a person retiring at age 67?

10. A defined benefit pension plan calls for the rollover of assets to individual retirement accounts upon retirement. The interest rate of the plan is 5%. For a benefit of $125,000 per year, what amount would be the rollover for a person retiring at age 67?

11. The assets of a defined contribution pension plan amount to $500,000. Determine the annual pension amount for a person who retires at age 65 if the interest rate is 6%.

12. The assets of a defined contribution pension plan amount to $1,000,000. Determine the annual pension amount for a person who retires at age 65 if the interest rate is 5%.

13. The assets of a defined contribution pension plan amount to $500,000. Determine the annual pension amount for a person who retires at age 66 if the interest rate is 5%.

14. The assets of a defined contribution pension plan amount to $1,000,000. Determine the annual pension amount for a person who retires at age 66 if the interest rate is 5%.

15. The assets of a defined contribution pension plan amount to $500,000. Determine the annual pension amount for a person who retires at age 67 if the interest rate is 6%.

16. The assets of a defined contribution pension plan amount to $500,000. Determine the annual pension amount for a person who retires at age 67 if the interest rate is 5%.

17. A person terminates employment at age 65 and assets of $750,000 are transferred into an individual retirement account on December 31 of that year. The money is invested at 6% effective. The person chooses to withdraw the annual accumulation of interest until age $70\frac{1}{2}$, at which time the minimum distribution rules take effect. The individual is married, and the spouse's age is 68 at that time. Generate a table of the first five years of minimum distributions.

18. A person terminates employment at age 65 and assets of $750,000 are transferred into an individual retirement account on December 31 of that year. The money is invested at 6% effective. The person continues working in another occupation and chooses to leave the annual accumulation of interest until age $70\frac{1}{2}$, at which time the minimum distribution rules take effect. The individual is married, and the spouse's age is 58 at that time. Generate a table of the first five years of minimum distributions.

19. A person terminates employment at age 65 and assets of $800,000 are transferred into an individual retirement account on December 31 of that year. The money is invested at 6% effective. The person chooses to withdraw the annual accumulation of interest until age $70\frac{1}{2}$, at which time the minimum distributions are to be withdrawn. Prior to the beginning of distributions the spouse dies and the person names a daughter who will be 35 at the time the distribution begins. Generate a table of the first five years of minimum distributions.

20. An unmarried participant of a defined contribution pension plan has assets of $1,000,000 but dies prior to the beginning of distributions. A trust, in which the oldest beneficiary is 35 at the time of the participant's death, has been named the beneficiary. The money is invested at 6% effective. Determine the first five years of minimum distributions from the trust.

21. After the first distribution from an IRA, at age $70\frac{1}{2}$, there are assets of $1,000,000. The unmarried owner of the IRA dies during the first year. A trust, in which the oldest beneficiary is 55 at the time of the owner's death, has been named the beneficiary. The money is invested at 6% effective. Determine the first five years of minimum distributions from the trust.

11.5 INTEGRATION WITH SOCIAL SECURITY

The fact that the law mandates employers to contribute 6.2% of each employee's compensation, up to the FICA wage base, to the Social Security system may be taken into account when private pension plans are designed. That is, private pension plans may be integrated with Social Security because it is perceived that employers would be contributing to the employees pension by way of the matching FICA contributions.

The integration rules for defined contribution plans specify:

1. The integration level a plan may use.
2. The maximum permitted disparity in employer contribution rates in excess of the base contribution in the plan.

The integration level may be set from 20% to 100% of the FICA taxable wage base for the plan-year. If the integration level is no more than 20% of the taxable wage base, then the maximum permitted disparity is 5.7%. If the integration level is between 20% and 80% of the taxable wage base, the maximum permitted disparity if 5.4%, and if the integration level is greater than 80% of the taxable wage base, the maximum permitted disparity is 4.3%. If the base rate of the plan is less than the permitted disparity rate, then the maximum excess is twice the base rate. If the base rate of the plan is at least the permitted disparity rate, then the maximum excess is the base rate plus the permitted disparity.

The integration rules for defined benefit plans require:

1. The integration/offset level a plan may use.
2. The maximum permitted disparity in employer benefit rates.
3. Any uniform disparity.
4. Reductions for early retirement.
5. Benefits, rights, and features.

The two methods by which defined benefit plans may be integrated with Social Security are Excess and Offset. In Excess plans, an integration level is established such that the rate of contributions below that level is less than the rate of contributions above that level. In Offset plans, the pension benefit will be offset (reduced), to some extent by the anticipated Social Security benefit. Beyond recognition of the concept, integration with Social Security is beyond the level of this book.

APPENDIX

I

Summary of Formulas

CHAPTER 1

1.1 SIMPLE INTEREST

$$I = Prt \qquad t_o = \frac{N}{360}$$

$$t_e = \frac{N}{365} \qquad N = 365 - (D_1 - D_2).$$

1.2 FUTURE VALUE AND PRESENT VALUE

$$S = P + I \qquad S = P(1 + rt)$$

$$P = \frac{S}{1 + rt} \qquad P = S(1 - dt)$$

1.3 BANK DISCOUNT

$$d \le \frac{r}{n}(360) \qquad d \le \frac{r}{n}(365) \qquad d = \frac{r}{1 + rt}$$

$$r = \frac{d}{1 - dt} \qquad (1 + rt)(1 - dt) = 1$$

CHAPTER 2

2.1 EQUATIONS OF VALUE

$$\sum_{j=1}^{k} \frac{S}{1 + rn_j} = \sum_{i=1}^{m} \frac{S_i}{1 + rt_i}$$

$$\sum_{j=1}^{k} S(1 + r[t_k - n_j]) = \sum_{i=1}^{m} S_i(1 + r[t_k - t_i])$$

$$\sum_{j=1}^{k} S_j(1 - dn_j) = \sum_{i=1}^{m} S_i(1 - dt_i)$$

$$\frac{S}{1 + rn} = \sum_{i=1}^{m} \frac{S_i}{1 + rt_i}$$

$$S(1 - dn) = \sum_{i-1}^{m} S_i(1 - dt_i) \qquad Sn = \sum_{i=1}^{m} S_i t_i$$

2.2 ADD-ON INTEREST AND APPROXIMATE TRUE INTEREST RATE

$$R_F = \frac{L(L + 1)}{N(N + 1)} \qquad P_f = PL - R_F I$$

$$r_t \cong \frac{2FI}{B_0(N + 1)} \qquad r_t \cong \frac{2rN}{N + 1}$$

2.3 SHORT-TERM INVESTMENT RATE OF INTEREST

$$r = \frac{E - (B + C - W)}{Bt + C(t - t_1) - W(t - t_2)}$$

$$r = \frac{E - B - \sum F_i}{Bt + \sum F_i(t - t_i)} \qquad 1 + j_k = \frac{B_k}{B_{k-1} + F_{k-1}(t - t_k)}$$

$$r = \prod_{k=i}^{n} (1 + j_k) - 1$$

CHAPTER 3

3.1 COMPOUNDING FREQUENCY AND PERIODIC INTEREST RATE

$$i = \frac{r}{f} \qquad n = ft$$

3.2 FUTURE AND PRESENT VALUES

$$S_n = P(1 + i)^n \qquad r_e = \left[\left(1 + \frac{r}{f}\right)^f - 1\right] \times 100\%$$

$$n = \frac{\text{Ln}\left(\frac{S}{P}\right)}{\text{Ln}(1 + i)} \qquad i = \left(\frac{S}{P}\right)^{\frac{1}{n}} - 1$$

$$P = \frac{S_n}{(1 + i)^n} \qquad S_t = Pe^{rt} \qquad S_t = Pe^{krt}$$

$$P = S_t e^{-rt} \qquad P = S_t e^{-krt}$$

$$t = \frac{\text{Ln}\left(\dfrac{S}{P}\right)}{kr} \qquad t_{2\,\text{times}} \approx \frac{72}{r\%}$$

$$t_{3\,\text{times}} \approx \frac{114}{r\%} \qquad t_{5\,\text{times}} \approx \frac{167}{r\%}$$

3.3 DISCOUNT AT COMPOUND INTEREST

$$P = M(1+i)^{-n} \qquad P = Me^{-krt}$$

$$P = \frac{M}{1+rt}$$

CHAPTER 4

4.1 FUTURE AMOUNT OF AN ANNUITY AND SINKING FUND

$$S = p\left[\frac{(1+i)^n - 1}{i}\right] = ps_{\overline{n}|}$$

4.2 PRESENT VALUE OF AN ANNUITY AND AMORTIZATION

$$A = p\left[\frac{1-(1+i)^{-n}}{i}\right] = pa_{\overline{n}|}$$

$$A = L\left(1 - \frac{N}{100}\right)$$

4.3 ANNUITY DUE

$$\ddot{S} = p\left[\frac{(1+i)^n - 1}{i}\right](1+i) = ps_{\overline{n}|}(1+i)$$

$$\ddot{S} = S(1+i)$$

$$\ddot{A} = p\left[\frac{1-(1+i)^{-n}}{i}\right](1+i) = pa_{\overline{n}|}(1+i)$$

$$\ddot{A} = A(1+i)$$

4.4 PRACTICAL APPLICATIONS

$$n = \frac{\text{Ln}\left(1 + \dfrac{Si}{p}\right)}{\text{Ln}(1+i)}$$

$$n = \frac{\text{Ln}\left((1 + \dfrac{Si}{p(1+i)}\right)}{\text{Ln}(1+i)} \quad \text{for annuity due}$$

$$n = -\frac{\text{Ln}\left(1 - \dfrac{Ai}{p}\right)}{\text{Ln}(1+i)}$$

$$n = -\frac{\text{Ln}\left(1 - \dfrac{Ai}{p(1+i)}\right)}{\text{Ln}(1+i)} \quad \text{for annuity due}$$

$$B_m = A(1+i)^m - p\left[\frac{(1+i)^m - 1}{i}\right]$$

$$p_n = B_{n-1}(1+i)$$
$$B_m = (A - p)(1+i)^{m-1}$$

$$-p\left[\frac{(1+i)^{m-1} - 1}{i}\right] \quad \text{for annuity due}$$

$$P = p(1+i)^{-(n-m+1)} \qquad I = p - P$$
$$I = p[1 - (1+i)^{-(n-m+1)}]$$

$$m = n + 1 + \frac{\text{Ln}(k)}{\text{Ln}(1+i)}$$

CHAPTER 5

5.1 DEFERRED ANNUITIES

$$d = ft_d - 1 \qquad A = A_d(1+i)^d$$
$$A_d = A(1+i)^{-d}$$

$$A_d = p\left[\frac{1-(1+i)^{-n}}{i}\right](1+i)^{-d}$$

5.2 GENERAL (COMPLEX) ANNUITIES

$$(1+j)^{f'} = (1+i)^f \qquad 1+j = (1+i)^{\frac{f}{f'}}$$

$$j = (1+i)^{\frac{f}{f'}} - 1 \qquad S = p\left[\frac{(1+j)^n - 1}{j}\right]$$

$$A = p\left[\frac{1-(1+j)^{-n}}{j}\right] \qquad \ddot{S} = S(1+j)$$

$$\ddot{S} = p\left[\frac{(1+j)^n - 1}{j}\right](1+j)$$

$$\ddot{A} = p\left[\frac{1-(1+j)^{-n}}{j}\right](1+j)$$

$$\ddot{A} = A(1+j) \qquad d = f't_d - 1$$

$$A_d = p\left[\frac{1-(1+j)^{-n}}{j}\right](1+j)^{-d}$$

5.3 ANNUITY IN PERPETUITY

$$A_\infty = \frac{p}{i} \qquad \ddot{A}_\infty = \frac{p}{i} + p \qquad A_\infty = \frac{p}{j}$$

$$\ddot{A}_\infty = \frac{p}{j} + p \qquad A_{d\infty} = A_\infty(1+j)^{-d}$$

5.4 GEOMETRICALLY VARYING ANNUITIES

$$A = p_1\left[\frac{1 - \left(\dfrac{1+k}{1+i}\right)^n}{i-k}\right] \qquad A = \frac{np_1}{1+i}$$

$$S = p_1 \left[\frac{(1+i)^n - (1+k)^n}{i-k} \right]$$

$$S = np_1(1+k)^{n-1} \qquad p_n = p_1(1+k)^{n-1}$$

CHAPTER 6

6.1 PRICE OF A BOND ON AN INTEREST PAYMENT DATE

$$i = \frac{R}{f} \qquad b = \frac{B}{f} \qquad I = bF \qquad n = ft$$

$$P = M(1+i)^{-n} + bF\left[\frac{1-(1+i)^{-n}}{i}\right]$$

$$r = \frac{I}{P} \qquad R = (1+r)^2 - 1$$

$$t = Y_2 - Y_1 + \text{Fraction of } Y_0$$

6.2 PRICE OF A BOND BETWEEN INTEREST PAYMENT DATES

$$P_0 = M(1+i)^{-(n+1)} + bF\left[\frac{1-(1+i)^{-(n+1)}}{i}\right]$$

$$P = P_0(1+i)^k \qquad C = Q + FBt$$

6.3 ZERO COUPON BONDS

$$P = F(1+i)^{-n} \qquad n = f\left(t + \frac{m}{y}\right)$$

6.4 DURATION AND VOLATILITY

$$\bar{d} = \frac{\displaystyle\sum_{t=1}^{n} t(1+i)^{-t}F_t + nM(1+i)^{-n}}{\displaystyle\sum_{t=1}^{n}(1+i)^{-t}F_t + M(1+i)^{-n}}$$

$$a_{\overline{n}|} = \frac{1-(1+i)^{-n}}{i}$$

$$\bar{d} = \left[\frac{bF\left(\dfrac{a_{\overline{n}|}(1+i) - n(1+i)^{-n}}{i}\right)}{bFa_{\overline{n}|} + M(1+i)^{-n}} + nM(1+i)^{-n}\right]$$

$$\bar{v} = \frac{\bar{d}}{1+i}$$

CHAPTER 7

Chapter 7 does not have any specific-use formulas.

CHAPTER 8

8.1 FUNDAMENTAL BREAKEVEN ANALYSIS

$$C(q) = cq + C_0 \qquad R(q) = sq$$
$$P(q) = R(q) - C(q) \qquad q = q't$$
$$C(t) = cq't + C_0 \qquad R(t) = sq'(t - t_R)$$

$$t_B = \frac{C_0 + sq't_R}{(s-c)q'} \qquad I = (C_0 + cq't_R)rt_L$$

$$t = \frac{(C_0 + cq't_R)rt_0 + sq't_R + C_0}{(s-c)q' - (C_0 + cq't_R)r}$$

8.2 BREAKEVEN BASED ON SUPPLY AND DEMAND

$$P_D = p'q + P_0 \qquad P_S = p'q + P_0$$

$$R(q) = p'q^2 + P_0 q \qquad q_{v(\text{Max}R)} = -\frac{P_0}{2p'}$$

$$P(q) = p'q^2 + (P_0 - c)q - C_0$$

$$q_{v(\text{Max}P)} = -\frac{P_0 - c}{2p'} \qquad x = \frac{-b \pm \sqrt{b^2 - 4ac}}{2a}$$

8.3 BREAKEVEN BASED ON FINANCE

$$C = (1 + r_c)cN \qquad R = (1 - r_s)sN$$

$$s = \left(\frac{1 + r_c}{1 - r_s}\right)c \qquad R_E = P_E t$$

$$R_N = P_N(t - t_N) \qquad t_B = \frac{P_L t_L}{P_L - P_0}$$

$$t_B = \frac{\text{Ln}\left[\dfrac{P_L - P_0}{P_L(1+r)^{-t_L} - P_0}\right]}{\text{Ln}(1 + r)}$$

CHAPTER 9

9.1 ELEMENTS OF PROBABILITY

$$P(E) = \frac{n}{N} \qquad P(A \cup B) = P(A) + P(B)$$

$$q_x = \frac{d_x}{l_x} \qquad P(A \cap B) = P(A) \times P(B)$$

9.2 MORTALITY

$$l_{x+1} = l_x - d_x \qquad d_x = l_x q_x \qquad q_x = \frac{d_x}{l_x}$$

$$d_x = l_x - l_{x+1} \qquad p_x = \frac{l_{x+1}}{l_x} \qquad p_x + q_x = 1$$

$$_n p_x = \frac{l_{x+n}}{l_x} \qquad _n p_x + {}_n q_x = 1$$

9.3 COMMUTATION FUNCTIONS

$$D_x = l_x v^x \qquad N_x = \sum_{t=x}^{\infty} D_t$$

$$C_x = d_x v^{x+1} \qquad M_x = \sum_{t=0}^{\infty} C_t$$

9.4 LIFE ANNUITIES

$$A_x = P \left[\frac{l_{x+n}}{l_x} \right] v^n \qquad A_x = P \left[\frac{D_{x+n}}{D_x} \right]$$

$$A_x = P \sum_{t=1}^{x+n} \left(\frac{l_{x+t}}{l_x} \right) v^t \qquad N_x = \sum_{t=x}^{\infty} D_t$$

$$A_{x:\overline{n}|} = P \left[\frac{N_{x+t} - N_{x+t+n}}{D_x} \right] \qquad A_x = P \left[\frac{N_{x+t}}{D_x} \right]$$

$$A_{x:\overline{n_2}|} = P_1 \left[\frac{N_{x+t} - N_{x+t+n_1}}{D_x} \right]_{r_1}$$

$$+ P_2 \left[\frac{N_{x+t+n_1} - N_{x+t+n_1+n_2}}{D_x} \right]_{r_2}$$

$$A_x = P_1 \left[\frac{N_{x+t} - N_{x+t+n_1}}{D_x} \right]_{r_1}$$

$$+ P_2 \left[\frac{N_{x+t+n_1}}{D_x} \right]_{r_2} \qquad \ddot{A}_x = A_x + P$$

$$A_x = P_1 \left[\frac{N_{x+t} - N_{x+t+n}}{D_x} \right] + P_2 \left[\frac{N_{x+t+n}}{D_x} \right]$$

9.5 LIFE INSURANCE: NET SINGLE PREMIUM

$$A_x = 1{,}000 \left(\frac{d_x v}{l_x} \right) \qquad A_x = B \left(\frac{C_x}{D_x} \right)$$

$$A_{x:\overline{n}|} = B \left[\frac{M_x - M_{x+n}}{D_x} \right] \qquad A_x = B \left[\frac{M_x}{D_x} \right]$$

$$A_x = B \left[\frac{D_{x+n}}{D_x} \right] + B \left[\frac{M_x - M_{x+n}}{D_x} \right]$$

$$A_x = B \left[\frac{M_x - M_{x+n} + D_{x+n}^E}{D_x} \right]$$

9.6 LIFE INSURANCE: NET ANNUAL PREMIUM

$${}_nP_x = B \left[\frac{M_x - M_{x+n}}{N_x - N_{x+n}} \right] \qquad P_x = B \left[\frac{M_x}{N_x} \right]$$

$${}_nP_x = B \left[\frac{M_x}{N_x - N_{x+n}} \right]$$

$${}_nP_x = B \left[\frac{M_x - M_{x+n} + D_{x+n}^E}{N_x - N_{x+n}} \right]$$

9.7 LIFE INSURANCE RESERVES

$${}_tV_x = \frac{(Pl_t + V_{t-1})(1+i) - Bd_t}{l_{t+1}}$$

$${}_tV_x = \frac{P[N_x - N_{x+t}] - B[M_x - M_{x+t}]}{D_{x+t}}$$

$$t \le n$$

$${}_tV_x = \frac{P[N_x - N_{x+n}] - B[M_x - M_{x+t}]}{D_{x+t}}$$

$$t \ge n \qquad A_{x+t} = {}_tV_x + {}_tA_x$$

$${}_tV_x = \frac{B[M_{x+t} - M_{x+n}] - P_x[N_{x+t} - N_{x+n}]}{D_{x+t}}$$

$${}_tV_x = \frac{B[M_{x+t} - M_{x+n} + D_{x+t}^E] - P_x[N_{x+t} - N_{x+n}]}{D_{x+t}}$$

$$M_{t+n} = M_t + \frac{{}_tV_x D_t}{B} \qquad B = {}_tV_x \left(\frac{D_t}{M_t} \right)$$

CHAPTER 10

10.1 A BRIEF HISTORY OF SOCIAL SECURITY

$$R_w = \frac{5}{9}(780 - M_a)\%$$

$$R_w = \frac{5}{9}(M_n - M_a)\%, \qquad M_n - M_a \le 36$$

$$R_w = 20\% + \frac{5}{12}(M_n - M_a - 36)\%,$$

$$M_n - M_a > 36 \qquad R_s = \frac{25}{36}(780 - M_a)\%$$

$$R_s = \frac{25}{36}(M_n - M_a)\%, \qquad M_n - M_a \le 36$$

$$R_s = 25\% + \frac{5}{12}(M_n - M_a - 36)\%,$$

$$M_n - M_a > 36$$

$$PIA_s = \left(1 - \frac{R_s}{100} \right)(0.50)PIA_w$$

10.2 DETERMINATION OF BENEFITS

$$B_d = \left[1 + \left(\frac{DRC}{12} \right)(M_d - M_n) \right] PIA_n$$

CHAPTER 11

11.1 NO SPECIFIC FORMULAS

11.2 DEFINED BENEFIT PENSION PLANS

$$\ddot{A}_x = a_x \left(\frac{S}{12} \right) \qquad s = \frac{(1+i)^{RA-(AA-1)} - 1}{i}$$

$$C = \frac{\ddot{A}_x}{s} \qquad L = \ddot{A}_x - B(1+i)^{RA-AA} \qquad C = \frac{L}{s}$$

$$C' = \frac{a_x}{s} \qquad C = C'PE \qquad N = E + C$$

$$E = \frac{N}{1 + PC'} \qquad \ddot{A}_x = a_x \left(\frac{E}{12} \right)$$

11.3 DEFINED CONTRIBUTION PENSION PLANS

$$C = PS \qquad C = P(N - C) \qquad C = \left(\frac{P}{1 + P} \right) N$$

$$P = \frac{C}{S} \qquad F = Ps \qquad P_e = \frac{F_k}{s_e}$$

$$C_e = P_e S_e \qquad A_x = a_x (1 + i)^{RA - AA}$$

$$A = A_x S \qquad B = \frac{T}{\sum A} \qquad C = BA$$

$$F = C(1 + i)^{RA - AA} \qquad M = \frac{F}{a_x}$$

11.4 DISTRIBUTIONS FROM PENSION PLANS

$$\ddot{A}_x = Ma_x \qquad M = \frac{\ddot{A}}{a_x}$$

$$R = \frac{\ddot{A}}{a_x} \text{ for minimum distributions}$$

APPENDIX

II

Tables

TABLE A-1 JULIAN DATES												
Day	Jan	Feb	Mar	April	May	June	July	Aug	Sept	Oct	Nov	Dec	Day
1	1	32	60	91	121	152	182	213	244	274	305	335	1
2	2	33	61	92	122	153	183	214	245	275	306	336	2
3	3	34	62	93	123	154	184	215	246	276	307	337	3
4	4	35	63	94	124	155	185	216	247	277	308	338	4
5	5	36	64	95	125	156	186	217	248	278	309	339	5
6	6	37	65	96	126	157	187	218	249	279	310	340	6
7	7	38	66	97	127	158	188	219	250	280	311	341	7
8	8	39	67	98	128	159	189	220	251	281	312	342	8
9	9	40	68	99	129	160	190	221	252	282	313	343	9
10	10	41	69	100	130	161	191	222	253	283	314	344	10
11	11	42	70	101	131	162	192	223	254	284	315	345	11
12	12	43	71	102	132	163	193	224	255	285	316	346	12
13	13	44	72	103	133	164	194	225	256	286	317	347	13
14	14	45	73	104	134	165	195	226	257	287	318	348	14
15	15	46	74	105	135	166	196	227	258	288	319	349	15
16	16	47	75	106	136	167	197	228	259	289	320	350	16
17	17	48	76	107	137	168	198	229	260	290	321	351	17
18	18	49	77	108	138	169	199	230	261	291	322	352	18
19	19	50	78	109	139	170	200	231	262	292	323	353	19
20	20	51	79	110	140	171	201	232	263	293	324	354	20
21	21	52	80	111	141	172	202	233	264	294	325	355	21
22	22	53	81	112	142	173	203	234	265	295	326	356	22
23	23	54	82	113	143	174	204	235	266	296	327	357	23
24	24	55	83	114	144	175	205	236	267	297	328	358	24
25	25	56	84	115	145	176	206	237	268	298	329	359	25
26	26	57	85	116	146	177	207	238	269	299	330	360	26
27	27	58	86	117	147	178	208	239	270	300	331	361	27
28	28	59	87	118	148	179	209	240	271	301	332	362	28
29	29		88	119	149	180	210	241	272	302	333	363	29
30	30		89	120	150	181	211	242	273	303	334	364	30
31	31		90		151		212	243		304		365	31

Note: Add one to each day after February 28 in leap years.

TABLE A-2 FUTURE AMOUNT OF $1.00 $(1 + i)^n$

n	$\frac{5}{12}$%	$\frac{1}{2}$%	$\frac{7}{12}$%	$\frac{2}{3}$%	$\frac{3}{4}$%	1%	2%	3%	n
1	1.00417	1.00500	1.00583	1.00667	1.00750	1.01000	1.02000	1.03000	1
2	1.00835	1.01003	1.01170	1.01338	1.01506	1.02010	1.04040	1.06090	2
3	1.01255	1.01508	1.01760	1.02013	1.02267	1.03030	1.06121	1.09273	3
4	1.01677	1.02015	1.02354	1.02693	1.03034	1.04060	1.08243	1.12551	4
5	1.02101	1.02525	1.02951	1.03378	1.03807	1.05101	1.10408	1.15927	5
6	1.02526	1.03038	1.03551	1.04067	1.04585	1.06152	1.12616	1.19405	6
7	1.02953	1.03553	1.04155	1.04761	1.05370	1.07214	1.14869	1.22987	7
8	1.03382	1.04071	1.04763	1.05459	1.06160	1.08286	1.17166	1.26677	8
9	1.03813	1.04591	1.05374	1.06163	1.06956	1.09369	1.19509	1.30477	9
10	1.04246	1.05114	1.05989	1.06870	1.07758	1.10462	1.21899	1.34392	10
11	1.04680	1.05640	1.06607	1.07583	1.08566	1.11567	1.24337	1.38423	11
12	1.05116	1.06168	1.07229	1.08300	1.09381	1.12683	1.26824	1.42576	12
13	1.05554	1.06699	1.07855	1.09022	1.10201	1.13809	1.29361	1.46853	13
14	1.05994	1.07232	1.08484	1.09749	1.11028	1.14947	1.31948	1.51259	14
15	1.06436	1.07768	1.09116	1.10480	1.11860	1.16097	1.34587	1.55797	15
16	1.06879	1.08307	1.09753	1.11217	1.12699	1.17258	1.37279	1.60471	16
17	1.07324	1.08849	1.10393	1.11958	1.13544	1.18430	1.40024	1.65285	17
18	1.07772	1.09393	1.11037	1.12705	1.14396	1.19615	1.42825	1.70243	18
19	1.08221	1.09940	1.11685	1.13456	1.15254	1.20811	1.45681	1.75351	19
20	1.08672	1.10490	1.12336	1.14213	1.16118	1.22019	1.48595	1.80611	20
21	1.09124	1.11042	1.12992	1.14974	1.16989	1.23239	1.51567	1.86029	21
22	1.09579	1.11597	1.13651	1.15740	1.17867	1.24472	1.54598	1.91610	22
23	1.10036	1.12155	1.14314	1.16512	1.18751	1.25716	1.57690	1.97359	23
24	1.10494	1.12716	1.14981	1.17289	1.19641	1.26973	1.60844	2.03279	24
25	1.10955	1.13280	1.15651	1.18071	1.20539	1.28243	1.64061	2.09378	25
26	1.11417	1.13846	1.16326	1.18858	1.21443	1.29526	1.67342	2.15659	26
27	1.11881	1.14415	1.17005	1.19650	1.22354	1.30821	1.70689	2.22129	27
28	1.12347	1.14987	1.17687	1.20448	1.23271	1.32129	1.74102	2.28793	28
29	1.12815	1.15562	1.18374	1.21251	1.24196	1.33450	1.77584	2.35657	29
30	1.13285	1.16140	1.19064	1.22059	1.25127	1.34785	1.81136	2.42726	30
31	1.13757	1.16721	1.19759	1.22873	1.26066	1.36133	1.84759	2.50008	31
32	1.14231	1.17304	1.20457	1.23692	1.27011	1.37494	1.88454	2.57508	32
33	1.14707	1.17891	1.21160	1.24517	1.27964	1.38869	1.92223	2.65234	33
34	1.15185	1.18480	1.21867	1.25347	1.28923	1.40258	1.96068	2.73191	34
35	1.15665	1.19073	1.22578	1.26182	1.29890	1.41660	1.99989	2.81386	35
36	1.16147	1.19668	1.23293	1.27024	1.30865	1.43077	2.03989	2.89828	36
37	1.16631	1.20266	1.24012	1.27871	1.31846	1.44508	2.08069	2.98523	37
38	1.17117	1.20868	1.24735	1.28723	1.32835	1.45953	2.12230	3.07478	38
39	1.17605	1.21472	1.25463	1.29581	1.33831	1.47412	2.16474	3.16703	39
40	1.18095	1.22079	1.26195	1.30445	1.34835	1.48886	2.20804	3.26204	40

TABLE A-2 *continued*

n	$\frac{5}{12}\%$	$\frac{1}{2}\%$	$\frac{7}{12}\%$	$\frac{2}{3}\%$	$\frac{3}{4}\%$	1%	2%	3%	n
41	1.18587	1.22690	1.26931	1.31315	1.35846	1.50375	2.25220	3.35990	41
42	1.19081	1.23303	1.27671	1.32190	1.36865	1.51879	2.29724	3.46070	42
43	1.19577	1.23920	1.28416	1.33071	1.37891	1.53398	2.34319	3.56452	43
44	1.20076	1.24539	1.29165	1.33959	1.38926	1.54932	2.39005	3.67145	44
45	1.20576	1.25162	1.29919	1.34852	1.39968	1.56481	2.43785	3.78160	45
46	1.21078	1.25788	1.30676	1.35751	1.41017	1.58046	2.48661	3.89504	46
47	1.21583	1.26417	1.31439	1.36656	1.42075	1.59626	2.53634	4.01190	47
48	1.22090	1.27049	1.32205	1.37567	1.43141	1.61223	2.58707	4.13225	48
49	1.22598	1.27684	1.32977	1.38484	1.44214	1.62835	2.63881	4.25622	49
50	1.23109	1.28323	1.33752	1.39407	1.45296	1.64463	2.69159	4.38391	50
51	1.23622	1.28964	1.34533	1.40336	1.46385	1.66108	2.74542	4.51542	51
52	1.24137	1.29609	1.35317	1.41272	1.47483	1.67769	2.80033	4.65089	52
53	1.24654	1.30257	1.36107	1.42214	1.48589	1.69447	2.85633	4.79041	53
54	1.25174	1.30908	1.36901	1.43162	1.49704	1.71141	2.91346	4.93412	54
55	1.25695	1.31563	1.37699	1.44116	1.50827	1.72852	2.97173	5.08215	55
56	1.26219	1.32221	1.38502	1.45077	1.51958	1.74581	3.03117	5.23461	56
57	1.26745	1.32882	1.39310	1.46044	1.53098	1.76327	3.09179	5.39165	57
58	1.27273	1.33546	1.40123	1.47018	1.54246	1.78090	3.15362	5.55340	58
59	1.27803	1.34214	1.40940	1.47998	1.55403	1.79871	3.21670	5.72000	59
60	1.28336	1.34885	1.41763	1.48985	1.56568	1.81670	3.28103	5.89160	60
61	1.28871	1.35559	1.42589	1.49978	1.57742	1.83486	3.34665	6.06835	61
62	1.29408	1.36237	1.43421	1.50978	1.58925	1.85321	3.41358	6.25040	62
63	1.29947	1.36918	1.44258	1.51984	1.60117	1.87174	3.48186	6.43791	63
64	1.30488	1.37603	1.45099	1.52997	1.61318	1.89046	3.55149	6.63105	64
65	1.31032	1.38291	1.45946	1.54017	1.62528	1.90937	3.62252	6.82998	65
66	1.31578	1.38982	1.46797	1.55044	1.63747	1.92846	3.69497	7.03488	66
67	1.32126	1.39677	1.47653	1.56078	1.64975	1.94774	3.76887	7.24593	67
68	1.32677	1.40376	1.48515	1.57118	1.66213	1.96722	3.84425	7.46331	68
69	1.33229	1.41078	1.49381	1.58166	1.67459	1.98689	3.92114	7.68721	69
70	1.33785	1.41783	1.50252	1.59220	1.68715	2.00676	3.99956	7.91782	70
71	1.34342	1.42492	1.51129	1.60282	1.69980	2.02683	4.07955	8.15536	71
72	1.34902	1.43204	1.52011	1.61350	1.71255	2.04710	4.16114	8.40002	72
73	1.35464	1.43920	1.52897	1.62426	1.72540	2.06757	4.24436	8.65202	73
74	1.36028	1.44640	1.53789	1.63509	1.73834	2.08825	4.32925	8.91158	74
75	1.36595	1.45363	1.54686	1.64599	1.75137	2.10913	4.41584	9.17893	75
76	1.37164	1.46090	1.55589	1.65696	1.76451	2.13022	4.50415	9.45429	76
77	1.37736	1.46821	1.56496	1.66801	1.77774	2.15152	4.59424	9.73792	77
78	1.38310	1.47555	1.57409	1.67913	1.79108	2.17304	4.68612	10.03006	78
79	1.38886	1.48292	1.58327	1.69032	1.80451	2.19477	4.77984	10.33096	79
80	1.39465	1.49034	1.59251	1.70159	1.81804	2.21672	4.87544	10.64089	80

TABLE A-2 *continued*

n	$\frac{5}{12}$%	$\frac{1}{2}$%	$\frac{7}{12}$%	$\frac{2}{3}$%	$\frac{3}{4}$%	1%	2%	3%	n
81	1.40046	1.49779	1.60180	1.71293	1.83168	2.23888	4.97295	10.96012	81
82	1.40629	1.50528	1.61114	1.72435	1.84542	2.26127	5.07241	11.28892	82
83	1.41215	1.51281	1.62054	1.73585	1.85926	2.28388	5.17386	11.62759	83
84	1.41804	1.52037	1.62999	1.74742	1.87320	2.30672	5.27733	11.97642	84
85	1.42394	1.52797	1.63950	1.75907	1.88725	2.32979	5.38288	12.33571	85
86	1.42988	1.53561	1.64907	1.77080	1.90141	2.35309	5.49054	12.70578	86
87	1.43584	1.54329	1.65869	1.78260	1.91567	2.37662	5.60035	13.08695	87
88	1.44182	1.55101	1.66836	1.79449	1.93003	2.40038	5.71235	13.47956	88
89	1.44783	1.55876	1.67809	1.80645	1.94451	2.42439	5.82660	13.88395	89
90	1.45386	1.56655	1.68788	1.81849	1.95909	2.44863	5.94313	14.30047	90
91	1.45992	1.57439	1.69773	1.83062	1.97379	2.47312	6.06200	14.72948	91
92	1.46600	1.58226	1.70763	1.84282	1.98859	2.49785	6.18324	15.17137	92
93	1.47211	1.59017	1.71759	1.85511	2.00350	2.52283	6.30690	15.62651	93
94	1.47824	1.59812	1.72761	1.86747	2.01853	2.54806	6.43304	16.09530	94
95	1.48440	1.60611	1.73769	1.87992	2.03367	2.57354	6.56170	16.57816	95
96	1.49059	1.61414	1.74783	1.89246	2.04892	2.59927	6.69293	17.07551	96
97	1.49680	1.62221	1.75802	1.90507	2.06429	2.62527	6.82679	17.58777	97
98	1.50303	1.63032	1.76828	1.91777	2.07977	2.65152	6.96333	18.11540	98
99	1.50930	1.63848	1.77859	1.93056	2.09537	2.67803	7.10259	18.65887	99
100	1.51558	1.64667	1.78897	1.94343	2.11108	2.70481	7.24465	19.21863	100
101	1.52190	1.65490	1.79940	1.95639	2.12692	2.73186	7.38954	19.79519	101
102	1.52824	1.66318	1.80990	1.96943	2.14287	2.75918	7.53733	20.38905	102
103	1.53461	1.67149	1.82046	1.98256	2.15894	2.78677	7.68808	21.00072	103
104	1.54100	1.67985	1.83108	1.99577	2.17513	2.81464	7.84184	21.63074	104
105	1.54742	1.68825	1.84176	2.00908	2.19145	2.84279	7.99867	22.27966	105
106	1.55387	1.69669	1.85250	2.02247	2.20788	2.87121	8.15865	22.94805	106
107	1.56035	1.70517	1.86331	2.03596	2.22444	2.89993	8.32182	23.63649	107
108	1.56685	1.71370	1.87418	2.04953	2.24112	2.92893	8.48826	24.34559	108
109	1.57338	1.72227	1.88511	2.06319	2.25793	2.95822	8.65802	25.07596	109
110	1.57993	1.73088	1.89611	2.07695	2.27487	2.98780	8.83118	25.82823	110
111	1.58651	1.73953	1.90717	2.09079	2.29193	3.01768	9.00781	26.60308	111
112	1.59312	1.74823	1.91829	2.10473	2.30912	3.04785	9.18796	27.40117	112
113	1.59976	1.75697	1.92948	2.11876	2.32644	3.07833	9.37172	28.22321	113
114	1.60643	1.76576	1.94074	2.13289	2.34388	3.10911	9.55916	29.06991	114
115	1.61312	1.77459	1.95206	2.14711	2.36146	3.14020	9.75034	29.94200	115
116	1.61984	1.78346	1.96345	2.16142	2.37917	3.17161	9.94535	30.84026	116
117	1.62659	1.79238	1.97490	2.17583	2.39702	3.20332	10.14425	31.76547	117
118	1.63337	1.80134	1.98642	2.19034	2.41500	3.23536	10.34714	32.71843	118
119	1.64018	1.81035	1.99801	2.20494	2.43311	3.26771	10.55408	33.69999	119
120	1.64701	1.81940	2.00966	2.21964	2.45136	3.30039	10.76516	34.71099	120

TABLE A-2 *continued*

n	4%	5%	6%	7%	8%	9%	10%	12%	n
1	1.04000	1.05000	1.06000	1.07000	1.08000	1.09000	1.10000	1.12000	1
2	1.08160	1.10250	1.12360	1.14490	1.16640	1.18810	1.21000	1.25440	2
3	1.12486	1.15763	1.19102	1.22504	1.25971	1.29503	1.33100	1.40493	3
4	1.16986	1.21551	1.26248	1.31080	1.36049	1.41158	1.46410	1.57352	4
5	1.21665	1.27628	1.33823	1.40255	1.46933	1.53862	1.61051	1.76234	5
6	1.26532	1.34010	1.41852	1.50073	1.58687	1.67710	1.77156	1.97382	6
7	1.31593	1.40710	1.50363	1.60578	1.71382	1.82804	1.94872	2.21068	7
8	1.36857	1.47746	1.59385	1.71819	1.85093	1.99256	2.14359	2.47596	8
9	1.42331	1.55133	1.68948	1.83846	1.99900	2.17189	2.35795	2.77308	9
10	1.48024	1.62889	1.79085	1.96715	2.15892	2.36736	2.59374	3.10585	10
11	1.53945	1.71034	1.89830	2.10485	2.33164	2.58043	2.85312	3.47855	11
12	1.60103	1.79586	2.01220	2.25219	2.51817	2.81266	3.13843	3.89598	12
13	1.66507	1.88565	2.13293	2.40985	2.71962	3.06580	3.45227	4.36349	13
14	1.73168	1.97993	2.26090	2.57853	2.93719	3.34173	3.79750	4.88711	14
15	1.80094	2.07893	2.39656	2.75903	3.17217	3.64248	4.17725	5.47357	15
16	1.87298	2.18287	2.54035	2.95216	3.42594	3.97031	4.59497	6.13039	16
17	1.94790	2.29202	2.69277	3.15882	3.70002	4.32763	5.05447	6.86604	17
18	2.02582	2.40662	2.85434	3.37993	3.99602	4.71712	5.55992	7.68997	18
19	2.10685	2.52695	3.02560	3.61653	4.31570	5.14166	6.11591	8.61276	19
20	2.19112	2.65330	3.20714	3.86968	4.66096	5.60441	6.72750	9.64629	20
21	2.27877	2.78596	3.39956	4.14056	5.03383	6.10881	7.40025	10.80385	21
22	2.36992	2.92526	3.60354	4.43040	5.43654	6.65860	8.14027	12.10031	22
23	2.46472	3.07152	3.81975	4.74053	5.87146	7.25787	8.95430	13.55235	23
24	2.56330	3.22510	4.04893	5.07237	6.34118	7.91108	9.84973	15.17863	24
25	2.66584	3.38635	4.29187	5.42743	6.84848	8.62308	10.83471	17.00006	25
26	2.77247	3.55567	4.54938	5.80735	7.39635	9.39916	11.91818	19.04007	26
27	2.88337	3.73346	4.82235	6.21387	7.98806	10.24508	13.10999	21.32488	27
28	2.99870	3.92013	5.11169	6.64884	8.62711	11.16714	14.42099	23.88387	28
29	3.11865	4.11614	5.41839	7.11426	9.31727	12.17218	15.86309	26.74993	29
30	3.24340	4.32194	5.74349	7.61226	10.06266	13.26768	17.44940	29.95992	30
31	3.37313	4.53804	6.08810	8.14511	10.86767	14.46177	19.19434	33.55511	31
32	3.50806	4.76494	6.45339	8.71527	11.73708	15.76333	21.11378	37.58173	32
33	3.64838	5.00319	6.84059	9.32534	12.67605	17.18203	23.22515	42.09153	33
34	3.79432	5.25335	7.25103	9.97811	13.69013	18.72841	25.54767	47.14252	34
35	3.94609	5.51602	7.68609	10.67658	14.78534	20.41397	28.10244	52.79962	35
36	4.10393	5.79182	8.14725	11.42394	15.96817	22.25123	30.91268	59.13557	36
37	4.26809	6.08141	8.63609	12.22362	17.24563	24.25384	34.00395	66.23184	37
38	4.43881	6.38548	9.15425	13.07927	18.62528	26.43668	37.40434	74.17966	38
39	4.61637	6.70475	9.70351	13.99482	20.11530	28.81598	41.14478	83.08122	39
40	4.80102	7.03999	10.28572	14.97446	21.72452	31.40942	45.25926	93.05097	40

TABLE A-2 *continued*

n	4%	5%	6%	7%	8%	9%	10%	12%	n
41	4.99306	7.39199	10.90286	16.02267	23.46248	34.23627	49.78518	104.21709	41
42	5.19278	7.76159	11.55703	17.14426	25.33948	37.31753	54.76370	116.72314	42
43	5.40050	8.14967	12.25045	18.34435	27.36664	40.67611	60.24007	130.72991	43
44	5.61652	8.55715	12.98548	19.62846	29.55597	44.33696	66.26408	146.41750	44
45	5.84118	8.98501	13.76461	21.00245	31.92045	48.32729	72.89048	163.98760	45
46	6.07482	9.43426	14.59049	22.47262	34.47409	52.67674	80.17953	183.66612	46
47	6.31782	9.90597	15.46592	24.04571	37.23201	57.41765	88.19749	205.70605	47
48	6.57053	10.40127	16.39387	25.72891	40.21057	62.58524	97.01723	230.39078	48
49	6.83335	10.92133	17.37750	27.52993	43.42742	68.21791	106.71896	258.03767	49
50	7.10668	11.46740	18.42015	29.45703	46.90161	74.35752	117.39085	289.00219	50
51	7.39095	12.04077	19.52536	31.51902	50.65374	81.04970	129.12994	323.68245	51
52	7.68659	12.64281	20.69689	33.72535	54.70604	88.34417	142.04293	362.52435	52
53	7.99405	13.27495	21.93870	36.08612	59.08252	96.29514	156.24723	406.02727	53
54	8.31381	13.93870	23.25502	38.61215	63.80913	104.96171	171.87195	454.75054	54
55	8.64637	14.63563	24.65032	41.31500	68.91386	114.40826	189.05914	509.32061	55
56	8.99222	15.36741	26.12934	44.20705	74.42696	124.70501	207.96506	570.43908	56
57	9.35191	16.13578	27.69710	47.30155	80.38112	135.92846	228.76156	638.89177	57
58	9.72599	16.94257	29.35893	50.61265	86.81161	148.16202	251.63772	715.55878	58
59	10.11503	17.78970	31.12046	54.15554	93.75654	161.49660	276.80149	801.42583	59
60	10.51963	18.67919	32.98769	57.94643	101.25706	176.03129	304.48164	897.59693	60
61	10.94041	19.61315	34.96695	62.00268	109.35763	191.87411	334.92980	1005.30857	61
62	11.37803	20.59380	37.06497	66.34286	118.10624	209.14278	368.42278	1125.94559	62
63	11.83315	21.62349	39.28887	70.98686	127.55474	227.96563	405.26506	1261.05906	63
64	12.30648	22.70467	41.64620	75.95595	137.75912	248.48253	445.79157	1412.38615	64
65	12.79874	23.83990	44.14497	81.27286	148.77985	270.84596	490.37073	1581.87249	65
66	13.31068	25.03190	46.79367	86.96196	160.68223	295.22210	539.40780	1771.69719	66
67	13.84311	26.28349	49.60129	93.04930	173.53681	321.79209	593.34858	1984.30085	67
68	14.39684	27.59766	52.57737	99.56275	187.41976	350.75338	652.68344	2222.41695	68
69	14.97271	28.97755	55.73201	106.53214	202.41334	382.32118	717.95178	2489.10699	69
70	15.57162	30.42643	59.07593	113.98939	218.60641	416.73009	789.74696	2787.79983	70
71	16.19448	31.94775	62.62049	121.96865	236.09492	454.23579	868.72165	3122.33581	71
72	16.84226	33.54513	66.37772	130.50646	254.98251	495.11702	955.59382	3497.01610	72
73	17.51595	35.22239	70.36038	139.64191	275.38111	539.67755	1051.15320	3916.65804	73
74	18.21659	36.98351	74.58200	149.41684	297.41160	588.24853	1156.26852	4386.65700	74
75	18.94525	38.83269	79.05692	159.87602	321.20453	641.19089	1271.89537	4913.05584	75
76	19.70306	40.77432	83.80034	171.06734	346.90089	698.89807	1399.08491	5502.62254	76
77	20.49119	42.81304	88.82836	183.04205	374.65296	761.79890	1538.99340	6162.93725	77
78	21.31083	44.95369	94.15806	195.85500	404.62520	830.36080	1692.89274	6902.48972	78
79	22.16327	47.20137	99.80754	209.56485	436.99522	905.09327	1862.18201	7730.78848	79
80	23.04980	49.56144	105.79599	224.23439	471.95483	986.55167	2048.40021	8658.48310	80

TABLE A-3 PRESENT VALUE OF $1.00 (1 + i)^{-n}$

n	$\frac{5}{12}\%$	$\frac{1}{2}\%$	$\frac{7}{12}\%$	$\frac{2}{3}\%$	$\frac{3}{4}\%$	1%	2%	3%	n
1	0.99585	0.99502	0.99420	0.99338	0.99256	0.99010	0.98039	0.97087	1
2	0.99172	0.99007	0.98843	0.98680	0.98517	0.98030	0.96117	0.94260	2
3	0.98760	0.98515	0.98270	0.98026	0.97783	0.97059	0.94232	0.91514	3
4	0.98351	0.98025	0.97700	0.97377	0.97055	0.96098	0.92385	0.88849	4
5	0.97942	0.97537	0.97134	0.96732	0.96333	0.95147	0.90573	0.86261	5
6	0.97536	0.97052	0.96570	0.96092	0.95616	0.94205	0.88797	0.83748	6
7	0.97131	0.96569	0.96010	0.95455	0.94904	0.93272	0.87056	0.81309	7
8	0.96728	0.96089	0.95453	0.94823	0.94198	0.92348	0.85349	0.78941	8
9	0.96327	0.95610	0.94900	0.94195	0.93496	0.91434	0.83676	0.76642	9
10	0.95927	0.95135	0.94350	0.93571	0.92800	0.90529	0.82035	0.74409	10
11	0.95529	0.94661	0.93802	0.92952	0.92109	0.89632	0.80426	0.72242	11
12	0.95133	0.94191	0.93258	0.92336	0.91424	0.88745	0.78849	0.70138	12
13	0.94738	0.93722	0.92717	0.91725	0.90743	0.87866	0.77303	0.68095	13
14	0.94345	0.93256	0.92180	0.91117	0.90068	0.86996	0.75788	0.66112	14
15	0.93954	0.92792	0.91645	0.90514	0.89397	0.86135	0.74301	0.64186	15
16	0.93564	0.92330	0.91114	0.89914	0.88732	0.85282	0.72845	0.62317	16
17	0.93175	0.91871	0.90585	0.89319	0.88071	0.84438	0.71416	0.60502	17
18	0.92789	0.91414	0.90060	0.88727	0.87416	0.83602	0.70016	0.58739	18
19	0.92404	0.90959	0.89538	0.88140	0.86765	0.82774	0.68643	0.57029	19
20	0.92020	0.90506	0.89018	0.87556	0.86119	0.81954	0.67297	0.55368	20
21	0.91639	0.90056	0.88502	0.86976	0.85478	0.81143	0.65978	0.53755	21
22	0.91258	0.89608	0.87989	0.86400	0.84842	0.80340	0.64684	0.52189	22
23	0.90880	0.89162	0.87479	0.85828	0.84210	0.79544	0.63416	0.50669	23
24	0.90503	0.88719	0.86971	0.85260	0.83583	0.78757	0.62172	0.49193	24
25	0.90127	0.88277	0.86467	0.84695	0.82961	0.77977	0.60953	0.47761	25
26	0.89753	0.87838	0.85965	0.84134	0.82343	0.77205	0.59758	0.46369	26
27	0.89381	0.87401	0.85467	0.83577	0.81730	0.76440	0.58586	0.45019	27
28	0.89010	0.86966	0.84971	0.83023	0.81122	0.75684	0.57437	0.43708	28
29	0.88640	0.86533	0.84478	0.82474	0.80518	0.74934	0.56311	0.42435	29
30	0.88273	0.86103	0.83988	0.81927	0.79919	0.74192	0.55207	0.41199	30
31	0.87906	0.85675	0.83501	0.81385	0.79324	0.73458	0.54125	0.39999	31
32	0.87542	0.85248	0.83017	0.80846	0.78733	0.72730	0.53063	0.38834	32
33	0.87178	0.84824	0.82536	0.80310	0.78147	0.72010	0.52023	0.37703	33
34	0.86817	0.84402	0.82057	0.79779	0.77565	0.71297	0.51003	0.36604	34
35	0.86456	0.83982	0.81581	0.79250	0.76988	0.70591	0.50003	0.35538	35
36	0.86098	0.83564	0.81108	0.78725	0.76415	0.69892	0.49022	0.34503	36
37	0.85740	0.83149	0.80638	0.78204	0.75846	0.69200	0.48061	0.33498	37
38	0.85385	0.82735	0.80170	0.77686	0.75281	0.68515	0.47119	0.32523	38
39	0.85030	0.82323	0.79705	0.77172	0.74721	0.67837	0.46195	0.31575	39
40	0.84677	0.81914	0.79243	0.76661	0.74165	0.67165	0.45289	0.30656	40

TABLE A-3 *continued*

n	$\frac{5}{12}\%$	$\frac{1}{2}\%$	$\frac{7}{12}\%$	$\frac{2}{3}\%$	$\frac{3}{4}\%$	1%	2%	3%	n
41	0.84326	0.81506	0.78783	0.76153	0.73613	0.66500	0.44401	0.29763	41
42	0.83976	0.81101	0.78326	0.75649	0.73065	0.65842	0.43530	0.28896	42
43	0.83628	0.80697	0.77872	0.75148	0.72521	0.65190	0.42677	0.28054	43
44	0.83281	0.80296	0.77420	0.74650	0.71981	0.64545	0.41840	0.27237	44
45	0.82935	0.79896	0.76971	0.74156	0.71445	0.63905	0.41020	0.26444	45
46	0.82591	0.79499	0.76525	0.73665	0.70913	0.63273	0.40215	0.25674	46
47	0.82248	0.79103	0.76081	0.73177	0.70385	0.62646	0.39427	0.24926	47
48	0.81907	0.78710	0.75640	0.72692	0.69861	0.62026	0.38654	0.24200	48
49	0.81567	0.78318	0.75201	0.72211	0.69341	0.61412	0.37896	0.23495	49
50	0.81229	0.77929	0.74765	0.71732	0.68825	0.60804	0.37153	0.22811	50
51	0.80892	0.77541	0.74331	0.71257	0.68313	0.60202	0.36424	0.22146	51
52	0.80556	0.77155	0.73900	0.70785	0.67804	0.59606	0.35710	0.21501	52
53	0.80222	0.76771	0.73472	0.70317	0.67300	0.59016	0.35010	0.20875	53
54	0.79889	0.76389	0.73046	0.69851	0.66799	0.58431	0.34323	0.20267	54
55	0.79557	0.76009	0.72622	0.69388	0.66301	0.57853	0.33650	0.19677	55
56	0.79227	0.75631	0.72201	0.68929	0.65808	0.57280	0.32991	0.19104	56
57	0.78899	0.75255	0.71782	0.68472	0.65318	0.56713	0.32344	0.18547	57
58	0.78571	0.74880	0.71366	0.68019	0.64832	0.56151	0.31710	0.18007	58
59	0.78245	0.74508	0.70952	0.67569	0.64349	0.55595	0.31088	0.17483	59
60	0.77921	0.74137	0.70541	0.67121	0.63870	0.55045	0.30478	0.16973	60
61	0.77597	0.73768	0.70131	0.66677	0.63395	0.54500	0.29881	0.16479	61
62	0.77275	0.73401	0.69725	0.66235	0.62923	0.53960	0.29295	0.15999	62
63	0.76955	0.73036	0.69320	0.65796	0.62454	0.53426	0.28720	0.15533	63
64	0.76635	0.72673	0.68918	0.65361	0.61989	0.52897	0.28157	0.15081	64
65	0.76317	0.72311	0.68519	0.64928	0.61528	0.52373	0.27605	0.14641	65
66	0.76001	0.71952	0.68121	0.64498	0.61070	0.51855	0.27064	0.14215	66
67	0.75685	0.71594	0.67726	0.64071	0.60615	0.51341	0.26533	0.13801	67
68	0.75371	0.71237	0.67333	0.63646	0.60164	0.50833	0.26013	0.13399	68
69	0.75058	0.70883	0.66943	0.63225	0.59716	0.50330	0.25503	0.13009	69
70	0.74747	0.70530	0.66555	0.62806	0.59272	0.49831	0.25003	0.12630	70
71	0.74437	0.70179	0.66169	0.62390	0.58830	0.49338	0.24513	0.12262	71
72	0.74128	0.69830	0.65785	0.61977	0.58392	0.48850	0.24032	0.11905	72
73	0.73820	0.69483	0.65403	0.61567	0.57958	0.48366	0.23561	0.11558	73
74	0.73514	0.69137	0.65024	0.61159	0.57526	0.47887	0.23099	0.11221	74
75	0.73209	0.68793	0.64647	0.60754	0.57098	0.47413	0.22646	0.10895	75
76	0.72905	0.68451	0.64272	0.60351	0.56673	0.46944	0.22202	0.10577	76
77	0.72603	0.68110	0.63899	0.59952	0.56251	0.46479	0.21766	0.10269	77
78	0.72302	0.67772	0.63529	0.59555	0.55832	0.46019	0.21340	0.09970	78
79	0.72002	0.67434	0.63160	0.59160	0.55417	0.45563	0.20921	0.09680	79
80	0.71703	0.67099	0.62794	0.58769	0.55004	0.45112	0.20511	0.09398	80

TABLE A-3 *continued*

n	$\frac{5}{12}\%$	$\frac{1}{2}\%$	$\frac{7}{12}\%$	$\frac{2}{3}\%$	$\frac{3}{4}\%$	1%	2%	3%	n
81	0.71405	0.66765	0.62430	0.58379	0.54595	0.44665	0.20109	0.09124	81
82	0.71109	0.66433	0.62068	0.57993	0.54188	0.44223	0.19715	0.08858	82
83	0.70814	0.66102	0.61708	0.57609	0.53785	0.43785	0.19328	0.08600	83
84	0.70520	0.65773	0.61350	0.57227	0.53385	0.43352	0.18949	0.08350	84
85	0.70227	0.65446	0.60994	0.56848	0.52987	0.42922	0.18577	0.08107	85
86	0.69936	0.65121	0.60640	0.56472	0.52593	0.42497	0.18213	0.07870	86
87	0.69646	0.64797	0.60289	0.56098	0.52201	0.42077	0.17856	0.07641	87
88	0.69357	0.64474	0.59939	0.55726	0.51813	0.41660	0.17506	0.07419	88
89	0.69069	0.64154	0.59591	0.55357	0.51427	0.41248	0.17163	0.07203	89
90	0.68782	0.63834	0.59246	0.54991	0.51044	0.40839	0.16826	0.06993	90
91	0.68497	0.63517	0.58902	0.54626	0.50664	0.40435	0.16496	0.06789	91
92	0.68213	0.63201	0.58561	0.54265	0.50287	0.40034	0.16173	0.06591	92
93	0.67930	0.62886	0.58221	0.53905	0.49913	0.39638	0.15856	0.06399	93
94	0.67648	0.62573	0.57883	0.53548	0.49541	0.39246	0.15545	0.06213	94
95	0.67367	0.62262	0.57548	0.53194	0.49172	0.38857	0.15240	0.06032	95
96	0.67088	0.61952	0.57214	0.52841	0.48806	0.38472	0.14941	0.05856	96
97	0.66809	0.61644	0.56882	0.52491	0.48443	0.38091	0.14648	0.05686	97
98	0.66532	0.61337	0.56552	0.52144	0.48082	0.37714	0.14361	0.05520	98
99	0.66256	0.61032	0.56224	0.51798	0.47724	0.37341	0.14079	0.05359	99
100	0.65981	0.60729	0.55898	0.51455	0.47369	0.36971	0.13803	0.05203	100
101	0.65707	0.60427	0.55574	0.51115	0.47016	0.36605	0.13533	0.05052	101
102	0.65435	0.60126	0.55252	0.50776	0.46666	0.36243	0.13267	0.04905	102
103	0.65163	0.59827	0.54931	0.50440	0.46319	0.35884	0.13007	0.04762	103
104	0.64893	0.59529	0.54613	0.50106	0.45974	0.35529	0.12752	0.04623	104
105	0.64624	0.59233	0.54296	0.49774	0.45632	0.35177	0.12502	0.04488	105
106	0.64355	0.58938	0.53981	0.49444	0.45292	0.34828	0.12257	0.04358	106
107	0.64088	0.58645	0.53668	0.49117	0.44955	0.34484	0.12017	0.04231	107
108	0.63822	0.58353	0.53357	0.48792	0.44620	0.34142	0.11781	0.04108	108
109	0.63558	0.58063	0.53047	0.48469	0.44288	0.33804	0.11550	0.03988	109
110	0.63294	0.57774	0.52740	0.48148	0.43959	0.33469	0.11324	0.03872	110
111	0.63031	0.57487	0.52434	0.47829	0.43631	0.33138	0.11101	0.03759	111
112	0.62770	0.57201	0.52130	0.47512	0.43307	0.32810	0.10884	0.03649	112
113	0.62509	0.56916	0.51827	0.47197	0.42984	0.32485	0.10670	0.03543	113
114	0.62250	0.56633	0.51527	0.46885	0.42664	0.32164	0.10461	0.03440	114
115	0.61992	0.56351	0.51228	0.46574	0.42347	0.31845	0.10256	0.03340	115
116	0.61734	0.56071	0.50931	0.46266	0.42031	0.31530	0.10055	0.03243	116
117	0.61478	0.55792	0.50636	0.45959	0.41718	0.31218	0.09858	0.03148	117
118	0.61223	0.55514	0.50342	0.45655	0.41408	0.30908	0.09665	0.03056	118
119	0.60969	0.55238	0.50050	0.45353	0.41100	0.30602	0.09475	0.02967	119
120	0.60716	0.54963	0.49760	0.45052	0.40794	0.30299	0.09289	0.02881	120

TABLE A-3 *continued*

n	4%	5%	6%	7%	8%	9%	10%	12%	n
1	0.96154	0.95238	0.94340	0.93458	0.92593	0.91743	0.90909	0.89286	1
2	0.92456	0.90703	0.89000	0.87344	0.85734	0.84168	0.82645	0.79719	2
3	0.88900	0.86384	0.83962	0.81630	0.79383	0.77218	0.75131	0.71178	3
4	0.85480	0.82270	0.79209	0.76290	0.73503	0.70843	0.68301	0.63552	4
5	0.82193	0.78353	0.74726	0.71299	0.68058	0.64993	0.62092	0.56743	5
6	0.79031	0.74622	0.70496	0.66634	0.63017	0.59627	0.56447	0.50663	6
7	0.75992	0.71068	0.66506	0.62275	0.58349	0.54703	0.51316	0.45235	7
8	0.73069	0.67684	0.62741	0.58201	0.54027	0.50187	0.46651	0.40388	8
9	0.70259	0.64461	0.59190	0.54393	0.50025	0.46043	0.42410	0.36061	9
10	0.67556	0.61391	0.55839	0.50835	0.46319	0.42241	0.38554	0.32197	10
11	0.64958	0.58468	0.52679	0.47509	0.42888	0.38753	0.35049	0.28748	11
12	0.62460	0.55684	0.49697	0.44401	0.39711	0.35553	0.31863	0.25668	12
13	0.60057	0.53032	0.46884	0.41496	0.36770	0.32618	0.28966	0.22917	13
14	0.57748	0.50507	0.44230	0.38782	0.34046	0.29925	0.26333	0.20462	14
15	0.55526	0.48102	0.41727	0.36245	0.31524	0.27454	0.23939	0.18270	15
16	0.53391	0.45811	0.39365	0.33873	0.29189	0.25187	0.21763	0.16312	16
17	0.51337	0.43630	0.37136	0.31657	0.27027	0.23107	0.19784	0.14564	17
18	0.49363	0.41552	0.35034	0.29586	0.25025	0.21199	0.17986	0.13004	18
19	0.47464	0.39573	0.33051	0.27651	0.23171	0.19449	0.16351	0.11611	19
20	0.45639	0.37689	0.31180	0.25842	0.21455	0.17843	0.14864	0.10367	20
21	0.43883	0.35894	0.29416	0.24151	0.19866	0.16370	0.13513	0.09256	21
22	0.42196	0.34185	0.27751	0.22571	0.18394	0.15018	0.12285	0.08264	22
23	0.40573	0.32557	0.26180	0.21095	0.17032	0.13778	0.11168	0.07379	23
24	0.39012	0.31007	0.24698	0.19715	0.15770	0.12640	0.10153	0.06588	24
25	0.37512	0.29530	0.23300	0.18425	0.14602	0.11597	0.09230	0.05882	25
26	0.36069	0.28124	0.21981	0.17220	0.13520	0.10639	0.08391	0.05252	26
27	0.34682	0.26785	0.20737	0.16093	0.12519	0.09761	0.07628	0.04689	27
28	0.33348	0.25509	0.19563	0.15040	0.11591	0.08955	0.06934	0.04187	28
29	0.32065	0.24295	0.18456	0.14056	0.10733	0.08215	0.06304	0.03738	29
30	0.30832	0.23138	0.17411	0.13137	0.09938	0.07537	0.05731	0.03338	30
31	0.29646	0.22036	0.16425	0.12277	0.09202	0.06915	0.05210	0.02980	31
32	0.28506	0.20987	0.15496	0.11474	0.08520	0.06344	0.04736	0.02661	32
33	0.27409	0.19987	0.14619	0.10723	0.07889	0.05820	0.04306	0.02376	33
34	0.26355	0.19035	0.13791	0.10022	0.07305	0.05339	0.03914	0.02121	34
35	0.25342	0.18129	0.13011	0.09366	0.06763	0.04899	0.03558	0.01894	35
36	0.24367	0.17266	0.12274	0.08754	0.06262	0.04494	0.03235	0.01691	36
37	0.23430	0.16444	0.11579	0.08181	0.05799	0.04123	0.02941	0.01510	37
38	0.22529	0.15661	0.10924	0.07646	0.05369	0.03783	0.02673	0.01348	38
39	0.21662	0.14915	0.10306	0.07146	0.04971	0.03470	0.02430	0.01204	39
40	0.20829	0.14205	0.09722	0.06678	0.04603	0.03184	0.02209	0.01075	40

TABLE A-3 *continued*

n	4%	5%	6%	7%	8%	9%	10%	12%	n
41	0.20028	0.13528	0.09172	0.06241	0.04262	0.02921	0.02009	0.00960	41
42	0.19257	0.12884	0.08653	0.05833	0.03946	0.02680	0.01826	0.00857	42
43	0.18517	0.12270	0.08163	0.05451	0.03654	0.02458	0.01660	0.00765	43
44	0.17805	0.11686	0.07701	0.05095	0.03383	0.02255	0.01509	0.00683	44
45	0.17120	0.11130	0.07265	0.04761	0.03133	0.02069	0.01372	0.00610	45
46	0.16461	0.10600	0.06854	0.04450	0.02901	0.01898	0.01247	0.00544	46
47	0.15828	0.10095	0.06466	0.04159	0.02686	0.01742	0.01134	0.00486	47
48	0.15219	0.09614	0.06100	0.03887	0.02487	0.01598	0.01031	0.00434	48
49	0.14634	0.09156	0.05755	0.03632	0.02303	0.01466	0.00937	0.00388	49
50	0.14071	0.08720	0.05429	0.03395	0.02132	0.01345	0.00852	0.00346	50
51	0.13530	0.08305	0.05122	0.03173	0.01974	0.01234	0.00774	0.00309	51
52	0.13010	0.07910	0.04832	0.02965	0.01828	0.01132	0.00704	0.00276	52
53	0.12509	0.07533	0.04558	0.02771	0.01693	0.01038	0.00640	0.00246	53
54	0.12028	0.07174	0.04300	0.02590	0.01567	0.00953	0.00582	0.00220	54
55	0.11566	0.06833	0.04057	0.02420	0.01451	0.00874	0.00529	0.00196	55
56	0.11121	0.06507	0.03827	0.02262	0.01344	0.00802	0.00481	0.00175	56
57	0.10693	0.06197	0.03610	0.02114	0.01244	0.00736	0.00437	0.00157	57
58	0.10282	0.05902	0.03406	0.01976	0.01152	0.00675	0.00397	0.00140	58
59	0.09886	0.05621	0.03213	0.01847	0.01067	0.00619	0.00361	0.00125	59
60	0.09506	0.05354	0.03031	0.01726	0.00988	0.00568	0.00328	0.00111	60
61	0.09140	0.05099	0.02860	0.01613	0.00914	0.00521	0.00299	0.00099	61
62	0.08789	0.04856	0.02698	0.01507	0.00847	0.00478	0.00271	0.00089	62
63	0.08451	0.04625	0.02545	0.01409	0.00784	0.00439	0.00247	0.00079	63
64	0.08126	0.04404	0.02401	0.01317	0.00726	0.00402	0.00224	0.00071	64
65	0.07813	0.04195	0.02265	0.01230	0.00672	0.00369	0.00204	0.00063	65
66	0.07513	0.03995	0.02137	0.01150	0.00622	0.00339	0.00185	0.00056	66
67	0.07224	0.03805	0.02016	0.01075	0.00576	0.00311	0.00169	0.00050	67
68	0.06946	0.03623	0.01902	0.01004	0.00534	0.00285	0.00153	0.00045	68
69	0.06679	0.03451	0.01794	0.00939	0.00494	0.00262	0.00139	0.00040	69
70	0.06422	0.03287	0.01693	0.00877	0.00457	0.00240	0.00127	0.00036	70
71	0.06175	0.03130	0.01597	0.00820	0.00424	0.00220	0.00115	0.00032	71
72	0.05937	0.02981	0.01507	0.00766	0.00392	0.00202	0.00105	0.00029	72
73	0.05709	0.02839	0.01421	0.00716	0.00363	0.00185	0.00095	0.00026	73
74	0.05490	0.02704	0.01341	0.00669	0.00336	0.00170	0.00086	0.00023	74
75	0.05278	0.02575	0.01265	0.00625	0.00311	0.00156	0.00079	0.00020	75
76	0.05075	0.02453	0.01193	0.00585	0.00288	0.00143	0.00071	0.00018	76
77	0.04880	0.02336	0.01126	0.00546	0.00267	0.00131	0.00065	0.00016	77
78	0.04692	0.02225	0.01062	0.00511	0.00247	0.00120	0.00059	0.00014	78
79	0.04512	0.02119	0.01002	0.00477	0.00229	0.00110	0.00054	0.00013	79
80	0.04338	0.02018	0.00945	0.00446	0.00212	0.00101	0.00049	0.00012	80

TABLE A-4 FUTURE AMOUNT OF $1.00 PER PERIOD $s_{\overline{n}|} = \dfrac{(1+i)^n - 1}{i}$

n	$\frac{5}{12}\%$	$\frac{1}{2}\%$	$\frac{7}{12}\%$	$\frac{2}{3}\%$	$\frac{3}{4}\%$	1%	2%	3%	n
1	1.00000	1.00000	1.00000	1.00000	1.00000	1.00000	1.00000	1.00000	1
2	2.00417	2.00500	2.00583	2.00667	2.00750	2.01000	2.02000	2.03000	2
3	3.01252	3.01502	3.01753	3.02004	3.02256	3.03010	3.06040	3.09090	3
4	4.02507	4.03010	4.03514	4.04018	4.04523	4.06040	4.12161	4.18363	4
5	5.04184	5.05025	5.05867	5.06711	5.07556	5.10101	5.20404	5.30914	5
6	6.06285	6.07550	6.08818	6.10089	6.11363	6.15202	6.30812	6.46841	6
7	7.08811	7.10588	7.12370	7.14157	7.15948	7.21354	7.43428	7.66246	7
8	8.11764	8.14141	8.16525	8.18918	8.21318	8.28567	8.58297	8.89234	8
9	9.15147	9.18212	9.21288	9.24377	9.27478	9.36853	9.75463	10.15911	9
10	10.18960	10.22803	10.26663	10.30540	10.34434	10.46221	10.94972	11.46388	10
11	11.23206	11.27917	11.32651	11.37410	11.42192	11.56683	12.16872	12.80780	11
12	12.27886	12.33556	12.39259	12.44993	12.50759	12.68250	13.41209	14.19203	12
13	13.33002	13.39724	13.46488	13.53293	13.60139	13.80933	14.68033	15.61779	13
14	14.38556	14.46423	14.54342	14.62315	14.70340	14.94742	15.97394	17.08632	14
15	15.44550	15.53655	15.62826	15.72063	15.81368	16.09690	17.29342	18.59891	15
16	16.50986	16.61423	16.71942	16.82544	16.93228	17.25786	18.63929	20.15688	16
17	17.57865	17.69730	17.81695	17.93761	18.05927	18.43044	20.01207	21.76159	17
18	18.65189	18.78579	18.92088	19.05719	19.19472	19.61475	21.41231	23.41444	18
19	19.72961	19.87972	20.03126	20.18424	20.33868	20.81090	22.84056	25.11687	19
20	20.81181	20.97912	21.14810	21.31880	21.49122	22.01900	24.29737	26.87037	20
21	21.89853	22.08401	22.27147	22.46093	22.65240	23.23919	25.78332	28.67649	21
22	22.98977	23.19443	23.40139	23.61066	23.82230	24.47159	27.29898	30.53678	22
23	24.08556	24.31040	24.53789	24.76807	25.00096	25.71630	28.84496	32.45288	23
24	25.18592	25.43196	25.68103	25.93319	26.18847	26.97346	30.42186	34.42647	24
25	26.29086	26.55912	26.83084	27.10608	27.38488	28.24320	32.03030	36.45926	25
26	27.40041	27.69191	27.98735	28.28678	28.59027	29.52563	33.67091	38.55304	26
27	28.51458	28.83037	29.15061	29.47536	29.80470	30.82089	35.34432	40.70963	27
28	29.63339	29.97452	30.32066	30.67187	31.02823	32.12910	37.05121	42.93092	28
29	30.75686	31.12439	31.49753	31.87634	32.26094	33.45039	38.79223	45.21885	29
30	31.88501	32.28002	32.68126	33.08885	33.50290	34.78489	40.56808	47.57542	30
31	33.01787	33.44142	33.87190	34.30945	34.75417	36.13274	42.37944	50.00268	31
32	34.15544	34.60862	35.06949	35.53818	36.01483	37.49407	44.22703	52.50276	32
33	35.29776	35.78167	36.27406	36.77510	37.28494	38.86901	46.11157	55.07784	33
34	36.44483	36.96058	37.48566	38.02026	38.56458	40.25770	48.03380	57.73018	34
35	37.59668	38.14538	38.70433	39.27373	39.85381	41.66028	49.99448	60.46208	35
36	38.75334	39.33610	39.93010	40.53556	41.15272	43.07688	51.99437	63.27594	36
37	39.91481	40.53279	41.16303	41.80579	42.46136	44.50765	54.03425	66.17422	37
38	41.08112	41.73545	42.40314	43.08450	43.77982	45.95272	56.11494	69.15945	38
39	42.25229	42.94413	43.65050	44.37173	45.10817	47.41225	58.23724	72.23423	39
40	43.42834	44.15885	44.90512	45.66754	46.44648	48.88637	60.40198	75.40126	40

TABLE A-4 *continued*

n	$\frac{5}{12}\%$	$\frac{1}{2}\%$	$\frac{7}{12}\%$	$\frac{2}{3}\%$	$\frac{3}{4}\%$	1%	2%	3%	n
41	44.60929	45.37964	46.16707	46.97199	47.79483	50.37524	62.61002	78.66330	41
42	45.79517	46.60654	47.43638	48.28514	49.15329	51.87899	64.86222	82.02320	42
43	46.98598	47.83957	48.71309	49.60704	50.52194	53.39778	67.15947	85.48389	43
44	48.18175	49.07877	49.99725	50.93775	51.90086	54.93176	69.50266	89.04841	44
45	49.38251	50.32416	51.28890	52.27734	53.29011	56.48107	71.89271	92.71986	45
46	50.58827	51.57578	52.58809	53.62585	54.68979	58.04589	74.33056	96.50146	46
47	51.79906	52.83366	53.89485	54.98336	56.09996	59.62634	76.81718	100.39650	47
48	53.01489	54.09783	55.20924	56.34992	57.52071	61.22261	79.35352	104.40840	48
49	54.23578	55.36832	56.53129	57.72558	58.95212	62.83483	81.94059	108.54065	49
50	55.46176	56.64516	57.86106	59.11042	60.39426	64.46318	84.57940	112.79687	50
51	56.69285	57.92839	59.19858	60.50449	61.84721	66.10781	87.27099	117.18077	51
52	57.92907	59.21803	60.54390	61.90785	63.31107	67.76889	90.01641	121.69620	52
53	59.17045	60.51412	61.89708	63.32057	64.78590	69.44658	92.81674	126.34708	53
54	60.41699	61.81669	63.25814	64.74271	66.27180	71.14105	95.67307	131.13749	54
55	61.66873	63.12577	64.62715	66.17433	67.76883	72.85246	98.58653	136.07162	55
56	62.92568	64.44140	66.00414	67.61549	69.27710	74.58098	101.55826	141.15377	56
57	64.18787	65.76361	67.38916	69.06626	70.79668	76.32679	104.58943	146.38838	57
58	65.45532	67.09243	68.78227	70.52670	72.32765	78.09006	107.68122	151.78003	58
59	66.72805	68.42789	70.18350	71.99688	73.87011	79.87096	110.83484	157.33343	59
60	68.00608	69.77003	71.59290	73.47686	75.42414	81.66967	114.05154	163.05344	60
61	69.28944	71.11888	73.01053	74.96670	76.98982	83.48637	117.33257	168.94504	61
62	70.57815	72.47448	74.43642	76.46648	78.56724	85.32123	120.67922	175.01339	62
63	71.87222	73.83685	75.87063	77.97626	80.15650	87.17444	124.09281	181.26379	63
64	73.17169	75.20603	77.31321	79.49610	81.75767	89.04619	127.57466	187.70171	64
65	74.47657	76.58206	78.76421	81.02607	83.37085	90.93665	131.12616	194.33276	65
66	75.78689	77.96497	80.22366	82.56625	84.99613	92.84602	134.74868	201.16274	66
67	77.10267	79.35480	81.69164	84.11669	86.63360	94.77448	138.44365	208.19762	67
68	78.42393	80.75157	83.16817	85.67747	88.28336	96.72222	142.21253	215.44355	68
69	79.75070	82.15533	84.65332	87.24865	89.94548	98.68944	146.05678	222.90686	69
70	81.08299	83.56611	86.14713	88.83031	91.62007	100.67634	149.97791	230.59406	70
71	82.42084	84.98394	87.64965	90.42251	93.30722	102.68310	153.97747	238.51189	71
72	83.76426	86.40886	89.16094	92.02533	95.00703	104.70993	158.05702	246.66724	72
73	85.11328	87.84090	90.68105	93.63883	96.71958	106.75703	162.21816	255.06726	73
74	86.46791	89.28010	92.21002	95.26309	98.44498	108.82460	166.46252	263.71928	74
75	87.82820	90.72650	93.74791	96.89817	100.18331	110.91285	170.79177	272.63086	75
76	89.19415	92.18014	95.29478	98.54416	101.93469	113.02198	175.20761	281.80978	76
77	90.56579	93.64104	96.85066	100.20112	103.69920	115.15220	179.71176	291.26407	77
78	91.94315	95.10924	98.41562	101.86913	105.47694	117.30372	184.30600	301.00200	78
79	93.32624	96.58479	99.98972	103.54826	107.26802	119.47675	188.99212	311.03206	79
80	94.71510	98.06771	101.57299	105.23858	109.07253	121.67152	193.77196	321.36302	80

TABLE A-4 *continued*

n	$\frac{5}{12}\%$	$\frac{1}{2}\%$	$\frac{7}{12}\%$	$\frac{2}{3}\%$	$\frac{3}{4}\%$	1%	2%	3%	n
81	96.10975	99.55805	103.16550	106.94017	110.89057	123.88824	198.64740	332.00391	81
82	97.51021	101.05584	104.76730	108.65310	112.72225	126.12712	203.62034	342.96403	82
83	98.91650	102.56112	106.37844	110.37746	114.56767	128.38839	208.69275	354.25295	83
84	100.32865	104.07393	107.99898	112.11331	116.42693	130.67227	213.86661	365.88054	84
85	101.74669	105.59430	109.62897	113.86073	118.30013	132.97900	219.14394	377.85695	85
86	103.17063	107.12227	111.26848	115.61980	120.18738	135.30879	224.52682	390.19266	86
87	104.60051	108.65788	112.91754	117.39060	122.08879	137.66187	230.01735	402.89844	87
88	106.03635	110.20117	114.57623	119.17320	124.00445	140.03849	235.61770	415.98539	88
89	107.47816	111.75217	116.24459	120.96769	125.93449	142.43888	241.33006	429.46496	89
90	108.92599	113.31094	117.92268	122.77414	127.87899	144.86327	247.15666	443.34890	90
91	110.37985	114.87749	119.61057	124.59264	129.83809	147.31190	253.09979	457.64937	91
92	111.83976	116.45188	121.30829	126.42326	131.81187	149.78502	259.16179	472.37885	92
93	113.30576	118.03414	123.01593	128.26608	133.80046	152.28287	265.34502	487.55022	93
94	114.77787	119.62431	124.73352	130.12118	135.80397	154.80570	271.65192	503.17672	94
95	116.25611	121.22243	126.46113	131.98866	137.82250	157.35376	278.08496	519.27203	95
96	117.74051	122.82854	128.19882	133.86858	139.85616	159.92729	284.64666	535.85019	96
97	119.23110	124.44268	129.94665	135.76104	141.90508	162.52657	291.33959	552.92569	97
98	120.72789	126.06490	131.70467	137.66611	143.96937	165.15183	298.16638	570.51346	98
99	122.23093	127.69522	133.47295	139.58389	146.04914	167.80335	305.12971	588.62887	99
100	123.74022	129.33370	135.25154	141.51445	148.14451	170.48138	312.23231	607.28773	100
101	125.25581	130.98037	137.04051	143.45788	150.25560	173.18620	319.47695	626.50636	101
102	126.77771	132.63527	138.83991	145.41426	152.38251	175.91806	326.86649	646.30156	102
103	128.30595	134.29845	140.64981	147.38369	154.52538	178.67724	334.40382	666.69060	103
104	129.84055	135.96994	142.47027	149.36625	156.68432	181.46401	342.09190	687.69132	104
105	131.38156	137.64979	144.30134	151.36202	158.85945	184.27865	349.93374	709.32206	105
106	132.92898	139.33804	146.14310	153.37110	161.05090	187.12144	357.93241	731.60172	106
107	134.48285	141.03473	147.99560	155.39358	163.25878	189.99265	366.09106	754.54977	107
108	136.04320	142.73990	149.85891	157.42954	165.48322	192.89258	374.41288	778.18627	108
109	137.61004	144.45360	151.73309	159.47907	167.72435	195.82151	382.90114	802.53185	109
110	139.18342	146.17587	153.61820	161.54226	169.98228	198.77972	391.55916	827.60781	110
111	140.76335	147.90675	155.51430	163.61921	172.25715	201.76752	400.39034	853.43604	111
112	142.34986	149.64628	157.42147	165.71000	174.54908	204.78519	409.39815	880.03913	112
113	143.94299	151.39451	159.33976	167.81474	176.85819	207.83304	418.58611	907.44030	113
114	145.54275	153.15148	161.26924	169.93350	179.18463	210.91137	427.95783	935.66351	114
115	147.14918	154.91724	163.20998	172.06639	181.52851	214.02049	437.51699	964.73341	115
116	148.76230	156.69183	165.16204	174.21350	183.88998	217.16069	447.26733	994.67542	116
117	150.38214	158.47529	167.12548	176.37492	186.26915	220.33230	457.21268	1025.51568	117
118	152.00873	160.26766	169.10038	178.55076	188.66617	223.53562	467.35693	1057.28115	118
119	153.64210	162.06900	171.08680	180.74109	191.08117	226.77098	477.70407	1089.99958	119
120	155.28228	163.87935	173.08481	182.94604	193.51428	230.03869	488.25815	1123.69957	120

TABLE A-4 *continued*

n	4%	5%	6%	7%	8%	9%	10%	12%	n
1	1.0000	1.0000	1.0000	1.0000	1.0000	1.0000	1.0000	1.0000	1
2	2.0400	2.0500	2.0600	2.0700	2.0800	2.0900	2.1000	2.1200	2
3	3.1216	3.1525	3.1836	3.2149	3.2464	3.2781	3.3100	3.3744	3
4	4.2465	4.3101	4.3746	4.4399	4.5061	4.5731	4.6410	4.7793	4
5	5.4163	5.5256	5.6371	5.7507	5.8666	5.9847	6.1051	6.3528	5
6	6.6330	6.8019	6.9753	7.1533	7.3359	7.5233	7.7156	8.1152	6
7	7.8983	8.1420	8.3938	8.6540	8.9228	9.2004	9.4872	10.0890	7
8	9.2142	9.5491	9.8975	10.2598	10.6366	11.0285	11.4359	12.2997	8
9	10.5828	11.0266	11.4913	11.9780	12.4876	13.0210	13.5795	14.7757	9
10	12.0061	12.5779	13.1808	13.8164	14.4866	15.1929	15.9374	17.5487	10
11	13.4864	14.2068	14.9716	15.7836	16.6455	17.5603	18.5312	20.6546	11
12	15.0258	15.9171	16.8699	17.8885	18.9771	20.1407	21.3843	24.1331	12
13	16.6268	17.7130	18.8821	20.1406	21.4953	22.9534	24.5227	28.0291	13
14	18.2919	19.5986	21.0151	22.5505	24.2149	26.0192	27.9750	32.3926	14
15	20.0236	21.5786	23.2760	25.1290	27.1521	29.3609	31.7725	37.2797	15
16	21.8245	23.6575	25.6725	27.8881	30.3243	33.0034	35.9497	42.7533	16
17	23.6975	25.8404	28.2129	30.8402	33.7502	36.9737	40.5447	48.8837	17
18	25.6454	28.1324	30.9057	33.9990	37.4502	41.3013	45.5992	55.7497	18
19	27.6712	30.5390	33.7600	37.3790	41.4463	46.0185	51.1591	63.4397	19
20	29.7781	33.0660	36.7856	40.9955	45.7620	51.1601	57.2750	72.0524	20
21	31.9692	35.7193	39.9927	44.8652	50.4229	56.7645	64.0025	81.6987	21
22	34.2480	38.5052	43.3923	49.0057	55.4568	62.8733	71.4027	92.5026	22
23	36.6179	41.4305	46.9958	53.4361	60.8933	69.5319	79.5430	104.6029	23
24	39.0826	44.5020	50.8156	58.1767	66.7648	76.7898	88.4973	118.1552	24
25	41.6459	47.7271	54.8645	63.2490	73.1059	84.7009	98.3471	133.3339	25
26	44.3117	51.1135	59.1564	68.6765	79.9544	93.3240	109.1818	150.3339	26
27	47.0842	54.6691	63.7058	74.4838	87.3508	102.7231	121.0999	169.3740	27
28	49.9676	58.4026	68.5281	80.6977	95.3388	112.9682	134.2099	190.6989	28
29	52.9663	62.3227	73.6398	87.3465	103.9659	124.1354	148.6309	214.5828	29
30	56.0849	66.4388	79.0582	94.4608	113.2832	136.3075	164.4940	241.3327	30
31	59.3283	70.7608	84.8017	102.0730	123.3459	149.5752	181.9434	271.2926	31
32	62.7015	75.2988	90.8898	110.2182	134.2135	164.0370	201.1378	304.8477	32
33	66.2095	80.0638	97.3432	118.9334	145.9506	179.8003	222.2515	342.4294	33
34	69.8579	85.0670	104.1838	128.2588	158.6267	196.9823	245.4767	384.5210	34
35	73.6522	90.3203	111.4348	138.2369	172.3168	215.7108	271.0244	431.6635	35
36	77.5983	95.8363	119.1209	148.9135	187.1021	236.1247	299.1268	484.4631	36
37	81.7022	101.6281	127.2681	160.3374	203.0703	258.3759	330.0395	543.5987	37
38	85.9703	107.7095	135.9042	172.5610	220.3159	282.6298	364.0434	609.8305	38
39	90.4091	114.0950	145.0585	185.6403	238.9412	309.0665	401.4478	684.0102	39
40	95.0255	120.7998	154.7620	199.6351	259.0565	337.8824	442.5926	767.0914	40

TABLE A-4 *continued*

n	4%	5%	6%	7%	8%	9%	10%	12%	n
41	99.8265	127.8398	165.0477	214.6096	280.7810	369.2919	487.8518	860.1424	41
42	104.8196	135.2318	175.9505	230.6322	304.2435	403.5281	537.6370	964.3595	42
43	110.0124	142.9933	187.5076	247.7765	329.5830	440.8457	592.4007	1081.0826	43
44	115.4129	151.1430	199.7580	266.1209	356.9496	481.5218	652.6408	1211.8125	44
45	121.0294	159.7002	212.7435	285.7493	386.5056	525.8587	718.9048	1358.2300	45
46	126.8706	168.6852	226.5081	306.7518	418.4261	574.1860	791.7953	1522.2176	46
47	132.9454	178.1194	241.0986	329.2244	452.9002	626.8628	871.9749	1705.8838	47
48	139.2632	188.0254	256.5645	353.2701	490.1322	684.2804	960.1723	1911.5898	48
49	145.8337	198.4267	272.9584	378.9990	530.3427	746.8656	1057.1896	2141.9806	49
50	152.6671	209.3480	290.3359	406.5289	573.7702	815.0836	1163.9085	2400.0182	50
51	159.7738	220.8154	308.7561	435.9860	620.6718	889.4411	1281.2994	2689.0204	51
52	167.1647	232.8562	328.2814	467.5050	671.3255	970.4908	1410.4293	3012.7029	52
53	174.8513	245.4990	348.9783	501.2303	726.0316	1058.8349	1552.4723	3375.2272	53
54	182.8454	258.7739	370.9170	537.3164	785.1141	1155.1301	1708.7195	3781.2545	54
55	191.1592	272.7126	394.1720	575.9286	848.9232	1260.0918	1880.5914	4236.0050	55
56	199.8055	287.3482	418.8223	617.2436	917.8371	1374.5001	2069.6506	4745.3257	56
57	208.7978	302.7157	444.9517	661.4506	992.2640	1499.2051	2277.6156	5315.7647	57
58	218.1497	318.8514	472.6488	708.7522	1072.6451	1635.1335	2506.3772	5954.6565	58
59	227.8757	335.7940	502.0077	759.3648	1159.4568	1783.2955	2758.0149	6670.2153	59
60	237.9907	353.5837	533.1282	813.5204	1253.2133	1944.7921	3034.8164	7471.6411	60
61	248.5103	372.2629	566.1159	871.4668	1354.4704	2120.8234	3339.2980	8369.2380	61
62	259.4507	391.8760	601.0828	933.4695	1463.8280	2312.6975	3674.2278	9374.5466	62
63	270.8288	412.4699	638.1478	999.8124	1581.9342	2521.8403	4042.6506	10500.4922	63
64	282.6619	434.0933	677.4367	1070.7992	1709.4890	2749.8059	4447.9157	11761.5513	64
65	294.9684	456.7980	719.0829	1146.7552	1847.2481	2998.2885	4893.7073	13173.9374	65
66	307.7671	480.6379	763.2278	1228.0280	1996.0279	3269.1344	5384.0780	14755.8099	66
67	321.0778	505.6698	810.0215	1314.9900	2156.7102	3564.3565	5923.4858	16527.5071	67
68	334.9209	531.9533	859.6228	1408.0393	2330.2470	3886.1486	6516.8344	18511.8080	68
69	349.3177	559.5510	912.2002	1507.6020	2517.6667	4236.9020	7169.5178	20734.2249	69
70	364.2905	588.5285	967.9322	1614.1342	2720.0801	4619.2232	7887.4696	23223.3319	70
71	379.8621	618.9549	1027.0081	1728.1236	2938.6865	5035.9533	8677.2165	26011.1317	71
72	396.0566	650.9027	1089.6286	1850.0922	3174.7814	5490.1891	9545.9382	29133.4675	72
73	412.8988	684.4478	1156.0063	1980.5987	3429.7639	5985.3061	10501.5320	32630.4836	73
74	430.4148	719.6702	1226.3667	2120.2406	3705.1450	6524.9836	11552.6852	36547.1417	74
75	448.6314	756.6537	1300.9487	2269.6574	4002.5566	7113.2321	12708.9537	40933.7987	75
76	467.5766	795.4864	1380.0056	2429.5334	4323.7612	7754.4230	13980.8491	45846.8545	76
77	487.2797	836.2607	1463.8059	2600.6008	4670.6620	8453.3211	15379.9340	51349.4771	77
78	507.7709	879.0738	1552.6343	2783.6428	5045.3150	9215.1200	16918.9274	57512.4143	78
79	529.0817	924.0274	1646.7924	2979.4978	5449.9402	10045.4808	18611.8201	64414.9040	79
80	551.2450	971.2288	1746.5999	3189.0627	5886.9354	10950.5741	20474.0021	72145.6925	80

TABLE A-5 PRESENT VALUE OF $1.00 PER PERIOD $a_{\overline{n}|} = \frac{1-(1+i)^{-n}}{i}$

n	$\frac{5}{12}\%$	$\frac{1}{2}\%$	$\frac{7}{12}\%$	$\frac{2}{3}\%$	$\frac{3}{4}\%$	1%	2%	3%	n
1	0.99585	0.99502	0.99420	0.99338	0.99256	0.99010	0.98039	0.97087	1
2	1.98757	1.98510	1.98264	1.98018	1.97772	1.97040	1.94156	1.91347	2
3	2.97517	2.97025	2.96534	2.96044	2.95556	2.94099	2.88388	2.82861	3
4	3.95868	3.95050	3.94234	3.93421	3.92611	3.90197	3.80773	3.71710	4
5	4.93810	4.92587	4.91368	4.90154	4.88944	4.85343	4.71346	4.57971	5
6	5.91346	5.89638	5.87938	5.86245	5.84560	5.79548	5.60143	5.41719	6
7	6.88478	6.86207	6.83948	6.81701	6.79464	6.72819	6.47199	6.23028	7
8	7.85206	7.82296	7.79402	7.76524	7.73661	7.65168	7.32548	7.01969	8
9	8.81533	8.77906	8.74302	8.70719	8.67158	8.56602	8.16224	7.78611	9
10	9.77460	9.73041	9.68651	9.64290	9.59958	9.47130	8.98259	8.53020	10
11	10.72989	10.67703	10.62454	10.57242	10.52067	10.36763	9.78685	9.25262	11
12	11.68122	11.61893	11.55712	11.49578	11.43491	11.25508	10.57534	9.95400	12
13	12.62860	12.55615	12.48430	12.41303	12.34235	12.13374	11.34837	10.63496	13
14	13.57205	13.48871	13.40609	13.32420	13.24302	13.00370	12.10625	11.29607	14
15	14.51159	14.41662	14.32254	14.22934	14.13699	13.86505	12.84926	11.93794	15
16	15.44722	15.33993	15.23368	15.12848	15.02431	14.71787	13.57771	12.56110	16
17	16.37898	16.25863	16.13953	16.02167	15.90502	15.56225	14.29187	13.16612	17
18	17.30687	17.17277	17.04013	16.90894	16.77918	16.39827	14.99203	13.75351	18
19	18.23090	18.08236	17.93551	17.79034	17.64683	17.22601	15.67846	14.32380	19
20	19.15111	18.98742	18.82569	18.66590	18.50802	18.04555	16.35143	14.87747	20
21	20.06749	19.88798	19.71071	19.53566	19.36280	18.85698	17.01121	15.41502	21
22	20.98008	20.78406	20.59060	20.39967	20.21121	19.66038	17.65805	15.93692	22
23	21.88887	21.67568	21.46539	21.25795	21.05331	20.45582	18.29220	16.44361	23
24	22.79390	22.56287	22.33510	22.11054	21.88915	21.24339	18.91393	16.93554	24
25	23.69517	23.44564	23.19977	22.95749	22.71876	22.02316	19.52346	17.41315	25
26	24.59270	24.32402	24.05942	23.79883	23.54219	22.79520	20.12104	17.87684	26
27	25.48651	25.19803	24.91409	24.63460	24.35949	23.55961	20.70690	18.32703	27
28	26.37660	26.06769	25.76380	25.46484	25.17071	24.31644	21.28127	18.76411	28
29	27.26301	26.93302	26.60858	26.28957	25.97589	25.06579	21.84438	19.18845	29
30	28.14573	27.79405	27.44847	27.10885	26.77508	25.80771	22.39646	19.60044	30
31	29.02480	28.65080	28.28348	27.92270	27.56832	26.54229	22.93770	20.00043	31
32	29.90021	29.50328	29.11365	28.73116	28.35565	27.26959	23.46833	20.38877	32
33	30.77200	30.35153	29.93901	29.53426	29.13712	27.98969	23.98856	20.76579	33
34	31.64016	31.19555	30.75958	30.33205	29.91278	28.70267	24.49859	21.13184	34
35	32.50473	32.03537	31.57539	31.12455	30.68266	29.40858	24.99862	21.48722	35
36	33.36570	32.87102	32.38646	31.91181	31.44681	30.10751	25.48884	21.83225	36
37	34.22311	33.70250	33.19284	32.69385	32.20527	30.79951	25.96945	22.16724	37
38	35.07695	34.52985	33.99454	33.47071	32.95808	31.48466	26.44064	22.49246	38
39	35.92725	35.35309	34.79159	34.24243	33.70529	32.16303	26.90259	22.80822	39
40	36.77403	36.17223	35.58401	35.00903	34.44694	32.83469	27.35548	23.11477	40

	TABLE A-5 *continued*							
n	$\frac{5}{12}\%$	$\frac{1}{2}\%$	$\frac{7}{12}\%$	$\frac{2}{3}\%$	$\frac{3}{4}\%$	1%	2%	3%	n
41	37.61729	36.98729	36.37184	35.77056	35.18307	33.49969	27.79949	23.41240	41
42	38.45705	37.79830	37.15511	36.52705	35.91371	34.15811	28.23479	23.70136	42
43	39.29333	38.60527	37.93383	37.27852	36.63892	34.81001	28.66156	23.98190	43
44	40.12614	39.40823	38.70803	38.02502	37.35873	35.45545	29.07996	24.25427	44
45	40.95549	40.20720	39.47774	38.76658	38.07318	36.09451	29.49016	24.51871	45
46	41.78140	41.00219	40.24299	39.50323	38.78231	36.72724	29.89231	24.77545	46
47	42.60388	41.79322	41.00380	40.23499	39.48617	37.35370	30.28658	25.02471	47
48	43.42296	42.58032	41.76020	40.96191	40.18478	37.97396	30.67312	25.26671	48
49	44.23863	43.36350	42.51221	41.68402	40.87820	38.58808	31.05208	25.50166	49
50	45.05092	44.14279	43.25986	42.40134	41.56645	39.19612	31.42361	25.72976	50
51	45.85983	44.91820	44.00318	43.11392	42.24958	39.79814	31.78785	25.95123	51
52	46.66539	45.68975	44.74218	43.82177	42.92762	40.39419	32.14495	26.16624	52
53	47.46761	46.45746	45.47690	44.52494	43.60061	40.98435	32.49505	26.37499	53
54	48.26650	47.22135	46.20736	45.22345	44.26860	41.56866	32.83828	26.57766	54
55	49.06208	47.98145	46.93358	45.91733	44.93161	42.14719	33.17479	26.77443	55
56	49.85435	48.73776	47.65559	46.60662	45.58969	42.71999	33.50469	26.96546	56
57	50.64334	49.49031	48.37341	47.29135	46.24287	43.28712	33.82813	27.15094	57
58	51.42905	50.23911	49.08707	47.97154	46.89118	43.84863	34.14523	27.33101	58
59	52.21150	50.98419	49.79659	48.64722	47.53467	44.40459	34.45610	27.50583	59
60	52.99071	51.72556	50.50199	49.31843	48.17337	44.95504	34.76089	27.67556	60
61	53.76668	52.46324	51.20331	49.98520	48.80732	45.50004	35.05969	27.84035	61
62	54.53943	53.19726	51.90055	50.64755	49.43654	46.03964	35.35264	28.00034	62
63	55.30898	53.92762	52.59376	51.30551	50.06109	46.57390	35.63984	28.15567	63
64	56.07533	54.65435	53.28294	51.95912	50.68098	47.10287	35.92141	28.30648	64
65	56.83850	55.37746	53.96813	52.60839	51.29626	47.62661	36.19747	28.45289	65
66	57.59851	56.09698	54.64934	53.25337	51.90695	48.14516	36.46810	28.59504	66
67	58.35536	56.81291	55.32660	53.89408	52.51311	48.65857	36.73343	28.73305	67
68	59.10907	57.52529	55.99993	54.53054	53.11475	49.16690	36.99356	28.86704	68
69	59.85966	58.23411	56.66936	55.16279	53.71191	49.67020	37.24859	28.99712	69
70	60.60713	58.93942	57.33491	55.79085	54.30462	50.16851	37.49862	29.12342	70
71	61.35150	59.64121	57.99660	56.41475	54.89293	50.66190	37.74374	29.24604	71
72	62.09278	60.33951	58.65444	57.03452	55.47685	51.15039	37.98406	29.36509	72
73	62.83098	61.03434	59.30848	57.65019	56.05643	51.63405	38.21967	29.48067	73
74	63.56612	61.72571	59.95872	58.26178	56.63169	52.11292	38.45066	29.59288	74
75	64.29821	62.41365	60.60519	58.86931	57.20267	52.58705	38.67711	29.70183	75
76	65.02727	63.09815	61.24791	59.47283	57.76940	53.05649	38.89913	29.80760	76
77	65.75329	63.77926	61.88690	60.07235	58.33191	53.52127	39.11680	29.91029	77
78	66.47631	64.45697	62.52219	60.66789	58.89023	53.98146	39.33019	30.00999	78
79	67.19633	65.13132	63.15379	61.25950	59.44440	54.43709	39.53940	30.10679	79
80	67.91335	65.80231	63.78173	61.84718	59.99444	54.88821	39.74451	30.20076	80

TABLE A-5 *continued*

n	$\frac{5}{12}\%$	$\frac{1}{2}\%$	$\frac{7}{12}\%$	$\frac{2}{3}\%$	$\frac{3}{4}\%$	1%	2%	3%	n
81	68.62741	66.46996	64.40603	62.43098	60.54039	55.33486	39.94560	30.29200	81
82	69.33850	67.13428	65.02671	63.01090	61.08227	55.77709	40.14275	30.38059	82
83	70.04663	67.79531	65.64379	63.58699	61.62012	56.21494	40.33603	30.46659	83
84	70.75183	68.45304	66.25729	64.15926	62.15396	56.64845	40.52552	30.55009	84
85	71.45411	69.10750	66.86723	64.72774	62.68384	57.07768	40.71129	30.63115	85
86	72.15347	69.75871	67.47363	65.29246	63.20976	57.50265	40.89342	30.70986	86
87	72.84993	70.40668	68.07652	65.85344	63.73177	57.92342	41.07198	30.78627	87
88	73.54350	71.05142	68.67591	66.41070	64.24990	58.34002	41.24704	30.86045	88
89	74.23419	71.69296	69.27182	66.96427	64.76417	58.75249	41.41867	30.93248	89
90	74.92201	72.33130	69.86428	67.51418	65.27461	59.16088	41.58693	31.00241	90
91	75.60698	72.96647	70.45330	68.06044	65.78125	59.56523	41.75189	31.07030	91
92	76.28911	73.59847	71.03891	68.60309	66.28412	59.96557	41.91362	31.13621	92
93	76.96841	74.22734	71.62112	69.14214	66.78324	60.36195	42.07218	31.20021	93
94	77.64489	74.85307	72.19995	69.67762	67.27865	60.75441	42.22762	31.26234	94
95	78.31856	75.47569	72.77543	70.20956	67.77038	61.14298	42.38002	31.32266	95
96	78.98944	76.09522	73.34757	70.73797	68.25844	61.52770	42.52943	31.38122	96
97	79.65753	76.71166	73.91639	71.26288	68.74287	61.90862	42.67592	31.43808	97
98	80.32286	77.32503	74.48191	71.78432	69.22369	62.28576	42.81953	31.49328	98
99	80.98542	77.93536	75.04415	72.30231	69.70093	62.65917	42.96032	31.54687	99
100	81.64523	78.54264	75.60314	72.81686	70.17462	63.02888	43.09835	31.59891	100
101	82.30230	79.14691	76.15888	73.32801	70.64479	63.39493	43.23368	31.64942	101
102	82.95665	79.74817	76.71139	73.83577	71.11145	63.75736	43.36635	31.69847	102
103	83.60828	80.34644	77.26071	74.34017	71.57464	64.11619	43.49642	31.74609	103
104	84.25721	80.94173	77.80683	74.84123	72.03438	64.47148	43.62394	31.79232	104
105	84.90345	81.53406	78.34979	75.33897	72.49070	64.82325	43.74896	31.83720	105
106	85.54700	82.12344	78.88960	75.83341	72.94363	65.17153	43.87153	31.88078	106
107	86.18788	82.70989	79.42628	76.32458	73.39318	65.51637	43.99170	31.92308	107
108	86.82611	83.29342	79.95985	76.81250	73.83938	65.85779	44.10951	31.96416	108
109	87.46168	83.87405	80.49032	77.29718	74.28226	66.19583	44.22501	32.00404	109
110	88.09462	84.45180	81.01772	77.77866	74.72185	66.53053	44.33824	32.04276	110
111	88.72494	85.02666	81.54206	78.25695	75.15816	66.86191	44.44926	32.08035	111
112	89.35263	85.59867	82.06335	78.73206	75.59123	67.19001	44.55810	32.11684	112
113	89.97773	86.16783	82.58163	79.20404	76.02107	67.51486	44.66480	32.15227	113
114	90.60023	86.73416	83.09690	79.67289	76.44771	67.83649	44.76941	32.18667	114
115	91.22014	87.29767	83.60918	80.13863	76.87118	68.15494	44.87197	32.22007	115
116	91.83748	87.85838	84.11849	80.60129	77.29149	68.47024	44.97252	32.25250	116
117	92.45227	88.41630	84.62484	81.06088	77.70868	68.78242	45.07110	32.28398	117
118	93.06450	88.97144	85.12826	81.51743	78.12276	69.09150	45.16775	32.31454	118
119	93.67419	89.52382	85.62876	81.97096	78.53376	69.39753	45.26250	32.34421	119
120	94.28135	90.07345	86.12635	82.42148	78.94169	69.70052	45.35539	32.37302	120

TABLE A-5 *continued*

n	4%	5%	6%	7%	8%	9%	10%	12%	n
1	0.96154	0.95238	0.94340	0.93458	0.92593	0.91743	0.90909	0.89286	1
2	1.88609	1.85941	1.83339	1.80802	1.78326	1.75911	1.73554	1.69005	2
3	2.77509	2.72325	2.67301	2.62432	2.57710	2.53129	2.48685	2.40183	3
4	3.62990	3.54595	3.46511	3.38721	3.31213	3.23972	3.16987	3.03735	4
5	4.45182	4.32948	4.21236	4.10020	3.99271	3.88965	3.79079	3.60478	5
6	5.24214	5.07569	4.91732	4.76654	4.62288	4.48592	4.35526	4.11141	6
7	6.00205	5.78637	5.58238	5.38929	5.20637	5.03295	4.86842	4.56376	7
8	6.73274	6.46321	6.20979	5.97130	5.74664	5.53482	5.33493	4.96764	8
9	7.43533	7.10782	6.80169	6.51523	6.24689	5.99525	5.75902	5.32825	9
10	8.11090	7.72173	7.36009	7.02358	6.71008	6.41766	6.14457	5.65022	10
11	8.76048	8.30641	7.88687	7.49867	7.13896	6.80519	6.49506	5.93770	11
12	9.38507	8.86325	8.38384	7.94269	7.53608	7.16073	6.81369	6.19437	12
13	9.98565	9.39357	8.85268	8.35765	7.90378	7.48690	7.10336	6.42355	13
14	10.56312	9.89864	9.29498	8.74547	8.24424	7.78615	7.36669	6.62817	14
15	11.11839	10.37966	9.71225	9.10791	8.55948	8.06069	7.60608	6.81086	15
16	11.65230	10.83777	10.10590	9.44665	8.85137	8.31256	7.82371	6.97399	16
17	12.16567	11.27407	10.47726	9.76322	9.12164	8.54363	8.02155	7.11963	17
18	12.65930	11.68959	10.82760	10.05909	9.37189	8.75563	8.20141	7.24967	18
19	13.13394	12.08532	11.15812	10.33560	9.60360	8.95011	8.36492	7.36578	19
20	13.59033	12.46221	11.46992	10.59401	9.81815	9.12855	8.51356	7.46944	20
21	14.02916	12.82115	11.76408	10.83553	10.01680	9.29224	8.64869	7.56200	21
22	14.45112	13.16300	12.04158	11.06124	10.20074	9.44243	8.77154	7.64465	22
23	14.85684	13.48857	12.30338	11.27219	10.37106	9.58021	8.88322	7.71843	23
24	15.24696	13.79864	12.55036	11.46933	10.52876	9.70661	8.98474	7.78432	24
25	15.62208	14.09394	12.78336	11.65358	10.67478	9.82258	9.07704	7.84314	25
26	15.98277	14.37519	13.00317	11.82578	10.80998	9.92897	9.16095	7.89566	26
27	16.32959	14.64303	13.21053	11.98671	10.93516	10.02658	9.23722	7.94255	27
28	16.66306	14.89813	13.40616	12.13711	11.05108	10.11613	9.30657	7.98442	28
29	16.98371	15.14107	13.59072	12.27767	11.15841	10.19828	9.36961	8.02181	29
30	17.29203	15.37245	13.76483	12.40904	11.25778	10.27365	9.42691	8.05518	30
31	17.58849	15.59281	13.92909	12.53181	11.34980	10.34280	9.47901	8.08499	31
32	17.87355	15.80268	14.08404	12.64656	11.43500	10.40624	9.52638	8.11159	32
33	18.14765	16.00255	14.23023	12.75379	11.51389	10.46444	9.56943	8.13535	33
34	18.41120	16.19290	14.36814	12.85401	11.58693	10.51784	9.60857	8.15656	34
35	18.66461	16.37419	14.49825	12.94767	11.65457	10.56682	9.64416	8.17550	35
36	18.90828	16.54685	14.62099	13.03521	11.71719	10.61176	9.67651	8.19241	36
37	19.14258	16.71129	14.73678	13.11702	11.77518	10.65299	9.70592	8.20751	37
38	19.36786	16.86789	14.84602	13.19347	11.82887	10.69082	9.73265	8.22099	38
39	19.58448	17.01704	14.94907	13.26493	11.87858	10.72552	9.75696	8.23303	39
40	19.79277	17.15909	15.04630	13.33171	11.92461	10.75736	9.77905	8.24378	40

TABLE A-5 *continued*

n	4%	5%	6%	7%	8%	9%	10%	12%	n
41	19.99305	17.29437	15.13802	13.39412	11.96723	10.78657	9.79914	8.25337	41
42	20.18563	17.42321	15.22454	13.45245	12.00670	10.81337	9.81740	8.26194	42
43	20.37079	17.54591	15.30617	13.50696	12.04324	10.83795	9.83400	8.26959	43
44	20.54884	17.66277	15.38318	13.55791	12.07707	10.86051	9.84909	8.27642	44
45	20.72004	17.77407	15.45583	13.60552	12.10840	10.88120	9.86281	8.28252	45
46	20.88465	17.88007	15.52437	13.65002	12.13741	10.90018	9.87528	8.28796	46
47	21.04294	17.98102	15.58903	13.69161	12.16427	10.91760	9.88662	8.29282	47
48	21.19513	18.07716	15.65003	13.73047	12.18914	10.93358	9.89693	8.29716	48
49	21.34147	18.16872	15.70757	13.76680	12.21216	10.94823	9.90630	8.30104	49
50	21.48218	18.25593	15.76186	13.80075	12.23348	10.96168	9.91481	8.30450	50
51	21.61749	18.33898	15.81308	13.83247	12.25323	10.97402	9.92256	8.30759	51
52	21.74758	18.41807	15.86139	13.86212	12.27151	10.98534	9.92960	8.31035	52
53	21.87267	18.49340	15.90697	13.88984	12.28843	10.99573	9.93600	8.31281	53
54	21.99296	18.56515	15.94998	13.91573	12.30410	11.00525	9.94182	8.31501	54
55	22.10861	18.63347	15.99054	13.93994	12.31861	11.01399	9.94711	8.31697	55
56	22.21982	18.69854	16.02881	13.96256	12.33205	11.02201	9.95191	8.31872	56
57	22.32675	18.76052	16.06492	13.98370	12.34449	11.02937	9.95629	8.32029	57
58	22.42957	18.81954	16.09898	14.00346	12.35601	11.03612	9.96026	8.32169	58
59	22.52843	18.87575	16.13111	14.02192	12.36668	11.04231	9.96387	8.32294	59
60	22.62349	18.92929	16.16143	14.03918	12.37655	11.04799	9.96716	8.32405	60
61	22.71489	18.98028	16.19003	14.05531	12.38570	11.05320	9.97014	8.32504	61
62	22.80278	19.02883	16.21701	14.07038	12.39416	11.05798	9.97286	8.32593	62
63	22.88729	19.07508	16.24246	14.08447	12.40200	11.06237	9.97532	8.32673	63
64	22.96855	19.11912	16.26647	14.09764	12.40926	11.06640	9.97757	8.32743	64
65	23.04668	19.16107	16.28912	14.10994	12.41598	11.07009	9.97961	8.32807	65
66	23.12181	19.20102	16.31049	14.12144	12.42221	11.07347	9.98146	8.32863	66
67	23.19405	19.23907	16.33065	14.13219	12.42797	11.07658	9.98315	8.32913	67
68	23.26351	19.27530	16.34967	14.14223	12.43330	11.07943	9.98468	8.32958	68
69	23.33030	19.30981	16.36762	14.15162	12.43825	11.08205	9.98607	8.32999	69
70	23.39451	19.34268	16.38454	14.16039	12.44282	11.08445	9.98734	8.33034	70
71	23.45626	19.37398	16.40051	14.16859	12.44706	11.08665	9.98849	8.33066	71
72	23.51564	19.40379	16.41558	14.17625	12.45098	11.08867	9.98954	8.33095	72
73	23.57273	19.43218	16.42979	14.18341	12.45461	11.09052	9.99049	8.33121	73
74	23.62762	19.45922	16.44320	14.19010	12.45797	11.09222	9.99135	8.33143	74
75	23.68041	19.48497	16.45585	14.19636	12.46108	11.09378	9.99214	8.33164	75
76	23.73116	19.50950	16.46778	14.20220	12.46397	11.09521	9.99285	8.33182	76
77	23.77996	19.53285	16.47904	14.20767	12.46664	11.09653	9.99350	8.33198	77
78	23.82689	19.55510	16.48966	14.21277	12.46911	11.09773	9.99409	8.33213	78
79	23.87201	19.57628	16.49968	14.21755	12.47140	11.09883	9.99463	8.33226	79
80	23.91539	19.59646	16.50913	14.22201	12.47351	11.09985	9.99512	8.33237	80

TABLE A-6 MORTALITY TABLE BASED ON IRS TABLE 80CNSMT									3%	
x	l_x	d_x	q_x	D_x	N_x	C_x	M_x	L.E.	x
0	100,000	1260	0.012600	100,000.00	2,969,462.56	1,223.30	13,510.80	74.4	0
1	98,740	92	0.000932	95,864.08	2,869,462.56	86.72	12,287.50	74.3	1
2	98,648	64	0.000649	92,985.20	2,773,598.48	58.57	12,200.78	73.4	2
3	98,584	49	0.000497	90,218.33	2,680,613.28	43.54	12,142.21	72.4	3
4	98,535	40	0.000406	87,547.07	2,590,394.96	34.50	12,098.67	71.5	4
5	98,495	36	0.000366	84,962.65	2,502,847.89	30.15	12,064.17	70.5	5
6	98,459	33	0.000335	82,457.86	2,417,885.23	26.83	12,034.02	69.5	6
7	98,426	30	0.000305	80,029.35	2,335,427.37	23.68	12,007.19	68.5	7
8	98,396	26	0.000264	77,674.71	2,255,398.03	19.93	11,983.51	67.6	8
9	98,370	23	0.000234	75,392.41	2,177,723.32	17.11	11,963.58	66.6	9
10	98,347	19	0.000193	73,179.40	2,102,330.90	13.73	11,946.47	65.6	10
11	98,328	19	0.000193	71,034.24	2,029,151.50	13.33	11,932.74	64.6	11
12	98,309	24	0.000244	68,951.95	1,958,117.26	16.34	11,919.41	63.6	12
13	98,285	37	0.000376	66,927.30	1,889,165.30	24.46	11,903.07	62.6	13
14	98,248	52	0.000529	64,953.50	1,822,238.00	33.38	11,878.61	61.7	14
15	98,196	67	0.000682	63,028.28	1,757,284.50	41.75	11,845.23	60.7	15
16	98,129	82	0.000836	61,150.75	1,694,256.22	49.61	11,803.48	59.7	16
17	98,047	94	0.000959	59,320.05	1,633,105.47	55.22	11,753.87	58.8	17
18	97,953	102	0.001041	57,537.06	1,573,785.43	58.17	11,698.65	57.8	18
19	97,851	110	0.001124	55,803.06	1,516,248.36	60.90	11,640.48	56.9	19
20	97,741	118	0.001207	54,116.82	1,460,445.30	63.43	11,579.58	56.0	20
21	97,623	124	0.001270	52,477.17	1,406,328.48	64.71	11,516.15	55.0	21
22	97,499	129	0.001323	50,884.00	1,353,851.31	65.36	11,451.43	54.1	22
23	97,370	130	0.001335	49,336.58	1,302,967.31	63.95	11,386.07	53.2	23
24	97,240	130	0.001337	47,835.64	1,253,630.74	62.09	11,322.12	52.2	24
25	97,110	128	0.001318	46,380.28	1,205,795.10	59.35	11,260.03	51.3	25
26	96,982	126	0.001299	44,970.04	1,159,414.82	56.72	11,200.68	50.4	26
27	96,856	126	0.001301	43,603.51	1,114,444.78	55.07	11,143.95	49.4	27
28	96,730	126	0.001303	42,278.43	1,070,841.27	53.47	11,088.88	48.5	28
29	96,604	127	0.001315	40,993.56	1,028,562.84	52.32	11,035.42	47.6	29
30	96,477	127	0.001316	39,747.25	987,569.28	50.80	10,983.09	46.6	30
31	96,350	130	0.001349	38,538.76	947,822.03	50.48	10,932.29	45.7	31
32	96,220	132	0.001372	37,365.79	909,283.27	49.77	10,881.81	44.7	32
33	96,088	137	0.001426	36,227.70	871,917.48	50.15	10,832.04	43.8	33
34	95,951	143	0.001490	35,122.37	835,689.79	50.82	10,781.89	42.9	34
35	95,808	153	0.001597	34,048.57	800,567.41	52.79	10,731.08	41.9	35
36	95,655	163	0.001704	33,004.08	766,518.84	54.60	10,678.29	41.0	36
37	95,492	175	0.001833	31,988.19	733,514.76	56.91	10,623.68	40.1	37
38	95,317	188	0.001972	30,999.58	701,526.57	59.36	10,566.77	39.1	38
39	95,129	203	0.002134	30,037.32	670,526.99	62.23	10,507.41	38.2	39
40	94,926	220	0.002318	29,100.21	640,489.67	65.48	10,445.18	37.3	40

TABLE A-6 *continued*									3%
x	l_x	d_x	q_x	D_x	N_x	C_x	M_x	L.E.	x
41	94,706	241	0.002545	28,187.16	611,389.46	69.64	10,379.70	36.4	41
42	94,465	264	0.002795	27,296.53	583,202.30	74.06	10,310.06	35.5	42
43	94,201	288	0.003057	26,427.43	555,905.77	78.44	10,236.00	34.6	43
44	93,913	314	0.003344	25,579.25	529,478.34	83.03	10,157.55	33.7	44
45	93,599	343	0.003665	24,751.19	503,899.09	88.06	10,074.52	32.8	45
46	93,256	374	0.004010	23,942.22	479,147.90	93.22	9,986.46	31.9	46
47	92,882	410	0.004414	23,151.65	455,205.68	99.22	9,893.23	31.0	47
48	92,472	451	0.004877	22,378.11	432,054.02	105.96	9,794.02	30.1	48
49	92,021	495	0.005379	21,620.36	409,675.91	112.91	9,688.05	29.3	49
50	91,526	540	0.005900	20,877.73	388,055.55	119.59	9,575.14	28.4	50
51	90,986	584	0.006419	20,150.05	367,177.82	125.57	9,455.55	27.6	51
52	90,402	631	0.006980	19,437.59	347,027.77	131.72	9,329.98	26.8	52
53	89,771	684	0.007619	18,739.72	327,590.19	138.63	9,198.26	26.0	53
54	89,087	739	0.008295	18,055.28	308,850.46	145.41	9,059.63	24.1	54
55	88,348	797	0.009021	17,383.99	290,795.18	152.26	8,914.22	24.4	55
56	87,551	856	0.009777	16,725.40	273,411.20	158.76	8,761.97	23.6	56
57	86,695	919	0.010600	16,079.49	256,685.80	165.48	8,603.20	22.8	57
58	85,776	987	0.011507	15,445.67	240,606.31	172.55	8,437.72	22.0	58
59	84,789	1063	0.012537	14,823.24	225,160.64	180.43	8,265.17	21.3	59
60	83,726	1145	0.013676	14,211.07	210,337.39	188.68	8,084.74	20.5	60
61	82,581	1233	0.014931	13,608.47	196,126.32	197.27	7,896.06	19.8	61
62	81,348	1324	0.016276	13,014.84	182,517.85	205.66	7,698.79	19.1	62
63	80,024	1415	0.017682	12,430.11	169,503.00	213.39	7,493.13	18.4	63
64	78,609	1502	0.019107	11,854.68	157,072.89	219.91	7,279.74	17.7	64
65	77,107	1587	0.020582	11,289.49	145,218.21	225.59	7,059.83	17.0	65
66	75,520	1674	0.022166	10,735.08	133,928.72	231.03	6,834.24	16.3	66
67	73,846	1764	0.023888	10,191.38	123,193.65	236.36	6,603.21	15.7	67
68	72,082	1864	0.025859	9,658.19	113,002.27	242.48	6,366.86	15.1	68
69	70,218	1970	0.028055	9,134.40	103,344.08	248.81	6,124.38	14.4	69
70	68,248	2083	0.030521	8,619.54	94,209.68	255.41	5,875.57	13.8	70
71	66,165	2193	0.033144	8,113.07	85,590.14	261.07	5,620.16	13.2	71
72	63,972	2299	0.035938	7,615.70	77,477.07	265.72	5,359.08	12.6	72
73	61,673	2394	0.038818	7,128.16	69,861.37	268.64	5,093.37	12.1	73
74	59,279	2480	0.041836	6,651.91	62,733.21	270.18	4,824.73	11.5	74
75	56,799	2560	0.045071	6,187.98	56,081.30	270.78	4,554.54	11.0	75
76	54,239	2640	0.048673	5,736.97	49,893.32	271.11	4,283.77	10.5	76
77	51,599	2721	0.052734	5,298.77	44,156.35	271.28	4,012.66	9.9	77
78	48,878	2807	0.057429	4,873.15	38,857.58	271.71	3,741.38	9.4	78
79	46,071	2891	0.062751	4,459.51	33,984.43	271.69	3,469.67	8.9	79
80	43,180	2972	0.068828	4,057.93	29,524.92	271.16	3,197.98	8.5	80

TABLE A-6 *continued*									3%
x	l_x	d_x	q_x	D_x	N_x	C_x	M_x	*L.E.*	x
81	40,208	3036	0.075507	3,668.57	25,466.99	268.94	2,926.82	8.0	81
82	37,172	3077	0.082777	3,292.79	21,798.42	264.63	2,657.88	7.6	82
83	34,095	3083	0.090424	2,932.25	18,505.63	257.42	2,393.25	7.2	83
84	31,012	3052	0.098414	2,589.42	15,573.38	247.41	2,135.83	6.8	84
85	27,960	2999	0.107260	2,266.59	12,983.96	236.03	1,888.42	6.5	85
86	24,961	2923	0.117103	1,964.54	10,717.37	223.35	1,652.38	6.1	86
87	22,038	2803	0.127189	1,683.97	8,752.83	207.94	1,429.03	5.8	87
88	19,235	2637	0.137094	1,426.98	7,068.86	189.93	1,221.09	5.5	88
89	16,598	2444	0.147247	1,195.48	5,641.89	170.90	1,031.15	5.2	89
90	14,154	2246	0.158683	989.76	4,446.40	152.48	860.25	4.9	90
91	11,908	2045	0.171733	808.45	3,456.65	134.79	707.77	4.7	91
92	9,863	1831	0.185643	650.11	2,648.20	117.17	572.97	4.4	92
93	8,032	1608	0.200199	514.00	1,998.09	99.90	455.80	4.2	93
94	6,424	1381	0.214975	399.12	1,484.10	83.30	355.90	4.0	94
95	5,043	1159	0.229824	304.20	1,084.97	67.88	272.59	3.8	95
96	3,884	945	0.243306	227.46	780.78	53.73	204.72	3.7	96
97	2,939	754	0.256550	167.10	553.32	41.62	150.99	3.5	97
98	2,185	587	0.268650	120.62	386.21	31.46	109.37	3.4	98
99	1,598	448	0.280350	85.64	265.60	23.31	77.91	3.3	99
100	1,150	335	0.291304	59.84	179.95	16.92	54.60	3.2	100
101	815	245	0.300613	41.17	120.12	12.02	37.67	3.1	101
102	570	177	0.310526	27.96	78.94	8.43	25.66	3.0	102
103	393	126	0.320611	18.71	50.99	5.83	17.23	2.8	103
104	267	88	0.329588	12.34	32.27	3.95	11.40	2.7	104
105	179	60	0.335196	8.03	19.93	2.61	7.45	2.6	105
106	119	41	0.344538	5.19	11.90	1.73	4.84	2.4	106
107	78	27	0.346154	3.30	6.71	1.11	3.10	2.1	107
108	51	18	0.352941	2.09	3.41	0.72	2.00	1.6	108
109	33	33	1.000000	1.32	1.32	1.28	1.28	1.0	109
110	0								110

TABLE A-6 *continued*   5%

x	l_x	d_x	q_x	D_x	N_x	C_x	M_x	L.E.	x
0	100,000	1260	0.012600	100,000.00	1,992,208.86	1,200.00	5,132.91	74.4	0
1	98,740	92	0.000932	94,038.10	1,892,208.86	83.45	3,932.91	74.3	1
2	98,648	64	0.000649	89,476.64	1,798,170.76	55.29	3,849.46	73.4	2
3	98,584	49	0.000497	85,160.57	1,708,694.12	40.31	3,794.18	72.4	3
4	98,535	40	0.000406	81,064.99	1,623,533.55	31.34	3,753.87	71.5	4
5	98,495	36	0.000366	77,173.41	1,542,468.56	26.86	3,722.53	70.5	5
6	98,459	33	0.000335	73,471.62	1,465,295.15	23.45	3,695.66	69.5	6
7	98,426	30	0.000305	69,949.52	1,391,823.53	20.31	3,672.21	68.5	7
8	98,396	26	0.000264	66,598.29	1,321,874.01	16.76	3,651.90	67.6	8
9	98,370	23	0.000234	63,410.18	1,255,275.73	14.12	3,635.14	66.6	9
10	98,347	19	0.000193	60,376.53	1,191,865.55	11.11	3,621.02	65.6	10
11	98,328	19	0.000193	57,490.35	1,131,489.02	10.58	3,609.92	64.6	11
12	98,309	24	0.000244	54,742.13	1,073,998.67	12.73	3,599.34	63.6	12
13	98,285	37	0.000376	52,122.63	1,019,256.54	18.69	3,586.61	62.6	13
14	98,248	52	0.000529	49,621.92	967,133.91	25.01	3,567.92	61.7	14
15	98,196	67	0.000682	47,233.95	917,511.99	30.69	3,542.91	60.7	15
16	98,129	82	0.000836	44,954.03	870,278.04	35.78	3,512.21	59.7	16
17	98,047	94	0.000959	42,777.58	825,324.01	39.06	3,476.44	58.8	17
18	97,953	102	0.001041	40,701.49	782,546.43	40.36	3,437.38	57.8	18
19	97,851	110	0.001124	38,722.96	741,844.94	41.46	3,397.01	56.9	19
20	97,741	118	0.001207	36,837.55	703,121.97	42.36	3,355.56	56.0	20
21	97,623	124	0.001270	35,041.03	666,284.42	42.39	3,313.20	55.0	21
22	97,499	129	0.001323	33,330.02	631,243.39	42.00	3,270.81	54.1	22
23	97,370	130	0.001335	31,700.88	597,913.37	40.31	3,228.81	53.2	23
24	97,240	130	0.001337	30,151.00	566,212.49	38.39	3,188.50	52.2	24
25	97,110	128	0.001318	28,676.85	536,061.49	36.00	3,150.11	51.3	25
26	96,982	126	0.001299	27,275.29	507,384.63	33.75	3,114.12	50.4	26
27	96,856	126	0.001301	25,942.72	480,109.35	32.14	3,080.37	49.4	27
28	96,730	126	0.001303	24,675.21	454,166.63	30.61	3,048.23	48.5	28
29	96,604	127	0.001315	23,469.59	429,491.42	29.38	3,017.61	47.6	29
30	96,477	127	0.001316	22,322.60	406,021.84	27.99	2,988.23	46.6	30
31	96,350	130	0.001349	21,231.64	383,699.23	27.28	2,960.24	45.7	31
32	96,220	132	0.001372	20,193.32	362,467.60	26.38	2,932.96	44.7	32
33	96,088	137	0.001426	19,205.35	342,274.28	26.08	2,906.58	43.8	33
34	95,951	143	0.001490	18,264.73	323,068.92	25.92	2,880.50	42.9	34
35	95,808	153	0.001597	17,369.06	304,804.19	26.42	2,854.57	41.9	35
36	95,655	163	0.001704	16,515.54	287,435.13	26.80	2,828.16	41.0	36
37	95,492	175	0.001833	15,702.29	270,919.58	27.41	2,801.35	40.1	37
38	95,317	188	0.001972	14,927.15	255,217.30	28.04	2,773.95	39.1	38
39	95,129	203	0.002134	14,188.30	240,290.14	28.84	2,745.91	38.2	39
40	94,926	220	0.002318	13,483.83	226,101.85	29.76	2,717.07	37.3	40

| TABLE A-6 *continued* | | | | | | | | | 5% |
x	l_x	d_x	q_x	D_x	N_x	C_x	M_x	L.E.	x
41	94,706	241	0.002545	12,811.98	212,618.02	31.05	2,687.31	36.4	41
42	94,465	264	0.002795	12,170.83	199,806.04	32.39	2,656.26	35.5	42
43	94,201	288	0.003057	11,558.88	187,635.20	33.66	2,623.87	34.6	43
44	93,913	314	0.003344	10,974.80	176,076.33	34.95	2,590.21	33.7	44
45	93,599	343	0.003665	10,417.24	165,101.53	36.36	2,555.26	32.8	45
46	93,256	374	0.004010	9,884.83	154,684.29	37.76	2,518.91	31.9	46
47	92,882	410	0.004414	9,376.36	144,799.46	39.42	2,481.15	31.0	47
48	92,472	451	0.004877	8,890.45	135,423.10	41.30	2,441.73	30.1	48
49	92,021	495	0.005379	8,425.80	126,532.64	43.17	2,400.44	29.3	49
50	91,526	540	0.005900	7,981.41	118,106.84	44.85	2,357.27	28.4	50
51	90,986	584	0.006419	7,556.49	110,125.43	46.19	2,312.43	27.6	51
52	90,402	631	0.006980	7,150.47	102,568.94	47.53	2,266.23	26.8	52
53	89,771	684	0.007619	6,762.44	95,418.47	49.07	2,218.70	26.0	53
54	89,087	739	0.008295	6,391.34	88,656.03	50.49	2,169.63	24.1	54
55	88,348	797	0.009021	6,036.50	82,264.69	51.86	2,119.13	24.4	55
56	87,551	856	0.009777	5,697.19	76,228.19	53.05	2,067.27	23.6	56
57	86,695	919	0.010600	5,372.84	70,531.00	54.24	2,014.22	22.8	57
58	85,776	987	0.011507	5,062.75	65,158.16	55.48	1,959.98	22.0	58
59	84,789	1063	0.012537	4,766.18	60,095.41	56.91	1,904.50	21.3	59
60	83,726	1145	0.013676	4,482.32	55,329.23	58.38	1,847.59	20.5	60
61	82,581	1233	0.014931	4,210.49	50,846.91	59.87	1,789.21	19.8	61
62	81,348	1324	0.016276	3,950.12	46,636.42	61.23	1,729.34	19.1	62
63	80,024	1415	0.017682	3,700.79	42,686.30	62.32	1,668.11	18.4	63
64	78,609	1502	0.019107	3,462.24	38,985.51	63.00	1,605.79	17.7	64
65	77,107	1587	0.020582	3,234.37	35,523.27	63.40	1,542.78	17.0	65
66	75,520	1674	0.022166	3,016.95	32,288.90	63.69	1,479.38	16.3	66
67	73,846	1764	0.023888	2,809.60	29,271.95	63.92	1,415.69	15.7	67
68	72,082	1864	0.025859	2,611.89	26,462.35	64.33	1,351.78	15.1	68
69	70,218	1970	0.028055	2,423.19	23,850.47	64.75	1,287.45	14.4	69
70	68,248	2083	0.030521	2,243.05	21,427.28	65.20	1,222.70	13.8	70
71	66,165	2193	0.033144	2,071.04	19,184.23	65.37	1,157.50	13.2	71
72	63,972	2299	0.035938	1,907.04	17,113.19	65.27	1,092.13	12.6	72
73	61,673	2394	0.038818	1,750.96	15,206.15	64.73	1,026.86	12.1	73
74	59,279	2480	0.041836	1,602.85	13,455.19	63.86	962.13	11.5	74
75	56,799	2560	0.045071	1,462.66	11,852.34	62.78	898.26	11.0	75
76	54,239	2640	0.048673	1,330.22	10,389.68	61.66	835.48	10.5	76
77	51,599	2721	0.052734	1,205.22	9,059.46	60.53	773.81	9.9	77
78	48,878	2807	0.057429	1,087.30	7,854.24	59.47	713.29	9.4	78
79	46,071	2891	0.062751	976.05	6,766.94	58.33	653.82	8.9	79
80	43,180	2972	0.068828	871.24	5,790.89	57.11	595.49	8.5	80

TABLE A-6 *continued*									5%
x	l_x	d_x	q_x	D_x	N_x	C_x	M_x	*L.E.*	x
81	40,208	3036	0.075507	772.64	4,919.65	55.56	538.37	8.0	81
82	37,172	3077	0.082777	680.29	4,147.00	53.63	482.81	7.6	82
83	34,095	3083	0.090424	594.26	3,466.72	51.18	429.18	7.2	83
84	31,012	3052	0.098414	514.79	2,872.45	48.25	378.00	6.8	84
85	27,960	2999	0.107260	442.02	2,357.66	45.15	329.76	6.5	85
86	24,961	2923	0.117103	375.82	1,915.64	41.91	284.60	6.1	86
87	22,038	2803	0.127189	316.01	1,539.82	38.28	242.69	5.8	87
88	19,235	2637	0.137094	262.68	1,223.81	34.30	204.41	5.5	88
89	16,598	2444	0.147247	215.88	961.12	30.27	170.11	5.2	89
90	14,154	2246	0.158683	175.32	745.24	26.50	139.84	4.9	90
91	11,908	2045	0.171733	140.48	569.92	22.98	113.34	4.7	91
92	9,863	1831	0.185643	110.81	429.44	19.59	90.36	4.4	92
93	8,032	1608	0.200199	85.94	318.63	16.39	70.77	4.2	93
94	6,424	1381	0.214975	65.47	232.68	13.40	54.39	4.0	94
95	5,043	1159	0.229824	48.94	167.22	10.71	40.98	3.8	95
96	3,884	945	0.243306	35.90	118.27	8.32	30.27	3.7	96
97	2,939	754	0.256550	25.87	82.37	6.32	21.95	3.5	97
98	2,185	587	0.268650	18.32	56.50	4.69	15.63	3.4	98
99	1,598	448	0.280350	12.76	38.18	3.41	10.94	3.3	99
100	1,150	335	0.291304	8.75	25.42	2.43	7.53	3.2	100
101	815	245	0.300613	5.90	16.67	1.69	5.11	3.1	101
102	570	177	0.310526	3.93	10.77	1.16	3.42	3.0	102
103	393	126	0.320611	2.58	6.84	0.79	2.26	2.8	103
104	267	88	0.329588	1.67	4.26	0.52	1.47	2.7	104
105	179	60	0.335196	1.07	2.59	0.34	0.94	2.6	105
106	119	41	0.344538	0.68	1.52	0.22	0.60	2.4	106
107	78	27	0.346154	0.42	0.85	0.14	0.38	2.1	107
108	51	18	0.352941	0.26	0.42	0.09	0.24	1.6	108
109	33	33	1.000000	0.16	0.16	0.15	0.15	1.0	109
110	0								110

TABLE A-7 PRESENT VALUE: LIFE ANNUITY DUE OF $1.00 PER MONTH

Annual Interest Rate = 5.00%   Annual Amount = $12.00
Based on Table H, IRS Publication 1457, Section 605

Age	Amount	Age	Amount	Age	Amount
0	227.37	39	192.39	78	78.62
1	229.71	40	190.43	79	75.21
2	229.42	41	188.40	80	71.86
3	229.04	42	186.31	81	68.59
4	228.61	43	184.16	82	65.41
5	228.13	44	181.94	83	62.34
6	227.63	45	179.66	84	59.36
7	227.09	46	177.31	85	56.48
8	226.51	47	174.90	86	53.71
9	225.90	48	172.44	87	51.08
10	225.25	49	169.92	88	48.57
11	224.55	50	167.34	89	46.15
12	223.83	51	164.72	90	43.79
13	223.07	52	162.03	91	41.52
14	222.31	53	159.29	92	39.39
15	221.55	54	156.49	93	37.42
16	220.78	55	153.64	94	35.62
17	220.01	56	150.74	95	34.00
18	219.22	57	147.78	96	32.57
19	218.42	58	144.76	97	31.26
20	217.59	59	141.70	98	30.08
21	216.74	60	138.60	99	28.98
22	215.86	61	135.46	100	27.94
23	214.94	62	132.30	101	26.93
24	213.99	63	129.12	102	25.86
25	212.98	64	125.90	103	24.68
26	211.91	65	122.66	104	23.31
27	210.79	66	119.37	105	21.55
28	209.61	67	116.05	106	19.02
29	208.37	68	112.68	107	15.31
30	207.07	69	109.30	108	9.41
31	205.70	70	105.90	109	0.00
32	204.27	71	102.51	110	
33	202.77	72	99.12		
34	201.20	73	95.73		
35	199.57	74	92.34		
36	197.87	75	88.92		
37	196.11	76	85.49		
38	194.29	77	82.05		

TABLE A-7 *continued*

*Annual Interest Rate = **6.00%**   Annual Amount = $12.00*
Based on Table H, IRS Publication 1457, Section 605

Age	Amount	Age	Amount	Age	Amount
0	192.51	39	169.38	78	74.45
1	194.59	40	167.91	79	71.36
2	194.45	41	166.39	80	68.30
3	194.25	42	164.81	81	65.31
4	194.00	43	163.17	82	62.39
5	193.73	44	161.47	83	59.56
6	193.42	45	159.71	84	56.81
7	193.09	46	157.89	85	54.14
8	192.74	47	156.02	86	51.57
9	192.36	48	154.09	87	49.11
10	191.95	49	152.10	88	46.77
11	191.50	50	150.07	89	44.50
12	191.03	51	147.98	90	42.28
13	190.54	52	145.83	91	40.14
14	190.04	53	143.63	92	38.13
15	189.55	54	141.37	93	36.26
16	189.06	55	139.05	94	34.55
17	188.56	56	136.68	95	33.01
18	188.06	57	134.26	96	31.65
19	187.55	58	131.77	97	30.40
20	187.02	59	129.23	98	29.28
21	186.47	60	126.65	99	28.23
22	185.90	61	124.03	100	27.25
23	185.31	62	121.37	101	26.29
24	184.68	63	118.68	102	25.27
25	184.02	64	115.96	103	24.14
26	183.31	65	113.20	104	22.84
27	182.55	66	110.38	105	21.16
28	181.75	67	107.52	106	18.72
29	180.90	68	104.62	107	15.11
30	180.00	69	101.68	108	9.32
31	179.04	70	98.72	109	0.00
32	178.03	71	95.75	110	
33	176.96	72	92.77		
34	175.84	73	89.77		
35	174.66	74	86.76		
36	173.43	75	83.72		
37	172.13	76	80.65		
38	170.79	77	77.55		

TABLE A-8 SINGLE LIFE DISTRIBUTION FACTORS—IRS PUBLICATION 939

Age	Years	Age	Years	Age	Years	Age	Years	Age	Years
5	77.7	28	55.3	51	33.3	74	14.1	97	3.6
6	76.7	29	54.3	52	32.3	75	13.4	98	3.4
7	75.7	30	53.3	53	31.4	76	12.7	99	3.1
8	74.8	31	52.4	54	30.5	77	12.1	100	2.9
9	73.8	32	51.4	55	29.6	78	11.4	101	2.7
10	72.8	33	50.4	56	28.7	79	10.8	102	2.5
11	71.8	34	49.4	57	27.9	80	10.2	103	2.3
12	70.8	35	48.5	58	27.0	81	9.7	104	2.1
13	69.9	36	47.5	59	26.1	82	9.1	105	1.9
14	68.9	37	46.5	60	25.2	83	8.6	106	1.7
15	67.9	38	45.6	61	24.4	84	8.1	107	1.5
16	66.9	39	44.6	62	23.5	85	7.6	108	1.4
17	66.0	40	43.6	63	22.7	86	7.1	109	1.2
18	65.0	41	42.7	64	21.8	87	6.7	110	1.1
19	64.0	42	41.7	65	21.0	88	6.3	111	1.0
20	63.0	43	40.7	66	20.2	89	5.9		
21	62.1	44	39.8	67	19.4	90	5.5		
22	61.1	45	38.8	68	18.6	91	5.2		
23	60.1	46	37.9	69	17.8	92	4.9		
24	59.1	47	37.0	70	17.0	93	4.6		
25	58.2	48	36.0	71	16.3	94	4.3		
26	57.2	49	35.1	72	15.5	95	4.1		
27	56.2	50	34.2	73	14.8	96	3.8		

TABLE A-9 JOINT LIFE MINIMUM DISTRIBUTION FACTORS—10-YEAR AGE DIFFERENCE—IRS PUBLICATION 939, TABLE VI

Age	Years	Age	Years	Age	Years	Age	Years	Age	Years
70	27.4	80	18.7	90	11.4	100	6.3	110	3.1
71	26.5	81	17.9	91	10.8	101	5.9	111	2.9
72	25.6	82	17.1	92	10.2	102	5.5	112	2.6
73	24.7	83	16.3	93	9.6	103	5.2	113	2.4
74	23.8	84	15.5	94	9.1	104	4.9	114	2.1
75	22.9	85	14.8	95	8.6	105	4.5	115	1.9
76	22.0	86	14.1	96	8.1	106	4.2		
77	21.2	87	13.4	97	7.6	107	3.9		
78	20.3	88	12.7	98	7.1	108	3.7		
79	19.5	89	12.0	99	6.7	109	3.4		

Ages	50	51	52	53	54	55	56	57	58	59
TABLE A-10 ORDINARY JOINT AND LAST SURVIVOR ANNUITIES—TWO LIVES: IRS PUBLICATION 939										
50	40.4	40.0	39.5	39.1	38.7	38.3	38.0	37.6	37.3	37.1
51	40.0	39.5	39.0	38.5	38.1	37.7	37.4	37.0	36.7	36.4
52	39.5	39.0	38.5	38.0	37.6	37.2	36.8	36.4	36.0	35.7
53	39.1	38.5	38.0	37.5	37.1	36.6	36.2	35.8	35.4	35.1
54	38.7	38.1	37.6	37.1	36.6	36.1	35.7	35.2	34.8	34.5
55	38.3	37.7	37.2	36.6	36.1	35.6	35.1	34.7	34.3	33.9
56	38.0	37.4	36.8	36.2	35.7	35.1	34.7	34.2	33.7	33.3
57	37.6	37.0	36.4	35.8	35.2	34.7	34.2	33.7	33.2	32.8
58	37.3	36.7	36.0	35.4	34.8	34.3	33.7	33.2	32.8	32.3
59	37.1	36.4	35.7	35.1	34.5	33.9	33.3	32.8	32.3	31.8
60	36.8	36.1	35.4	34.8	34.1	33.5	32.9	32.4	31.9	31.3
61	36.6	35.8	35.1	34.5	33.8	33.2	32.6	32.0	31.4	30.9
62	36.3	35.6	34.9	34.2	33.5	32.9	32.2	31.6	31.1	30.5
63	36.1	35.4	34.6	33.9	33.2	32.6	31.9	31.3	30.7	30.1
64	35.9	35.2	34.4	33.7	33.0	32.3	31.6	31.0	30.4	29.8
65	35.8	35.0	34.2	33.5	32.7	32.0	31.4	30.7	30.0	29.4
66	35.6	34.8	34.0	33.3	32.5	31.8	31.1	30.4	29.8	29.1
67	35.5	34.7	33.9	33.1	32.3	31.6	30.9	30.2	29.5	28.8
68	35.3	34.5	33.7	32.9	32.1	31.4	30.7	29.9	29.2	28.6
69	35.2	34.4	33.6	32.8	32.0	31.2	30.5	29.7	29.0	28.3
70	35.1	34.3	33.4	32.6	31.8	31.1	30.3	29.5	28.8	28.1
71	35.0	34.2	33.3	32.5	31.7	30.9	30.1	29.4	28.6	27.9
72	34.9	34.1	33.2	32.4	31.6	30.8	30.0	29.2	28.4	27.7
73	34.8	34.0	33.1	32.3	31.5	30.6	29.8	29.1	28.3	27.5
74	34.8	33.9	33.0	32.2	31.4	30.5	29.7	28.9	28.1	27.4
75	34.7	33.8	33.0	32.1	31.3	30.4	29.6	28.8	28.0	27.2
76	34.6	33.8	32.9	32.0	31.2	30.3	29.5	28.7	27.9	27.1
77	34.6	33.7	32.8	32.0	31.1	30.3	29.4	28.6	27.8	27.0
78	34.5	33.6	32.8	31.9	31.0	30.2	29.3	28.5	27.7	26.9
79	34.5	33.6	32.7	31.8	31.0	30.1	29.3	28.4	27.6	26.8
80	34.5	33.6	32.7	31.8	30.9	30.1	29.2	28.4	27.5	26.7
81	34.4	33.5	32.6	31.8	30.9	30.0	29.2	28.3	27.5	26.6
82	34.4	33.5	32.6	31.7	30.8	30.0	29.1	28.3	27.4	26.6
83	34.4	33.5	32.6	31.7	30.8	29.9	29.1	28.2	27.4	26.5
84	34.3	33.4	32.5	31.7	30.8	29.9	29.0	28.2	27.3	26.5
85	34.3	33.4	32.5	31.6	30.7	29.9	29.0	28.1	27.3	26.4
86	34.3	33.4	32.5	31.6	30.7	29.8	29.0	28.1	27.2	26.4
87	34.3	33.4	32.5	31.6	30.7	29.8	28.9	28.1	27.2	26.4
88	34.3	33.4	32.5	31.6	30.7	29.8	28.9	28.0	27.2	26.3
89	34.3	33.3	32.4	31.5	30.7	29.8	28.9	28.0	27.2	26.3
90	34.2	33.3	32.4	31.5	30.6	29.8	28.9	28.0	27.1	26.3
91	34.2	33.3	32.4	31.5	30.6	29.7	28.9	28.0	27.1	26.3
92	34.2	33.3	32.4	31.5	30.6	29.7	28.8	28.0	27.1	26.2
93	34.2	33.3	32.4	31.5	30.6	29.7	28.8	28.0	27.1	26.2
94	34.2	33.3	32.4	31.5	30.6	29.7	28.8	27.9	27.1	26.2
95	34.2	33.3	32.4	31.5	30.6	29.7	28.8	27.9	27.1	26.2

TABLE A-10 *continued*

Ages	60	61	62	63	64	65	66	67	68	69
60	30.9	30.4	30.0	29.6	29.2	28.8	28.5	28.2	27.9	27.6
61	30.4	29.9	29.5	29.0	28.6	28.3	27.9	27.6	27.3	27.0
62	30.0	29.5	29.0	28.5	28.1	27.7	27.3	27.0	26.7	26.4
63	29.6	29.0	28.5	28.1	27.6	27.2	26.8	26.4	26.1	25.7
64	29.2	28.6	28.1	27.6	27.1	26.7	26.3	25.9	25.5	25.2
65	28.8	28.3	27.7	27.2	26.7	26.2	25.8	25.4	25.0	24.6
66	28.5	27.9	27.3	26.8	26.3	25.8	25.3	24.9	24.5	24.1
67	28.2	27.6	27.0	26.4	25.9	25.4	24.9	24.4	24.0	23.6
68	27.9	27.3	26.7	26.1	25.5	25.0	24.5	24.0	23.5	23.1
69	27.6	27.0	26.4	25.7	25.2	24.6	24.1	23.6	23.1	22.6
70	27.4	26.7	26.1	25.4	24.8	24.3	23.7	23.2	22.7	22.2
71	27.2	26.5	25.8	25.2	24.5	23.9	23.4	22.8	22.3	21.8
72	27.0	26.3	25.6	24.9	24.3	23.7	23.1	22.5	22.0	21.4
73	26.8	26.1	25.4	24.7	24.0	23.4	22.8	22.2	21.6	21.1
74	26.6	25.9	25.2	24.5	23.8	23.1	22.5	21.9	21.3	20.8
75	26.5	25.7	25.0	24.3	23.6	22.9	22.3	21.6	21.0	20.5
76	26.3	25.6	24.8	24.1	23.4	22.7	22.0	21.4	20.8	20.2
77	26.2	25.4	24.7	23.9	23.2	22.5	21.8	21.2	20.6	19.9
78	26.1	25.3	24.6	23.8	23.1	22.4	21.7	21.0	20.3	19.7
79	26.0	25.2	24.4	23.7	22.9	22.2	21.5	20.8	20.1	19.5
80	25.9	25.1	24.3	23.6	22.8	22.1	21.3	20.6	20.0	19.3
81	25.8	25.0	24.2	23.4	22.7	21.9	21.2	20.5	19.8	19.1
82	25.8	24.9	24.1	23.4	22.6	21.8	21.1	20.4	19.7	19.0
83	25.7	24.9	24.1	23.3	22.5	21.7	21.0	20.2	19.5	18.8
84	25.6	24.8	24.0	23.2	22.4	21.6	20.9	20.1	19.4	18.7
85	25.6	24.8	23.9	23.1	22.3	21.6	20.8	20.1	19.3	18.6
86	25.5	24.7	23.9	23.1	22.3	21.5	20.7	20.0	19.2	18.5
87	25.5	24.7	23.8	23.0	22.2	21.4	20.7	19.9	19.2	18.4
88	25.5	24.6	23.8	23.0	22.2	21.4	20.6	19.8	19.1	18.3
89	25.4	24.6	23.8	22.9	22.1	21.3	20.5	19.8	19.0	18.3
90	25.4	24.6	23.7	22.9	22.1	21.3	20.5	19.7	19.0	18.2
91	25.4	24.5	23.7	22.9	22.1	21.3	20.5	19.7	18.9	18.2
92	25.4	24.5	23.7	22.9	22.0	21.2	20.4	19.6	18.9	18.1
93	25.4	24.5	23.7	22.8	22.0	21.2	20.4	19.6	18.8	18.1
94	25.3	24.5	23.6	22.8	22.0	21.2	20.4	19.6	18.8	18.0
95	25.3	24.5	23.6	22.8	22.0	21.1	20.3	19.6	18.8	18.0
96	25.3	24.5	23.6	22.8	21.9	21.1	20.3	19.5	18.8	18.0
97	25.3	24.5	23.6	22.8	21.9	21.1	20.3	19.5	18.7	18.0
98	25.3	24.4	23.6	22.8	21.9	21.1	20.3	19.5	18.7	17.9
99	25.3	24.4	23.6	22.7	21.9	21.1	20.3	19.5	18.7	17.9
100	25.3	24.4	23.6	22.7	21.9	21.1	20.3	19.5	18.7	17.9
101	25.3	24.4	23.6	22.7	21.9	21.1	20.2	19.4	18.7	17.9
102	25.3	24.4	23.6	22.7	21.9	21.0	20.2	19.4	18.6	17.9
103	25.3	24.4	23.6	22.7	21.9	21.0	20.2	19.4	18.6	17.9
104	25.3	24.4	23.5	22.7	21.9	21.0	20.2	19.4	18.6	17.8
105	25.3	24.4	23.5	22.7	21.9	21.0	20.2	19.4	18.6	17.8

TABLE A-10 *continued*

Ages	70	71	72	73	74	75	76	77	78	79
70	21.8	21.3	20.9	20.6	20.2	19.9	19.6	19.4	19.1	18.9
71	21.3	20.9	20.5	20.1	19.7	19.4	19.1	18.8	18.5	18.3
72	20.9	20.5	20.0	19.6	19.3	18.9	18.6	18.3	18.0	17.7
73	20.6	20.1	19.6	19.2	18.8	18.4	18.1	17.8	17.5	17.2
74	20.2	19.7	19.3	18.8	18.4	18.0	17.6	17.3	17.0	16.7
75	19.9	19.4	18.9	18.4	18.0	17.6	17.2	16.8	16.5	16.2
76	19.6	19.1	18.6	18.1	17.6	17.2	16.8	16.4	16.0	15.7
77	19.4	18.8	18.3	17.8	17.3	16.8	16.4	16.0	15.6	15.3
78	19.1	18.5	18.0	17.5	17.0	16.5	16.0	15.6	15.2	14.9
79	18.9	18.3	17.7	17.2	16.7	16.2	15.7	15.3	14.9	14.5
80	18.7	18.1	17.5	16.9	16.4	15.9	15.4	15.0	14.5	14.1
81	18.5	17.9	17.3	16.7	16.2	15.6	15.1	14.7	14.2	13.8
82	18.3	17.7	17.1	16.5	15.9	15.4	14.9	14.4	13.9	13.5
83	18.2	17.5	16.9	16.3	15.7	15.2	14.7	14.2	13.7	13.2
84	18.0	17.4	16.7	16.1	15.5	15.0	14.4	13.9	13.4	13.0
85	17.9	17.1	16.5	15.8	15.2	14.6	14.1	13.5	13.0	12.5
87	17.7	17.0	16.4	15.7	15.1	14.5	13.9	13.4	12.9	12.4
88	17.6	16.9	16.3	15.6	15.0	14.4	13.8	13.2	12.7	12.2
89	17.6	16.9	16.2	15.5	14.9	14.3	13.7	13.1	12.6	12.0
90	17.5	16.8	16.1	15.4	14.8	14.2	13.6	13.0	12.4	11.9
91	17.4	16.7	16.0	15.4	14.7	14.1	13.5	12.9	12.3	11.8
92	17.4	16.7	16.0	15.3	14.6	14.0	13.4	12.8	12.2	11.7
93	17.3	16.6	15.9	15.2	14.6	13.9	13.3	12.7	12.1	11.6
94	17.3	16.6	15.9	15.2	14.5	13.9	13.2	12.6	12.0	11.5
95	17.3	16.5	15.8	15.1	14.5	13.8	13.2	12.6	12.0	11.4
96	17.2	16.5	15.8	15.1	14.4	13.8	13.1	12.5	11.9	11.3
97	17.2	16.5	15.8	15.1	14.4	13.7	13.1	12.5	11.9	11.3
98	17.2	16.4	15.7	15.0	14.3	13.7	13.0	12.4	11.8	11.2
99	17.2	16.4	15.7	15.0	14.3	13.6	13.0	12.4	11.8	11.2
100	17.1	16.4	15.7	15.0	14.3	13.6	12.9	12.3	11.7	11.1
101	17.1	16.4	15.6	14.9	14.2	13.6	12.9	12.3	11.7	11.1
102	17.1	16.4	15.6	14.9	14.2	13.5	12.9	12.2	11.6	11.0
103	17.1	16.3	15.6	14.9	14.2	13.5	12.9	12.2	11.6	11.0
104	17.1	16.3	15.6	14.9	14.2	13.5	12.8	12.2	11.6	11.0
105	17.1	16.3	15.6	14.9	14.2	13.5	12.8	12.2	11.5	10.9
106	17.1	16.3	15.6	14.8	14.1	13.5	12.8	12.2	11.5	10.9
107	17.0	16.3	15.6	14.8	14.1	13.4	12.8	12.1	11.5	10.9
108	17.0	16.3	15.5	14.8	14.1	13.4	12.8	12.1	11.5	10.9
109	17.0	16.3	15.5	14.8	14.1	13.4	12.8	12.1	11.5	10.9
110	17.0	16.3	15.5	14.8	14.1	13.4	12.7	12.1	11.5	10.9
111	17.0	16.3	15.5	14.8	14.1	13.4	12.7	12.1	11.5	10.8
112	17.0	16.3	15.5	14.8	14.1	13.4	12.7	12.1	11.5	10.8
113	17.0	16.3	15.5	14.8	14.1	13.4	12.7	12.1	11.4	10.8
114	17.0	16.3	15.5	14.8	14.1	13.4	12.7	12.1	11.4	10.8
115	17.0	16.3	15.5	14.8	14.1	13.4	12.7	12.1	11.4	10.8

TABLE A-10 *continued*

Ages	80	81	82	83	84	85	86	87	88	89
80	13.8	13.4	13.1	12.8	12.6	12.3	12.1	11.9	11.7	11.5
81	13.4	13.1	12.7	12.4	12.2	11.9	11.7	11.4	11.3	11.1
82	13.1	12.7	12.4	12.1	11.8	11.5	11.3	11.0	10.8	10.6
83	12.8	12.4	12.1	11.7	11.4	11.1	10.9	10.6	10.4	10.2
84	12.6	12.2	11.8	11.4	11.1	10.8	10.5	10.3	10.1	9.9
85	12.3	11.9	11.5	11.1	10.8	10.5	10.2	9.9	9.7	9.5
86	12.1	11.7	11.3	10.9	10.5	10.2	9.9	9.6	9.4	9.2
87	11.9	11.4	11.0	10.6	10.3	9.9	9.6	9.4	9.1	8.9
88	11.7	11.3	10.8	10.4	10.1	9.7	9.4	9.1	8.8	8.6
89	11.5	11.1	10.6	10.2	9.9	9.5	9.2	8.9	8.6	8.3
90	11.4	10.9	10.5	10.1	9.7	9.3	9.0	8.6	8.3	8.1
91	11.3	10.8	10.3	9.9	9.5	9.1	8.8	8.4	8.1	7.9
92	11.2	10.7	10.2	9.8	9.3	9.0	8.6	8.3	8.0	7.7
93	11.1	10.6	10.1	9.6	9.2	8.8	8.5	8.1	7.8	7.5
94	11.0	10.5	10.0	9.5	9.1	8.7	8.3	8.0	7.6	7.3
95	10.9	10.4	9.9	9.4	9.0	8.6	8.2	7.8	7.5	7.2
96	10.8	10.3	9.8	9.3	8.9	8.5	8.1	7.7	7.4	7.1
97	10.7	10.2	9.7	9.2	8.8	8.4	8.0	7.6	7.3	6.9
98	10.7	10.1	9.6	9.2	8.7	8.3	7.9	7.5	7.1	6.8
99	10.6	10.1	9.6	9.1	8.6	8.2	7.8	7.4	7.0	6.7
100	10.6	10.0	9.5	9.0	8.5	8.1	7.7	7.3	6.9	6.6
101	10.5	10.0	9.4	9.0	8.5	8.0	7.6	7.2	6.9	6.5
102	10.5	9.9	9.4	8.9	8.4	8.0	7.5	7.1	6.8	6.4
103	10.4	9.9	9.4	8.8	8.4	7.9	7.5	7.1	6.7	6.3
104	10.4	9.8	9.3	8.8	8.3	7.9	7.4	7.0	6.6	6.3
105	10.4	9.8	9.3	8.8	8.3	7.8	7.4	7.0	6.6	6.2
106	10.3	9.8	9.2	8.7	8.2	7.8	7.3	6.9	6.5	6.2
107	10.3	9.8	9.2	8.7	8.2	7.7	7.3	6.9	6.5	6.1
108	10.3	9.7	9.2	8.7	8.2	7.7	7.3	6.8	6.4	6.1
109	10.3	9.7	9.2	8.7	8.2	7.7	7.2	6.8	6.4	6.0
110	10.3	9.7	9.2	8.6	8.1	7.7	7.2	6.8	6.4	6.0
111	10.3	9.7	9.1	8.6	8.1	7.6	7.2	6.8	6.3	6.0
112	10.2	9.7	9.1	8.6	8.1	7.6	7.2	6.7	6.3	5.9
113	10.2	9.7	9.1	8.6	8.1	7.6	7.2	6.7	6.3	5.9
114	10.2	9.7	9.1	8.6	8.1	7.6	7.1	6.7	6.3	5.9
115	10.2	9.7	9.1	8.6	8.1	7.6	7.1	6.7	6.3	5.9

Bibliography

1. Cissell, R., H. Cissell, and D. C. Flaspohler. *Mathematics of Finance*. 8th ed. Boston: Houghton Mifflin Company, 1990.

2. Kellison, S. G. *The Theory of Interest*. 2nd ed. Boston: Irwin, 1991.

3. Reilly, F. K. *Investment Analysis and Portfolio Management*. 3rd ed. Chicago: The Dryden Press, 1989.

4. Shao, S. P. and L. P. Shao. *Mathematics for Management and Finance*. 6th ed. Dallas: Southwestern, 1990.

5. Trowbridge, C. L. "Magic Numbers." *The Actuary Newsletter of the Society of Actuaries*. (1985).

6. Workman, L. *Mathematical Foundations of Life Insurance*. Atlanta: Life Office Management Association, 1987.

Answers

Chapter 1 Set 1.1

1a)	16.67	1b)	29.17	1c)	37.50	1d)	62.50	1e)	112.50
1f)	150.00	1g)	250.00	1h)	500.00	3a)	Aug. 31	3b)	Nov. 29
3c)	Feb. 27	5a)	112.50	5b)	225.00	5c)	337.50	7)	97.70
9)	7.5%	11)	8 months						

Set 1.2

1a)	16.67	1b)	29.17	1c)	37.50	1d)	62.50	1e)	112.50
1f)	150.00	1g)	250.00	1h)	500.00	3a)	112.50	3b)	225.00
3c)	337.50	5)	7,700.00	7)	37.037	9)	10,000.00	11)	3,485.08
13)	1,941.75	15)	5,016.95						

Set 1.3

1)	8,200.00	3)	2,275.00	5)	9,750.00	7)	2,475.00	9)	9,625.00
	1,800.00		225.00		250.00		25.00		
11a)	19,629.01	11b)	370.99	13)	9,100.00	15)	2,375.00	17)	9,933.33
					900.00				183.33
19)	2,486.11	21a)	39,600.00	21b)	600.00	21c)	400.00	23)	7.4%
	11.11								
25)	10.714%	27)	9.89%	29)	13.636%	31)	21.951%	33)	7.254%
35)	9.424%								

37) $(1 + rt)(1 - dt) = 1$ $1 - dt = \dfrac{1}{1 + rt} \rightarrow dt = 1 - \dfrac{1}{1 + rt} \rightarrow dt = \dfrac{1 + rt - 1}{1 + rt} = \dfrac{rt}{1 + rt}$

$$d = \dfrac{r}{1 + rt}$$

$$1 + rt = \dfrac{1}{1 - dt} \rightarrow rt = \dfrac{1}{1 - dt} - 1 \rightarrow rt = \dfrac{1 - 1 + dt}{1 - dt} = \dfrac{dt}{1 - dt}$$

$$r = \dfrac{d}{1 - dt}$$

Chapter 2 Set 2.1

1)	7,458.46	3)	7,416.67	5)	4,841.02	7)	16,260.27	9)	7,409.77
	7,893.54		7,876.11		5,831.58				7,957.89
11)	9,161.11	13)	14,942.51	15)	15,726.01	17)	8.71 months	19)	8.75 months
	9,903.90								
21a)	Sept. 18	21b)	Sept. 18	23a)	13,155.56	23b)	26,311.11		

Set 2.2

1)	120.83	3)	123.33	5)	152.50	7)	189.88	9)	14.54%	11)	3.15%
							9,114.21				

13) 63.333% 15) 832.63 17) 2,479.50 19) 19.6% 21) 23.4% 23a) 47.50
23b) 181.03 23c) 14.00% 23d) 14.00% 23e) 25.80%

Set 2.3

1) 9.23% 3) 11.78% 5) 24.36% 7a) 11.75% 7b) Time weighting overstates the performance

Chapter 3 Set 3.1

	f	i	n			f	i	n
1)	2	4.0000%	12		3)	12	0.6667%	72
5)	4	2.5000%	28		7)	12	0.7500%	240
9)	4	1.5000%	20		11)	2	5.0000%	8
13)	12	0.4583%	24		15)	4	1.7500%	32
17)	4	2.1250%	60					

Set 3.2

1) 1,410.60 3) 10,576.43 5) 9,861.72 7-1) 7.123% 7-3) 7.229%
7-5) 5.095% 8-1) 708.92 8-3) 6,051.19 8-5) 3,650.48 9) 15.5 years
11) 1,819.40 13) 6,022.58 15) 302.10 17-3a) 10,585.04 17-3b) 10,626.28
17-5a) 9,892.33 17-5b) 9,961.26 19) 28.57 years 21) 2,225.55 23) 11.294%
25) 5.790% 27) 19 29) 24

Set 3.3

1) 8,385.61 3) 8,358.31 5) 5,926.55 7) 6,472.74
9) 8,243.61 11) 18,096.75 13) 14,699.42

Chapter 4 Set 4.1

1) 12.335562 3) 2.03 5) 104.07393 7) 12.006107
9) 1,233.56 11) 1,218.00 13) 10,407.39 15) 7,203.66
17) 72,851.75 19) 174.72 21) 658.20 23) 69,362.88
25) 1,615.38 27) 62,745.39 29a) 132.04 29b) 662.16

Set 4.2

1) 51.7255608 3) 10.5631229 5) 118.503515 7) 21.4821846
9) 5,172.56 11) 6,337.87 13) 84.39 15) 465.50
17) 1,732.34 19) 1,890.68

21)
Year	Payment	Interest	Balance
1	2,000	0.00	2,000.00
2	2,000	140.00	4,140.00
3	2,000	289.80	6,429.80
4	2,000	450.09	8,879.89
5	2,000	621.59	11,501.48

23)

Year	Payment	Interest	Balance
1	2,000	0.00	2,000.00
2	2,000	160.00	4,160.00
3	2,000	332.80	6,492.80
4	2,000	519.42	9,012.24
5	2,000	720.98	11,733.20

25) $p = 2,805.11$

Year	Interest	Principal	Balance
0			11,501.48
1	805.10	2,000.05	9,501.47
2	665.10	2,140.01	7,361.47
3	515.30	2,289.81	5,071.66
4	355.02	2,450.09	2,621.57
5	183.51	2,621.57	0.00

27) $p = 2,938.66$

Year	Interest	Principal	Balance
0			11,733.20
1	938.66	2,000.00	9,733.20
2	778.66	2,160.00	7,573.19
3	605.86	2,332.80	5,240.39
4	419.23	2,519.43	2,720.96
5	217.68	2,720.96	0.00

29) 10.263%

Set 4.3

1) 10,647.16   3) 29,269.47   5) 50,132.82   7) 54,670.56   9) 9,175.44
   8,720.78   14,709.84   33,931.87

Set 4.4

1) 9   3) 20   5) 8   7) 0.00   9) 0.00   11) 13.00
13) 36.00   15) 16.00   17) 156.59   19) 89.07   21) 8.00   23) 19.00
25) 8.00   27) 227.91   29) 94.44   31) 15   33) 38   35) 18
39) 97.58   41a) 997.96   41b) 990.33

Chapter 5 Set 5.1

1) 2,784.11   3) 1,804.04   5) 13,427.99   7) 634.33   9) 25,977,055.14

Set 5.2

1a)	6,971.67	1b)	5,176.26	3a) 27,070.41	3b) 14,988.23	5a) 162,045.97
5b)	37,493.78	7a)	7,006.36	7b) 5,202.01	9a) 27,473.47	9b) 15,211.39
11a)	178,655.68	11b)	41,336.90	13) 3,862.39	15) 6,266.95	17) 3,604.73
19)	30,681.10	21)	44,222.90	23a) 6,518.67	23b) 1,576.27	25) 613,287.43
27)	$A = 35,250.56$		$B = 36,300.67$	29) 13,777.70		

Set 5.3

1)	25,000.00	3)	20,000.00	5)	34,333.33	7)	65,000.00	9)	24,834.07
11)	70,200.07	13)	17,254.84	15)	53,780.49	17)	60,555.56	19)	58,715.32
21)	370,428.85	23)	455,986.80						

Set 5.4

1a) 117,538.11   1b) 123,415.01   3a) 120,000.04   3b) 127,200.04
                 13,194.79   13,727.86

5a) 119,435.91   5b) 127,796.43   7a) 436,476.91   7b) 458,300.75
                 14,282.46   4,780.11

9a) 324,814.96   9b) 344,303.85   11a) 151,473.07   11b) 162,076.18
                 5,463.81   3,216.87

13) 412,207.14
     32,641.81

15) $A = p_1 \left[\dfrac{1 - \left(\dfrac{1+k}{1+i}\right)^n}{i - k} \right] \rightarrow 1 - \left(\dfrac{1+k}{1+i}\right)^n = \dfrac{A(i-k)}{p_1} \rightarrow \left(\dfrac{1+k}{1+i}\right)^n = 1 - \dfrac{A(i-k)}{p_1}$

$n = \dfrac{\text{Ln}\left[1 - \dfrac{A(i-k)}{p_1}\right]}{\text{Ln}\left(\dfrac{1+k}{1+i}\right)} = -\dfrac{\text{Ln}\left[1 - \dfrac{A(i-k)}{p_1}\right]}{\text{Ln}\left(\dfrac{1+i}{1+k}\right)}$

Chapter 6   Set 6.1

1)	9,530.75	3)	1,077.95	5)	8,984.79	7)	969.19	9)	10,000.00	11)	9,185.56
13)	10.30%	15)	7.772%	17)	10.46%						

Set 6.2

1)	9,752.41	3)	953.00	5)	4,724.03	7)	9,350.00	9)	930.00	11)	4,823.97
13)	9,582.81	15)	9,538.08								

Set 6.3

1)	6,245.97	3)	610.27	5)	5,583.95	7)	675.56	9)	712.89	11)	6,164.84
13)	566.87	15)	1,129.01								

Set 6.4

1a) 4.29   1b) 4.12   3a) 7.76   3b) 7.57   5a) 4.01   5b) 3.78   7a) 4.29   7b) 4.12
9a) 7.57   9b) 7.24

Chapter 7   Set 7.1

	Objective	x_1	x_2
1)	45	5	0
5)	7	1	3
9)	57.50	2	3

	Objective	x_1	x_2
3)	27	7	3
7)	150,000	1,000,000	1,000,000

Set 7.2

	Objective	x_1	x_2	x_3
1)	42	0	7	0
3)	3,200	40	0	0
5)	270,000	1,000,000	500,000	500,000
7)	270,000	1,000,000	500,000	500,000
9)	33,750	750,000	1,000,000	250,000

Set 7.3

The Excel Solver Reports for Problems 1–10 will be for Problems 1–10 in Section 7.2. The Reports for Problems 11–13 are compiled onto one page each and begin on page 450.

7.3-11)		A	B	C	D		
		15	16	10	10		
	1,554	40	24	36	23		
	Machining	120	120				
	Assembly	88	160				
	Admin.	365	1,000				
	Limit on	15	20				
	Limit on	16	16				
	Req. on C	10	10				
	Req. on D	10	10				

Answer Report

Target Cell (Max)

Cell	Name	Original Value	Final Value
B3	Problem 11	0	1,554

Adjustable Cells

Cell	Name	Original Value	Final Value
C2	A	0	15
D2	B	0	16
E2	C	0	10
F2	D	0	10

Constraints

Cell	Name	Cell Value	Formula	Status	Slack
C8	Limit on B	16	C8 <= D8	Binding	0
C5	Assembly	88	C5 <= D5	Not Binding	72.5
C6	Admin	365	C6 <= D6	Not Binding	635
C7	Limit on A	15	C7 <= D7	Not Binding	5.5
C4	Machining	120	C4 <= D4	Binding	0
C9	Req. on C	10	C9 >= D9	Binding	0
C10	Req. on D	10	C10 >= D1	Binding	0

Sensivity Report
Adjustable Cells

Cell	Name	Final Value	Reduced Cost	Objective Coefficient	Allowable Increase	Allowable Decrease
C2	A	15	0	40	8	11.2
D2	B	16	0	24	1E + 30	4
E2	C	10	0	36	14	1E + 30
F2	D	10	0	23	77	1E + 30

Constraints

Cell	Name	Final Value	Shadow Price	Constraint R.H. Side	Allowable Increase	Allowable Decrease
C8	Limit on B	16	4	16	29	11
C5	Assembly	88	0	160	1E + 30	72.5
C6	Admin	365	0	1000	1E + 30	635
C7	Limit on A	15	0	20	1E + 30	5.5
C4	Machining	120	20	120	11	29
C9	Req. on C	10	−14	10	11.6	4.4
C10	Req. on D	10	−77	10	5.8	2.2

Limits Report

Cell	Target Name	Value				
B3	Problem 11	1,554				

Cell	Adjustable Name	Value	Lower Limit	Target Result	Upper Limit	Target Result
C2	A	15	0	974	14	1,554
D2	B	16	0	1,170	16	1,554
E2	C	10	10	1,554	10	1,554
F2	D	10	10	1,554	10	1,554

7.3-13)		Women	Men	Cosmetics		
		55,000	27,500	27,500		
3,850,000		40	24	36		
Total Area		110,000	110,000			
Women Area		55,000	55,000			
Men Area		27,500	33,000			
Cosmetics Ar		27,500	27,500			

Answer Report
Target Cell (Max)

Cell	Name	Original Value	Final Value			
B3	Problem 13	0	3,850,000			

Adjustable Cells

Cell	Name	Original Value	Final Value			
C2	Women	0	55,000			
D2	Men	0	27,500			
E2	Cosmetics	0	27,500			

Constraints

Cell	Name	Cell Value	Formula	Status	Slack	
C4	Total Area	110,000	C4 <= D4	Binding	0	
C5	Women Area	55,000	C5 <= D5	Binding	0	
C6	Men Area	27,500	C6 <= D6	Non Binding	5,500	
C7	Cosmetics Are	27,500	C7 <= D7	Binding	0	

Sensitivity Report						
Adjustable Cells						
Cell	Name	Final Value	Reduced Cost	Objective Coefficient	Allowable Increase	Allowable Decrease
C2	Women	55,000	0	40	1E + 30	16
D2	Men	27,500	0	24	12	24
E2	Cosmetics	27,500	0	36	1E + 30	12
Constraints						
Cell	Name	Final Value	Shadow Price	Constraint R.H. Side	Allowable Increase	Allowable Decrease
C4	Total Area	110,000	24	110,000	5,500	27,500
C5	Women Area	55,000	16	55,000	27500	5,500
C6	Men Area	27,500	0	33,000	1E + 30	5,500
C7	cosmetics Are	27,500	12	27,500	27,500	5,500

Limits Report							
Cell	Target Name	Value					
B3	Problem 13	3,850,000					
Cell	Adjustable Name	Value	Lower Limit	Target Result	Upper Limit	Target Result	
C2	Women	55,000	0	1,650,000	55,000	3,849,999	
D2	Men	27,500	0	3,190,000	27,500	3,850,000	
E2	Cosmetics	27,500	0	2,860,000	27,500	3,850,000	

Set 7.4

1. The maximum total net present value of $61 million can be obtained when we select a combination of x_2, x_3, and $0.5x_4$.

3. The maximum total net present value of $58.02 million can be obtained when we select the combination of x_1, $0.66x_2$, x_3, and $0.2x_5$.

5. The maximum total net present value of $65.61 million can be obtained if the company choose x_1, $0.66x_2$, and x_3.

7. To reach the maximum net present value of $25.26 million, Hot Cross Buns should select x_2, $0.69x_4$, and x_5.

9. The new outcomes stay exactly the same as Problem 8.

Set 7.5

1. The profit of this company is maximized at $2,277.78 when 3.89 units of x_2 and 2.23 units of x_3 are produced.

3. The optimal output mix is $y_1 = 55.556$, $y_2 = 44.444$, and $y_3 = 138.889$. Note that the units of y_1, y_2, y_3 are one thousand cans. This output mix should generate a profit of $11,777.78 for Vineyard Chia.

5. The output mix is 0.1 units of x_1, 7.27 units of x_2, 0.22 units of x_3 and 7.43 units of x_4. This output mix brings in a profit of $4,374.40.

7. Widget Warehouse has a maximum profit of $71,235.09 and an output mix of 3.97 units of x_2, 1.796 units of x_3 and 13.29 units of x_4.

9. Ye Olde Cheese Shoppe is able to reach the maximum profit of $239.39 if the combination of outputs is 2.93 units of x_1 and 1.34 units of x_3.

Chapter 8 Set 8.1

1)
q	C, R	t
10,000	$150,000	100

3)
q	C, R	t
60,000	$24,000	600

5) 145 days   7) 664 days   9) $9,200   11) $1,264   13) 149.3 days

15) 811.4 days   17) Sinking Fund Payment $= \$6,274,539.49$
Coupon Interest Payment $= \$6,000,000$
Annual Fixed Cost $= \$12,274,539.49$
Breakeven Quantity $= \$2,045.76$ units per year

Set 8.2

1)
q	p
7.9	9.54

3)
q	p
2	18

5)
q	p
7.5	75

7)
q_v	R
5.5	$60.50

9)
q_1	q_2	q_v	R
3.8	10	6.9	$13.16

11)
q_1	q_2	q_v	R
3.2	8.8	6.0	$12.12

13)
q	p
1,873.6	75.04

14)
q	p
1,714.7	116.43

Set 8.3

1) 32.7869   3) 52.4962   5) 102.4346   7) 55.4552 9) 109.0706
                                                                68.9541

11)   26.67   13)   6.60   15a)   10.84   15b)

     81.67   68.60   72.84 $t_B = \dfrac{\text{Ln}\left[\dfrac{B_N - B_E}{B_N(1+r)^{-t_N} - B_E}\right]}{\text{Ln}(1+r)}$
                                                $= 11.56$ years at age 73.56

Chapter 9 Set 9.1

1) 0.1667   3) 0.0278   5) 0.1667   7) 0.0007992   9) 0.9980
11) 0.0480

Set 9.2

1) 0.013676   3)   797   5) 2,092   7) Probability of age x living n years
9) 0.957623   11) 0.632693

13)

Age	l_x	d_x	q_x
50	92,637	696	0.00757
51	91,941	746	0.00818
52	91,195	714	0.00789
52	90,481		

15) 0.915742   17) 0.0005758

Set 9.3

1) 96,477   3) 127   5) 127   7) 50,884,000   9) 145,218,210   11) Verification

Set 9.4

1) 233,345.23   3) 245,149.12   5) 513,010.93   7) 549,152.85
9) 188,170.72   11) 294,237.31   13) 542,446.31   15) 261,808.84
17) 225,454.13   19) 289,190.35   21) 241,883.04   23) 32,141.22
25) 28,479.73   27) 24,366.00   29) 22,202.10   31) 79,714.85
33) 84,965.43   35) 92,175.02   37) 47,744.76   39) 103,135.64
41) 81,138.45   43) 125,932.80   45) 60,904.83   47) 53,908.85
49) 81,890.31

Set 9.5

1) 9.31   3) 464.33   5) 45,862.94   7) 56,890.44   9) 56,392.53
11) 9.13   13) 450.98   15) 29,534.54   17) 41,219.50   19) 38,988.52

Set 9.6

1) 9.31   3) 235.86   5) 2,105.07   7) 21.99   9) 4,861.53
11) 727.36   13) 2,282.87   15) 1,749.91   17) 65.98   19) 1,060.07
21) 1,995.88   23) 3,339.27   25) 3,043.03

Set 9.7

1) 229   3) 8,710   5) 45,293   7) 56,891   9) 41,874
11) 229   13) 8,710   15) 45,293   17) 56,891   19) 41,558
21) $M_{x+n} = 1,946.58$ 23) 72,426.93
19 yrs, 277 days

Chapter 10 Set 10.1

1) 20.00%   3) 21.67%   5) 23.33%   7) 25.00%   9) 26.67%
11) 28.33%   13) 30.00%   15) 16.67%   17) 13.33%   19) 25.00%
21a) 25.00%   21b) 0.375 PIAw   23a) 12.5%   23b) 0.4375 PIAw
25a) 35.00%   25b) 0.325 PIAw

Set 10.2

1) $PIA = 1,774.60$ 3) $PIA = 936.50$ 5) $PIA = 1,684.90$ 7) $PIA = 1,526.00$
Benefit = 1,382   Benefit = 936   Benefit = 1,469   Benefit = 1,660
9) $PIA = 1,388.30$ 11) $PIA = 1,311.80$ 13) Benefit = 503   15) Benefit = 341
Delayed PIA   Delayed PIA   17) Benefit = 596   19) Benefit = 830
1,554.80   1,600.30
Benefit = 1,750   Initial Benefit   21) Benefit = 875   23) Benefit = 950
1,901

Chapter 11 Set 11.1

1. The employer could provide for selected employees only and exclude other employees arbitrarily. If the employer wishes to qualify for income tax advantages, all employees must be treated equitably. The laws and regulations are intended to insure this equitable treatment in order to be qualified for income tax consideration.

Set 11.2
1) 21,527.69   5) 3,744.81   3) 27,409.74   7) 2,268.65
9) 46,478.66   11) 24,997.92   13) 27,123.72   15) 16,586.10

Set 11.3

1)	1,341,310	3)	3,049,074	5)	1,052,929	7)	257,336	9)	555,935
11)	1,639,430	13)	2,736,902	15)	C1 = 25,000 C2 = 2,240 C3 = 1,306	17)	C1 = 18,643 C2 = 1,945 C3 = 1,163	19)	1,305,282

Set 11.4
1) 471,667   3) 1,277,708   5) 746,063   7) 448,000
9) 725,313   13) 50,264/yr   11) 53,400/yr   15) 55,800/yr

17)

Age	Predistribution Balance	Distribution Factor	Distribution	Postdistribution Balance
70	750,000	27.4	27,372	722,628
71	765,985	26.5	28,905	737,080
72	781,305	25.6	30,520	750,785
73	795,833	24.7	32,220	763,613
74	809,429	23.8	34,010	775,420

19)

Age	Predistribution Balance	Distribution Factor	Distribution	Postdistribution Balance
70	800,000	27.4	29,197	770,803
71	817,051	26.5	30,832	786,219
72	833,392	25.6	32,554	800,838
73	848,888	24.7	34,368	814,520
74	863,391	23.8	36,277	827,114

21)

Age	Predistribution Balance	Distribution Factor	Distribution	Postdistribution Balance
70	1,000,000	42.7	23,419	976,581
36	1,035,176	47.5	21,793	1,013,382
37	1,074,185	46.5	23,101	1,051,085
38	1,114,150	45.5	24,487	1,089,663
39	1,155,043	44.5	25,956	1,129,087

Index